The Caspian Region
at a Crossroad

Challenges of a New Frontier

of Energy and Development

Edited by Hooshang Amirahmadi

Macmillan

© Hooshang Amirahmadi, 2000

First published 2000 by
MACMILLAN PRESS LTD
Houndmills, Basingstoke, Hampshire RG21 6XS
and London
Companies and representatives
throughout the world

ISBN 0-333-91380-9

A catalogue record for this book is available
from the British Library.

10 9 8 7 6 5 4 3 2 1
09 08 07 06 05 04 03 02 01 00

Printed in the United States of America by
Haddon Craftsmen
Bloomsburg, PA

Contents

Preface and Acknowledgments

For centuries, the Caspian region was the strategic crossroads for the different empires of Europe, Asia, and Africa. Control of vital land and water routes, particularly for trade in spices, assured great power and wealth; it also invited rivalry. As technological changes diminished the significance of these routes, the control of the region's rich hydrocarbon resources has become the focus of international business and political players. During the cold war, the area remained in the crossfire of different interests, as a fault line of East-West struggle.

The demise of the Soviet Union since the late 1980s once again changed the regional dynamics. A power vacuum was created, with lines of control less certain. Notwithstanding lesser hegemonic control, there has been no corresponding abatement of interest. Approximately 70 percent of the world's known oil reserves and 40 percent of its gas reserves are found in a sphere stretching from southern Russia in the north to Saudi Arabia in the south. With a growing dependence on this resource, the region has once again become a frontier. Those who sense a loss of control in the area, those who want to protect newly acquired interest and power, and those who want to gain a foothold and expand their interest in the area jostle for position and influence.

It is, however, a frontier vastly different from that of the past. At a time when the hegemonic powers and influences of the United States and Russia appear to be diminishing, other powers are emerging. The area is inviting new external interest, particularly from the oil-thirsty East Asia and Western oil companies. The Soviet collapse led to the formation of fifteen independent nation-states in the region. Today, the area is home to some 300 million people, predominantly young, many of them poor, and it is under severe environmental stress. In past scrambles, the needs of citizens and the quality of their environment were often forgotten. This can no longer be the case; the aspirations of the people and the demand for a precautionary approach to the environment cannot be wished away. The new frontier contains more—and increasingly varied—interests than ever before.

Old and new players are engaged in various interest-seeking games, under different banners, with different agendas. What dominates these games is gross shortsightedness, a state of affairs that can exacerbate the brutal circle of power, money, and corruption evident in the region, and is sure to become a recipe for disaster. For example, the Russian energy sector seeks to secure partnerships in investments made in the oil and gas industry of the newly independent states; these states themselves seek to break structural dependence on Russia; and Western oil, gas, and engineering companies seek major contracts. A mixture of myth and reality, arrogance and humility, despair and hope characterizes this engagement. The people and environment do not always figure high on the agendas of the players, and reflection on the long-term future of the region is sacrificed for immediate gains. Problems are confronted on the basis of existing practice and approaches, often from afar and in a manner ill suited to the new and rapidly changing context. Behavior is rationalized in terms of past metaphors and rhetoric, born under different circumstances. Yet, the forces of transformation have created a labyrinth in the region, affecting all actors and sectors. Both newcomers and those who have been around for centuries have to adjust to the new realities of the Caspian frontier; no one escapes the impact, whether good or bad.

The forces of transformation emerge from larger processes of global change and circumstances internal to the region, as well as from specific countries and sectors. The United States is subject to a two-fold adjustment: it is a relative newcomer to the area and a self-promoted sole superpower, following the demise of the cold war bipolar world order. As such, it has to learn "new rules of doing" in an unfamiliar area, at a time of growing Third World reassertion, rising strength of nations and regions, declining American economic power, and diminishing utility of offensive force.[1] As U.S. National Security Adviser Samuel R. Berger remarked after an Iraqi–U.S. standoff: "We have a strong tradition of non-engagement and self-reliance in the world; yet here we are in a position where other countries look to us to lead. But if we lead with too heavy a hand, they resent it."[2] At the time of the renewed crisis over the Iraqi weapons of mass destruction early in 1989, some senior American politicians were dismayed at Kofi Annan for "calling the shots" in defusing a very volatile situation.

Russia has to come to terms with the fact that their old republics are now independent and cannot be engaged in the games of the past. It also has to manage significant conservative and retrogressive forces within its cadre, seeking a mindless re-unification of the Soviet Union. Iran is faced

by a new set of neighbors along its northern borders. Internally, it is negotiating a difficult transition from an early pan-Islamic position to an even-
tual pan-Iranism. Its revolution is still an incomplete project. The new
republics in the Caucasus and Central Asia face a daunting task: a transformation from being merely a supply house to Moscow to generating
internal development and nation building. For most, independence came
unexpectedly. These countries are experiencing the problems associated
with a "collusion" between a fixation with elements of the old system and
a desire to find a "third way," integrating certain traditional values with
those that are emerging or subdued for many years. Multinationals are
trying to do business in an environment in which economic sense and
political reality are often at odds and institutional frameworks are lacking
or evolving. The fundamental transformation from centralized control to
independence creates loopholes and vacuums, inviting corruption and
other criminal activities.

The transformation is also manifested in unsettling shifts between
exclusionary and inclusionary politics, between the sovereignty of nation-
states and regionalism, between ethnocentrism and pluralism, between
integration and self-centered development, between sectoral and holistic
growth, between anthropocentric and environcentric development, and
between traditional and emerging values. History teaches us that new
frontiers bring significant dangers and risks for all involved. In the
Caspian region, the mixture of global change and the rapid opening of a
new frontier have created a particularly volatile situation as values clash.
The presence of rich and strategically significant hydrocarbon resources,
the gradual slipping of Soviet control over its "near-abroad," and the rising aspirations of citizens are all among contributing factors to this situation. The clash is reflected in numerous conflicts and postures, military
and non-military, which continue to beset the region. How these will be
managed represents a new challenge for the Caspian Region. The potential is rich: overall there could be a transformation from a contested and
exploited region to one that is stable and sustainable, from broken and
strained relationships to mended and new solidarity. Thus, from a region
in the crossfire, the Caspian could emerge as a bastion of security for
energy and development in the twenty-first century. The key challenge is
how to engage and direct the vast potential that can be released if new
thinking and collective action can be adopted.

This book attempts to bring some order to the current environment of
disorder in the study of the region, indicating commonalties and threats,
as well as suggesting ways in which existing and emerging challenges
could be addressed. It aims at stimulating thinking about the challenges
of the region and facilitating the constructive engagement of all players.
The overall message of the book is that there is a need for reflection prior
to action in the area, reflection aimed not only at the specific context, but

also at the nature of the relationships between the larger world and the
frontier area. To achieve this, the book focuses on five major themes: the
people and resources, development and environment, pipelines and outlets, security and geopolitics, and legal regimes of the Caspian Sea. These
are themes that are currently either too hastily explained or their study is
focused on serving particular players or interests. A rich analysis of the
themes emerges in this book as the contributors, who come from different backgrounds, present different perspectives and prospects. There has
been no attempt to impose methodological, ideological, or stylistic uni-
formity and orthodoxy on the contributors—it is believed that a more
prosperous and secure future for the region can only be achieved through
airing and respecting the contrasting perspectives of all actors involved.

The book does, however, have an Iranian focus. Iran has emerged as a
major regional player from the U.S.-Iraq conflict. In the region, Iran is
also one of the "oldest" and, arguably, the most politically stable country;
it possesses considerable resources and skills and enjoys a unique geographic position. Despite the "dual containment" policy of the United
States, Iran has maintained a strong influence in the region. In the recent
past, it has also displayed pragmatism in regional policy and manifested
increasing political maturity. There is little doubt that Iran will play a
significant if not pivotal role in future regional politics and development.
This, however, does detract from a primary conclusion of all the contributors, namely that regional cooperation is the key for future stability and
prosperity in the Caspian region. Indeed, Iran, and particularly the nature
of its relations with the United States, may be considered the key in
achieving such cooperation. The Iranian focus of the book also follows
from the editor's long-standing relation with the region. Born and raised
on the southern shore of the Caspian Sea, he grew up enjoying the natural beauty of the Talish region in Iran's Gilan Province. His hope is that
the Caspian's environment will be preserved for the enjoyment of future
generations.

Organized into five parts, an introduction, and sixteen chapters, the
book provides the most up-to-date statement currently available about
the emerging Caspian frontier. In the introduction, the editor ties
together the many disparate aspects of the new frontier in an attempt to
offer a more conceptual and holistic understanding of its challenges and
prospects. In the first chapter of Part I, on people and resources, Schoeberlein and Ilkhamov explore the region's geographical and population
diversity, as well as the origins of national identities and their role in current politics. In the following chapter, Skagen analyzes the extent of the
Caspian region's oil and gas reserves and the potential of different countries to become significant producers. Focusing on Iran, Khajehpour-
Khouei in the next chapter discusses the significance of non-petroleum

resources and other forms of economic activity in the Caspian Sea area, an aspect often ignored in the scramble for oil and gas.

Part II, on development and environment, begins with Mostashari's contribution, in which she explores the history of the Azerbaijan's early oil industry with a view to draw parallels with current approaches. The author suggests that unwise policies today may contain in them the seeds of social and political instability, similar to that which gripped Baku at the turn of the century. In his chapter, Seznec warns that the export focus of the region's hydrocarbon industry will not necessarily make a significant contribution to the welfare of the Caucasus and Central Asia. Other, more sustainable development options warrant consideration, particularly using oil and gas resources to fuel the region's industrialization and the development of value-added economic activities. In their respective chapters, Namazi and Goodarzy discuss the varied environmental challenges and concomitant management initiatives of the Caspian Sea—the focal area of international hydrocarbon interest.

The countries of the Caucasus and Central Asia are landlocked. They must therefore rely on expensive pipelines through other countries to transport their oil and gas to maritime terminals and foreign markets. Part III, on pipelines and outlets, analyzes the pros and cons of the varied pipeline proposals devised to facilitate the export of oil and gas from the region. In his chapter, Ghorban analyzes the economic reasons why Iran provides the best possible option as the conduit for oil and gas from the Caucasus and Central Asia. He emphasizes the significance of swap arrangements with Iran as an economic alternative to more conventional pipeline proposals. Kovalev makes the case for Russia as being the most expedient route for exporting Caspian oil and gas. In the last chapter of this part, Amirahmadi gives an overview of pipeline politics and suggests that pros and cons of various lines call for a multiple pipeline approach where most nations in the region will benefit from the economic and strategic advantages that new oil and gas pipelines offer.

In the first chapter of Part IV, on security and geopolitics, Mojtahed-Zadeh provides a general perspective on the geopolitical factors that have been influencing international relations in the Caspian region. In particular, he discusses the role of Iran in any regional alignment as a matter of geo-economic and strategic necessity. In their chapter, McGuin and Mesbahi examine the United States' drive for influence in the region and the role that Turkey and Azerbaijan have played in that effort. In the last chapter of this part, Ansari examines the potential for military conflict in the region, given a new environment of economic and resource competition in the post-Soviet era.

The challenges of coexistence in a competitive environment will demand Caspian countries to adopt national laws and international regulations to govern their relations in areas such as overlapping resource exploitation, fishing, sanitation, pollution control, contraband, immigration, taxation, and so on. This struggle is illustrated in the search for an appropriate legal regime for the Caspian Sea. In Part V, on legal perspectives, chapters by Mirfendereski, Horton and Mamedov, and Movahed present a broad framework for developing a legal regime for the Caspian Sea, as well as the different perspectives of Azerbaijan and Iran on the issue. Russia's perspective is covered in the chapter by Kovalev.

In editing this book, the editor has received encouragement and support from many of his students, colleagues, and friends. Among them, Dr. Guive Mirfendereski and Stephen Boshoff deserve special mention. Guive's contribution was especially important as he helped rewrite and edit the chapters, while Stephen helped in the research and drafting of the introduction. I am also indebted to Siamak Namazi for helping with the initial setup of the project, particularly in communicating with the key contributors to the book. Caspian Associates, Inc., a strategic research and consulting firm specializing in the Caspian region, provided administrative support and sponsorship for the project. Needless to say, the individual authors and the editor remain accountable for any shortcomings and errors of the volume.

Hooshang Amirahmadi

New Brunswick, January 2000

Notes

1. See Hooshang Amirahmadi, "Global Restructuring, the Persian Gulf War, and the U.S. Quest for World Leadership," in *The United States and the Middle East: A Search for New Perspectives* (Albany: State University of New York Press, 1993).
2. See Steven Erlanger, "America, the Lone Wolf with a Following," *New York Times* (March 1, 1998).

INTRODUCTION

Challenges of the Caspian Region

Hooshang Amirahmadi

The purpose of this introduction is to tie together the issues raised in the present volume and to project these concerns into a possible scenario for the region in the next decade. Most writings on the area deal with specific and isolated topics, such as oil and gas reserves, pipelines, the Caspian Sea legal regime, and U.S.–Iranian relations. Although these issues are individually important, there is a need to analyze their common basis. Once this is done, it will be easier to better understand the current circumstances and speculate about the future trajectories of the region.

It will be argued that the Caspian region suffers from a primitive state of research and analysis, a fictional perspective of its problems and prospects, a partial view of its development requirements, and a dominance of global interests over local needs. Similarly troubling is the dominance of rhetoric, obliviousness to history, the zero-sum mindset of most major regional players, deep-rooted sources of regional instability, insensibleness to the need for regional cooperation, and the entrenched leadership of the largely undemocratic personalities.

For the Caspian region to prosper, these problems have to be addressed in a substantive manner. This will require developing new academic programs, educational facilities, and research institutions. Leaders must realize the need for development of open societies and public participation. But the most important change has to be initiated at the international policy level. The Caspian players must come together in common purpose and for the good of the region's people as well as for their own private selfish interests. Regional cooperation must be the guiding principle for these players to reconstruct their relations on a sustainable basis. Much can be learned from history and experiences elsewhere in the world.

The Caspian Region Defined

Notwithstanding the increasing recognition of its significance, there is no clear definition of what constitutes the Caspian region. Does it comprise the newly independent states, the littoral states, states within the catchment area of the Caspian Sea, states with noteworthy hydrocarbon reserves, or states that contain the primary network of hydrocarbon reserves and routes to transport oil and gas? Despite the changing configuration of empires in the region, significant population movement over time, and the focus of activity around the Caspian Sea, the area is not viewed as an integrated whole, but rather as a collection of isolated geographic fragments. A more robust methodological approach to defining the region is of primary importance for preparing a regional vision and for planning and policy formulation in individual countries.

The new emphasis on the Caspian Sea area results from its regional development potentials. Few countries, perhaps with the exception of Iran, can hope to achieve much development without regional cooperation. Most are landlocked and depend on other countries to transport oil and gas to world markets. Also, given their long history of exploitation, most are devoid of sufficient infrastructure, modern technology, appropriate expertise, consumer products, and domestic markets. Such a state of affairs does not fare well with a world where economic forces dominate international relations.

The accumulation of capital, agglomeration of production, and expansion of marketing opportunities is important for autonomy of the new Caspian economies. In today's world, regions form because of the need for transnational synergy and the concomitant prospect for development. This is different from the past, when regions were defined on the basis of religion, culture, colonial legacy, and like criteria. While economic institutions are key to regional development, non-economic networks are also needed for managing challenges that may threaten balance and growth.

The Caspian region may be defined as comprising Iran, Afghanistan, Pakistan, Turkey, Georgia, part of the Russian Federation, Armenia, Azerbaijan, Turkmenistan, Uzbekistan, Tajikistan, Kyrgyzstan, and Kazakhstan. This excludes India and the Arab world. Both political and economic factors underlie this proposed grouping of countries.[1] Even though the Caspian region is predominantly Muslim, Islam is not viewed as an important political force in the area. Rather, the proposed constellation "constitute(s) an economically viable assemblage of states with common developmental interests and an awareness of their potential for development synergy."[2]

Three interrelated bases for the development potential of the region may be identified: capital, transportation, and economic reciprocity. Growing world demand for hydrocarbon fuel guarantees capital, which could be distributed among all countries to an extent because of the land-locked nature of most littoral states and the need for pipelines that traverse several countries. Rail networks could be expanded to link countries in the region and the network could be integrated with other transport systems to facilitate interaction, particularly via maritime access. As most economies are in a state of sectoral imbalance, though not in the same way, significant potential for intra-regional trade exists. Iran, Turkey, Russia, and China form a ring of outlets to the rest of the world for the region.

The functioning of the Economic Cooperation Organization (ECO), established to facilitate economic affiliation and cooperation in the region, has been inhibited because it has not been formed with a clear regional focus. Turkey, for example, has tended to subordinate ECO to its ambition to join the European Union (EU) and to its relationship with the United States, Israel, and Azerbaijan. On the other hand, the Caspian focus of the Organization for Regional Cooperation of the Caspian States (ORCCS) may be too narrow to represent the varied interdependencies of countries in the region.

There are also political obstacles to regional cooperation. For example, Pakistan is not on good terms with India, while Russia has tenuous relations with the countries in its "near abroad." Similarly, the current policy of the United States toward the Caspian region and its so-called East-West directional strategy excludes Iran as a player in the region. Instead, the United States emphasizes relations with certain so-called pivotal states, particularly Turkey, Azerbaijan, Georgia, and Armenia, which are not economically linked or politically allied. The U.S. view of the region negates the important development interdependencies that exist between Iran and its northern neighbors. It is not also conducive to regional stability, which is a prerequisite for development in the region.

Within the larger region, there are other definitional issues. For example, in relation to the environment, how should the region be defined? Does it only include the Caspian Sea area and the rivers that feed it, or also the Black Sea, which will have to carry the risk of acting as conduit for Caspian oil? Apart from concerns about the adequacy of present navigational systems, the Turkish government, in particular, is concerned about the environmental risk to Istanbul owing to increased tanker traffic through the straits of Bosphorus and the Dardanelles. With this in mind, how should the Caspian Sea legal regime be resolved? For example, a legal regime based on the so-called donut principle of common resource use opens the opportunity for joint development while a division of the sea will reduce such possibility. From an environmental perspective, the first option is certainly more desirable, as it increases the chance for a region-wide management of resource extraction and pollution control.

Russia's role in the region should not be underestimated. It is a littoral state controlling a significant stretch of the Caspian shore. For example, it is difficult to envisage how the Caspian Sea legal regime can be resolved without Russia's cooperation, not to mention meeting its challenges of resource management and environmental protection. Further, Russia cannot be ignored as an outlet for some oil and gas from the region, and as a partner in intra-regional oil and gas swaps, transport, and trade. Sidelining Russia in regional developmental programs could also unleash its considerable "spoiling" potential, illustrated by its control of pipeline access for Central Asian oil and gas. It will also strengthen the hand of conservative and retrogressive elements within Russia's as yet fragile new power structure.

The Primitive State of Caspian Studies

There is a dearth of empirical study of the Caspian region, a factor that underlies the significant misunderstanding of the area's resources, problems, and needs. For example, despite severe stress on the Caspian's natural environment, relevant information in the field is scarce. As a consequence, aspects of the environmental regime are poorly understood. In recent years, a significant rise in sea level has caused loss of life, damage to infrastructure, and an increase in pollution. What causes the rise in Caspian's water level? A generally acceptable answer is yet to be formulated.

More empirical studies are required for theory building and conceptualization if a deeper, more analytical understanding of the region as a whole and in terms of its various building blocks is to be achieved. The initiative of Iran and Russia to establish a center for Caspian studies in Moscow in 1992 was aimed at addressing the need for environmental data collection and analysis. Unfortunately, inadequate resources and cooperation have inhibited this venture. Furthermore, disagreement over the legal regime of the Caspian Sea represents a major impediment to regional studies. For example, current debate on the issue focuses largely on past legal arrangements and understandings of use rights. What explicit implications the legal regime of the Caspian Sea would have for the environment, for example, remains copiously devoid of study because certain states do not wish to entertain the idea.

There is no doubt that the Caspian basin contains significant hydrocarbon resources. However, estimates of the region's oil and gas reserves vary, owing to a lack of reliable information. In particular, individual governments, desperate to attract foreign investment and to manage their own fragility in the face of high expectations and socio-economic hardship, tend to present a rosy picture of the extent of resources. Analysts and consultants, hoping to secure contracts, fuel this optimistic and romantic image of the area's riches. Western companies stand to benefit from exaggerating the extent of the region's reserves, as they seek favorable trading conditions with oil producers in and outside the region. The U.S. Government has also tended to ignore the problem because such exaggerations serve to justify its growing involvement in the region. Yet, in order to secure the socio-economic and political stability of the Caspian frontier, realistic rather than fictitious estimates of its resources are required.

More significantly, we are reminded that judgement is still out as to whether the oil and gas reserves of the region, and particularly the export of such resources, will make a significant contribution to the welfare of the Caucasian and Central Asian states and their inhabitants. As compared to the Persian Gulf, the oil and gas export industry in these countries face severe inhibiting factors. For example, gas from the Caspian region may not be as easy to be placed on world markets as many would like to believe. No doubt, more thorough studies are required about how best to utilize the Caspian's hydrocarbon resources for sustainable long-term socio-economic development.

At a more general level, the lack of empirical studies is also illustrated in external actors' perceptions and views of the Caspian Region as a whole. It is unfortunate that the Caspian Region, home to one of the world's oldest civilizations, and an ethnically diverse community, continues to be seen in the West in terms of certain stereotypes, namely hydrocarbon wealth, geopolitical importance, internal conflicts, and Islamic fundamentalism. The concomitant diplomacy and policies of the Western players toward the area are Anglo-centric and chauvinistic, ignoring the new reality of a world where different worldviews and cultural forces are being reasserted.

The need to change these outdated stereotypes forms the basis for future positive engagement in the region by the outside players. For example, it should be recognized that Islam and Islamic fundamentalism are not of similar significance as cultural and political forces throughout the Caspian region. Besides, as Ali Mazrui remarks: "against Western claims that Islamic 'fundamentalism' feeds terrorism, one powerful paradox of the twentieth century is often overlooked. While Islam may generate more political violence than Western culture, Western culture generates more street crime than Islam. . . . Western liberal democracy has enabled societies to enjoy openness, government accountability, popular participation, and high economic productivity, but Western pluralism has also been a breeding ground for racism, fascism, exploitation, and genocide. If history is to end in arrival at the ultimate political order, it will require more than the West's message on how to maximize the best in human nature."[3]

Students in Western elementary and high schools learn very little about the geography, cultures, or history of the Caspian area. At the same time, centers of Caspian studies at U.S. universities largely function as isolated enclaves, divorced from the internal dynamism and policy processes of the Caspian region. No wonder that this educational system should mold the current misunderstanding of the area, particularly in the United States.

The Dominance of Fictional Prospects

The dismal state of Caspian studies has produced more fiction than fact about the region, negatively effecting a more realistic assessment of regional problems and prospects. For example, current and projected medium-term production of hydrocarbon resources in the Caucasus and Central Asia is relatively meager. Yet, governments keen on attracting foreign investment and assisted by analysts in search of short-term opportunities often inflate estimates of resources. This could lead to an unrealistic estimation of the extent to which oil exports can boost development, guarantee foreign assistance, and improve the stability of governments for managing national affairs.

A more balanced view of resources and constraints to extraction would facilitate a deeper understanding of the interdependence of countries, as well as the interconnectedness of different facets of development. It would also temper the regional hegemonic aspirations of some countries. For example, inflation of Caspian resources and the consequent increase in U.S. activities there gives an outward expression of increasingly deeper American involvement in the region. Yet, it is not for certain that the United States, given the real size of regional resources, the likelihood of reaction from Moscow, and the minimal investments by the West thus far in the area, will engage in the long-term "hegemonic management" of the area.

Fictional notions also underlie popular geopolitical and developmental strategies for the region, particularly those originating in the West. The so-called New Silk Road proposal aimed at a western-oriented infrastructure and pipeline corridor stretching from Central Asia to Turkey to markets in the West is a case in point. It fails to consider the significance of Russian and Iranian political interests in the region, as well as their capacity to assist in its development. Further, it is oblivious to the underdevelopment of the Caucasus and Central Asia as a market for Western goods and to the structural dependence of these countries on its northern and southern neighbors. Ignored are also the serious geographic and political constraints that could impede the viability of a Western-oriented development corridor and the limited extent of the European market as an outlet for Caspian hydrocarbon resources. Worse still, it probably underestimates the tenacity of Caspian countries to resist any form of new hegemonic control and integration.

Simplistic Notions of a Complex Reality

Simplistic notions of a complex and interconnected phenomena such as the Caspian region abound. Views of both external and internal actors suffer from this problem. One reason why the complexity is overlooked is that commercial interests in the region preceded academic research. This relates to the rapid opening up of the region, expectations of its resource riches, and the work of consultants relating to the exploration of these resources. Generally, the area is viewed as a good piece of real estate—up for grabs by the toughest and highest bidder. Very little attention is focused on the people and environment of the place, and its long-term socio-economic and political stability. This mindset has led to the neglect by the littoral states to properly define their national interests. Leaders have been concerned primarily with securing the best deal in relation to the selling of hydrocarbon resources.

As an example, there is an almost reckless inattention to the legal regime of the Caspian Sea, while at the same time, international contract after contract is secured for the exploration and extraction of its oil and gas. The issue of delimitation of the Caspian Sea persists. It impedes implementation of development projects. The debate on the Caspian Sea legal regime is largely restricted to legal issues, instead of including important environmental and development imperatives. Legal perspectives can serve as a resource for determining an appropriate legal regime, but they cannot serve as the point of departure for regional development. The political and environmental context has changed significantly since the standing legal regime was introduced. Not only has the number of littoral states increased, but so has the extent of pressure on the Caspian Sea as a natural resource, both as a container of pollutants and a source of development and livelihood.

Simplistic notions of inherently complex phenomena also prevail in relation to perceptions of the internal dynamics of countries. For example, studies of post-revolutionary Iran have focused on the Islamic government, largely ignoring the non-state sector. As a consequence, the understanding of Iran has been largely reduced to one of the regime in power.[4] What is ignored is, for example, that with the republican revolution, Iranians entered the political scene en masse as participants. It was they who overthrew the Shah; it was they who fought Saddam Hussein while his war machine was backed by the West and the wealthy Persian Gulf monarchies; and it was they who forced the Islamic system to retreat from many restrictive cultural and social policies. If there is anything that the Iranian clergy are concerned with, it is not the United States or Israel, but the latent power of their own people.

Expediency versus Historical Perspective

Policy and action in the region show a dearth of historically based reflection. Life and governance in the Caspian region has been intertwined with the exploitation of oil for centuries. A richly textured history exists for individual places and for the region as a whole. This represents an important source of experience and knowledge, often forgotten in the rapidly changing development context of today, where decisions have to be made fast, and time spent on reflection is not viewed as an asset. For newly independent states, history is often associated with times best forgotten, as opposed to being viewed as a source of guidance and insight for the future. There is an ill-informed belief that the events of the past will not be repeated.

Perhaps, now more than ever, there is a need for reflection, and a view of history as a means to make sense of today's increasingly complex world. The scale and nature of the Caspian problematic is vastly different to what it once was. There are many more actors involved, more people with pressing needs to be met, and governments with less power to act unilaterally and no longer able to rely on force to achieve their aims. They are also more dependent on external sources of finance. Although the present situation is substantially more complex than before, we have the benefit of history to teach us precedent, to serve as both a guiding light and a warning.

Despite the curtailment of state powers today, many facets of the role played by the state in earlier times are important. First, the state's view of a valuable natural resource as a means to increase government revenue rather than one that forms the basis for more sustainable development warrants careful consideration. Second, the state's intense internal delays in decision making impeded the potential constructive development of the oil industry. Recent developments to consolidate the various ministries engaged in energy matters in the newly independent states should be welcomed in this regard. Also important is the kind of support the state provides in resources over which it has substantial control, from infrastructure to education, to allow for a more beneficial exploitation of the resource.

Finally, the state has a role to play in protecting the interests of smaller industries so that opportunities associated with hydrocarbon resources can be spread more widely. The state should not favor foreign capital to the detriment of local development. Instead, it has a responsibility to establish rules or conditions for foreign investment, which further support rather than inhibit local opportunities. Less dependence on foreign capital and a few multinationals would reduce dependency on the whims and woes of the international market. We are reminded of the harmful consequences of over-dependence by the incidents of the Baku oil crisis in the early twentieth century, and the control of transportation routes by first the Noble and later the Rothschild companies.

History offers important lessons for foreign companies. Even today, the tides can turn unexpectedly. Companies should not contribute to the vulnerability of governments through exploitative practices. When the tide turns, as it did in the Baku crisis, they cannot necessarily rely on the support of the state, and may lose everything. The institutional impasse, which was previously a result of a lack of coordination among government departments, could now conceivably occur between nation-states. The impasse leads to problems that currently relate to fundamental developmental and environmental issues.

History also suggests the wisdom of stronger states bearing responsibility, when possible, to secure against the perils faced by weaker states in the region. States could gain from respecting and assisting each other in developing their comparative advantages. In a region where the advantages of interdependence are prominent, focusing primarily on competitive advantage is inappropriate. The approach should be to gauge the competitive advantage of the region as a whole, and within that, to address the comparative advantage of its parts. The interdependent destinies of the Caspian region and the oil industry means that both will suffer if problems are addressed unilaterally. Regional cooperation is necessary not only for environmental management, but also for the installation of pipelines, among other development projects.

There are, however, contradictions in the extent to which history determines present-day action. The past may be forgotten in adherence to an export-driven development view, but it has been revived in shaping the foreign policy and the worldviews of key actors in the region. U.S. sanctions against Iran is a case in point, based on historic events and premises that may no longer be relevant. Current pipeline preferences have tended to take current U.S.-Iran relations as a given, not recognizing that much can change within the period of construction of these pipelines. Perhaps more time should be spent defining the preconditions for viable pipeline routes rather than speculating about routes within an uncertain context. The future development of the Caspian region is too grand a topic to be viewed through the lens of the current U.S. administration, the leaders of the Islamic Republic, or of present U.S.-Iran relations. Long after the current political players have left the scene, there still will be two great regions that must one day mend their fences and make the world a better place for posterity.

The Need for Comprehensive Development

Development perspectives for the area are partial rather than comprehensive. The governments of the newly independent states generally define national development in terms of the energy sector, and particularly one aspect of this sector: the export of hydrocarbon resources. People, the environment, and exactly how hydrocarbon wealth will relate to development, appear to feature relatively low on their agendas. The personal involvement by leaders of littoral states in energy matters clearly illustrates this point. In Azerbaijan, Turkmenistan, and Kazakhstan, leaders personally negotiate, ratify and sign contacts with foreign oil consortia.

Nowhere is the partial approach to development in the Caspian region better illustrated than in the response to the environment. The potential for environmental disaster is often forgotten, as consultants, analysts, leaders, and multinationals focus on the hydrocarbon resources and related commercial opportunity. This occurs despite the environmental devastation of the nearby Aral Sea, representing a stark reminder of the consequences of such inattention. Worse yet, environmental issues are allowed to become a part of political disputes or perspectives. A conference on the Caspian environment in March 1999 in Vienna, sponsored by NATO, did not permit Iran's participation.

The Caspian Sea is an apt metaphor for the development dynamics of the region. The physiographic and ecological characteristics of the Caspian as a natural system are clearly at odds with present political boundaries and institutions. Its proper workings and protection demand a management response different from the individual responsibilities of littoral states. Yet, no legal and institutional structures that reflect the essence of the natural system exist. Attempts to create such structures have not been successful, lacking funding, cooperation, and reliable information. At the same time, significant funding is poured into hydrocarbon exploration.

International institutions are also ill equipped to deal with the Caspian's environmental problems. This is partly due to their global focus and procedural characteristics. They are also handicapped in helping the region financially. Moreover, none has as yet developed a region-wide cooperative scheme for environmental protection. Funding and procedural difficulties have already neutralized one major initiative: the Caspian Sea Environmental Initiative, established in 1993 by the United Nations Development Program, the United Nations Environmental Program, and the World Bank. The Initiative was to arrest the deteriorating state of the Caspian Sea environment, but experienced funding problems because not all littoral states were members of the Global Environmental Facility, the funding agency of the United Nations Environmental Programs.

Rampant poverty in the region, augmented by a lack of appropriate laws and capacity for enforcement, make the unsustainable exploitation of the Caspian's natural resources a very real threat to a unique ecology. The rise of the Caspian Sea water level contributes not only to economic loss by damaging agricultural land, infrastructure, and buildings, but also increases pollution as some of the most polluted lands of the former Soviet Union are in the immediate vicinity. Thus far the causes of the high pollution levels of the Caspian waters have been industrial wastes and untreated sewerage poured into the sea. The extensive development of hydrocarbon resources promises greater problems, particularly as production moves increasingly offshore.

The economic impact of environmental degradation should not be underestimated, particularly because the Caspian shores offer significant non-oil economic potential. For example, although oil exports constitute Iran's major source of foreign exchange, agriculture leads all sectors with a 25 percent contribution to GDP, followed by oil with 15.5 percent. Iran's Caspian region, occupying only 4 percent of the country's land area, is responsible for 40 percent of the country's agricultural activity. As the government is emphasizing self-sufficiency in the food sector, the region's role in agriculture could increase. Throughout the region, in the Iranian and Russian sides, in particular, more people make their living by fishing and caviar production than by any other means. Moreover, the Caspian shores are a major attraction for tourists from all over the region, an industry that has grown significantly since the collapse of the Soviet Union.

Given the large and rapidly growing population of the Caspian region, it can ill afford to neglect any one of its resources offering potential for growth and development. Thus, it is not clear whether the hydrocarbon riches of the Caspian, and particularly an emphasis on its export, is a blessing or a curse to the region's newly independent states. It can also be argued that within the energy sector, there is an overriding emphasis on oil. This perspective underplays the significance of gas as an important fuel in the twenty-first century and is oblivious to the fact that oil fields require massive investments to keep them going. For example, Iran, which holds the second largest gas reserves in the world, is an importer of gas. To date, despite the availability of the required technology, it has not significantly explored the conversion of gas to liquid fuel, a strategy that overcomes the problems endemic to the gas industry such as the need for fixed outlets and long-term contracts.

Global Interests and Local Needs

Global interests in relation to geo-political strategy and the security of hydrocarbon resources, rather than local developmental needs, appear to drive engagement in the Caspian region. The conflicts between these forces represent a major impediment to future regional prosperity. At this stage, American corporations are by far the lead players in the business development debate of the Caspian region. How to get oil out of the region, in the quickest way possible, is their key interest. Governments will have to devise frameworks to engage these actors while also providing for the fulfillment of local needs.

The current emphasis on the extraction and export of Caspian oil and gas may not be in the long-term economic and social interest of the littoral states. In terms of proven reserves of hydrocarbon resources, the Caucasus and Central Asia are not as significant as the Persian Gulf or even Iran alone. Further, the Gulf benefits from the lowest marginal cost of production in the world for pumping crude, as well as low transit costs since the required pipelines to export terminals are short. Geographic and political factors impede the competitiveness of the Caspian oil and gas industry.

The required length of pipelines and demanding physical terrain will result in high capital and maintenance costs. Further, transit fees have to be paid to countries traversed by pipelines as well as management fees at maritime terminals. Added to these costs are the royalties and management costs to be paid to foreign companies responsible for financing and developing the oilfields and pipelines. All of these factors will reduce the share of the Caspian Sea countries to one third of the actual sale price of their crude oil. Under these circumstances, a lower crude price can have detrimental effects on the prospects of oil and gas development in the region.

Since hydrocarbon resources are not renewable, their export could compel governments to diversify rapidly and engage in not-so-viable industrialization programs. An influx of foreign exchange from oil exports can lead to over-valued local currencies, and a preference for the importation of goods and labor, thereby inhibiting the development of local industries and employment—a problem already faced by the Persian Gulf oil-exporting states. Thus, alternatives should be sought to development based on the export of oil and gas. However, given that the littoral states will depend on their oil and gas reserves for some time, any alternative must utilize and benefit from these natural resources.

One such alternative is development of energy-intensive industries such as the production of petrochemicals, aluminum, and direct reduction steel, which could be easily transported via a variety of routes to Europe and the Far East. Although these industries also demand considerable investment, it can be argued that returns on such investment could be three to four times higher than revenues from the export of oil and gas. Such a development focus has the added advantage of increasing the range of sources for foreign funding, thereby creating a more favorable and competitive investment environment. Substantial value could be added to a non-renewable resource, more jobs could be created, and economic independence can increase as countries become less reliant on the goodwill of adjoining states for the transit of resources. In the long term, more diverse economies would facilitate social and political stability as well as the containment of environmental degradation.

Zero-Sum and Win-Win Strategies

Zero-sum, as opposed to win-win, strategies characterize engagement by different actors in the region. The "great game" metaphor may be too simplistic a notion to describe the dynamics of the Caspian region. Yet elements of it remain in the policies of the United States and Russia, popularly viewed as the main players in the region. To a lesser extent perhaps, the involvement of multi-national oil, and non-oil, companies also exhibit elements of a game. The difference is, while the former actors are largely interested in strategic positioning of their interests, the latter are focused on economic matters.

The United States, in wanting to integrate Russia into Western oriented market and security arrangements, clearly illustrates a hegemonic tendency. A similar tendency motivates the U.S. policy to contain and exclude Iranian interests in the Caspian region. Similarly, Russia's "near abroad" policy toward the Caucasus and Central Asia is an assertion of its past hegemonic control. In contrast, Iran has followed a far more pragmatic strategy toward its new neighbors to the north, focusing on economic issues and underplaying ideological differences. The next major player, Turkey, has yet to strike a balance between its pragmatic interests in the region and Turkish nationalism.

Within the littoral states themselves, it is important to remember that the composition of the population is extremely varied. In only a few countries are members of the "titular" nationality—the group for which a particular territory was "designed"—a significant majority. Therefore, a major challenge of the region is to devise systems of inclusionary governance, which provide for meaningful participation of minorities. It is especially at this national level that the countries in the region need to devise win-win strategies and then coordinate them with similar approaches to international relations.

Rhetoric versus Reality

Rhetoric abounds in relations between actors engaged in the Caspian region. U.S.-Iran relations since the 1979 revolution illustrate this point. The Islamic regime has viewed the United States as "the Great Satan," the origin of all evil intentions. The United States has been equally guilty of inflammatory rhetoric that describes Iran as a "rogue" or "outlaw " state, and "the world's leading sponsor of terrorism." Rhetoric is to a large extent a relic of the region's past, its geo-political dynamics and power relations. It is also a product of the determinants of political and economic influences and control predating the collapse of the Soviet Union. But the post-Soviet era has witnessed a new form of rhetoric, one designed to demonize and incriminate the adverse party.

Today, rhetoric is sustained because players have not internalized the limits of what they can practically achieve individually in the current world order, as well as the extent to which development of the region is dependent on regional cooperation. For example, regional realities have made it difficult for U.S. policy and rhetoric to effectively influence their targets. In early 1998, the mismatch between rhetoric and practical action was illustrated by the United States conceding to limited swap arrangements for oil and gas between Iran and Central Asia. The gap was again demonstrated when the United States relaxed ILSA (secondary, extraterritorial sanctions against foreign firms) in relation to Iran's oil investment deal with Total, Gasprom, and Petronas companies. The failed "dual containment" policy was another instance of mismatch between rhetoric and reality.

Capitalizing on the potential of the Caspian for energy utilization and development in the twenty-first century will largely depend on regional cooperation. In this environment, hostile rhetoric confuses issues and destroys the atmosphere of mutual respect and understanding required for achieving cooperation. At the same time, it can represent a positive force. The deep-seated origins of rhetorical issues should not be underestimated. Rhetoric can serve to prevent societal change and changing relations from erupting in the face of those who promote change. All the players in the Caspian Region have to contend with conservative and retrogressive forces within their midst, forces that are sensitive to runaway change.

Sources of Regional Instability

Different actors with different priorities are playing different roles in the region. The emphasis on the export of hydrocarbon resources could be attributed in part to the central role that multinational oil companies, analysts, financiers, and consultants play in the region. Newly independent states have spared no effort to create favorable investment environments for these actors, though what constitutes such an environment over the long term is not always fully understood. This is not to say that these actors should not play a prominent role. Rather, there should be a clear understanding of their specific role as opposed to those of others. As long as oil companies' interests dominate in the region, more sustainable development scenarios will not be actively pursued.

This lopsided emphasis on a single sector could be attributed also to the legacy of centralized planning and lack of political development in these countries. Moreover, as the collapse of the Soviet Union was not inevitable, Central Asia and the Caucasus were caught off guard; they were clearly not prepared to deal with unexpected independence and the challenges of nation-building. Thus, the realities of structural interdependence between Russia and its "near abroad" cut through the economic and political domains, as well as through the areas of security and socialization. In short, the legacy of the Soviet era pervasively survives and should not be underestimated.

The policies of major players in the region, notably Russia and the United States, do not always facilitate a more rational understanding and development of the roles and involvement of different actors. For example, Moscow views its "near abroad" in Central Asia and the Caucasus as a primary area of historical influence and interest, demanding a foreign policy distinct from that of Russia's interstate relations with the rest of the world. It is an area where Russians want to be the sole guarantor of security and a key arbiter of major geopolitical shifts. Importantly, this view is broadly shared among various political groups in Russia, and the Russian government has in various occasions illustrated its ability and commitment to protecting its interests in the region.

The declared interest of the United States in the area similarly impedes a broader understanding and respect for the roles of different actors. Its East-West orientation, hegemonic tendency, and preoccupation with the "Islamic threat" exclude key players in the region. The shared interests of the Russians and Americans in the area, built around nuclear non-proliferation and containment of Islam, is likely to come under increasing pressure from other major regional or extra-regional players as competition for resources grows. Clearly, a return to the "cold war" mentality could further impede the engagement of all actors in the development process of the region.

The Caspian region is signified by a catalog of sources of tension and security issues. Existing conflicts include the Karabagh War between Armenia and Azerbaijan, the Chechen conflict involving Chechnya and Russia, the Turko-Kurdish hostility, and the war in Afghanistan. Potential sources of conflict also exist between Azerbaijan and Iran, as well as in Lezgin and Dagestan. Current conflicts, some longstanding and largely unresolved, have inhibited implementation of development proposals, drained resources, and exacerbated poor relations among the regional states and nationalities. The devastating impact of the Karabagh conflict on the economies of Armenia and Azerbaijan is illustrative of the kind of negative consequences that such conflicts have on regional development.

Moreover, low-intensity warfare by various ethnic and sectarian militias represents a threat to the long-term stability of the Caspian region. Although these are likely to originate more from intra-state instability than inter-state rivalries, the potential for conflicts to spill over political boundaries and draw in external actors cannot be ruled out. Most of these conflicts will not simply disappear as deep-rooted grievances and legitimate demands or claims fuel them. It is thus critical that they are recognized and drawn into nation-building and regional cooperation and development schemes.

The Need for Regional Cooperation

Regional cooperation on a range of issues is crucial to overcome the challenges and unlock the development potential of the Caspian. Yet most actors and countries emphasize self-interest over communal interests, seriously impeding regional cooperation and development prospects. The states in the Caucasus and Central Asia are landlocked, requiring cooperation with their neighbors for pipeline outlets for their hydrocarbon exports. The significance of pipelines in Caspian oil production needs no emphasis, as it is now a global issue. Caspian oil production can reach five million barrels per day (mbls/d) by the year 2010. Domestic consumption could surpass one mbls/d by the same year. However, existing pipeline capacity is some 400,000 bls/d, only sufficient to carry the so-called early oil.

Self-interest and the battle for power and political influence are most evident in the rivalry for the routes of these pipelines required for transporting "big oil" from the region. Pipelines provide investment, jobs, royalties, long-term access to energy sources and, critically, considerable political and economic leverage. Russia hopes to maintain its traditional control and influence by advocating oil and gas export through its own territory. On the other hand, the newly independent states are eager to shed their dependence on Russia and consider other options through Georgia, Turkey, and Iran. The international oil companies are also eager to find the most economical export outlets to ensure profitability of their ventures.

Yet, owing to U.S. sanctions against Iran, the most attractive route options through that country are not fully explored, thereby depriving the regional states of maximizing their oil revenues. Iran's option is particularly attractive in an environment of declining oil prices. The exclusion of Iran has made other alternatives so untenable that the United States has been forced to propose and defend pipeline routes across the Caspian Sea, an alternative that is environmentally hazardous given that the sea is earthquake prone.

The large size of some Caspian countries and the wide distribution of hydrocarbon resources, refineries, and population demand regional cooperation as a prerequisite for development. For example, owing to the distance between Kazakhstan's Tengiz oil field in the west and its main consumption centers in the north and southeastern parts of the country, crude has been exported to and via Russia, while Russian crude is imported to refineries in the east. In addition, under Soviet rule the function of the Caucasus and Central Asia was seen only in relation to that of

Moscow: each country was assigned a role in relation to serving the overall Russian economy. As a result, the economies, infrastructure, services, and facilities of the new republics are in a state of imbalance. Cooperation between states is simply the fastest and cheapest way to overcome the consequences of this imbalance.

Given the enormous cost of new pipelines, swap arrangements with Iran is a viable and economically practical alternative for exporting some oil from the Caucasus and Central Asia. The pipeline network and port infrastructure on the Caspian coast are largely in place or can be provided at reasonable cost and time, to carry both "early oil" and some later production. Iran's existing refineries at Tabriz, Teheran, Arak, and Isfahan could receive oil from the north rather than from the Persian Gulf. Other advantages of this arrangement would be the bypassing of politically unstable areas, avoiding environmental and navigational concerns related to the Caspian Sea and the Bosphorus, and promoting better exposure to the dynamic and growing Eastern markets.

The newly independent states lack money for investment in infrastructure required for development. However, much could be achieved through regional cooperation. One example of what could be done is Iran-Turkmen cooperation on gas exports, where Iran is financing pipeline infrastructure while Turkmenistan pays back its share in-kind with gas. Further examples of expanded regional cooperation include Iran's ambitious project to construct a canal linking the Caspian Sea with the Persian Gulf. Obviously, constructing the canal will require international cooperation, but if constructed, it would significantly reduce the transport cost of oil and other commodities from the Caspian Sea to Southeast Asia. It would also allow the possibility of Iran making available one of its Persian Gulf ports to the countries of the Caucasus and Central Asia.

The need for regional cooperation is not restricted solely to the transport of hydrocarbon resources. Equally significant is the cooperation over the unresolved legal regime of the Caspian Sea, an issue that threatens regional development and environmental management. In the absence of a definite legal regime, information about the environmental status of the Caspian, the nature of threats to the environment, and a joint management response remain fragmented. This includes the threat of rising sea levels, pollution, and exploitation of marine resources—notably sturgeon and seal. The ambiguity regarding the legal status of the Caspian has not deterred international investment in exploration and exploitation of oil and gas reserves. However, one must warn against ignoring the differences in character between contracts with predominantly political rather than practical economic purposes.

Different players are establishing a presence in the region rather than pumping large sums of money into the area. The extent of investment required is huge, given the aging state of existing facilities and infrastructure. A stable legal regime agreed to by all countries is necessary for such investment to occur. Claims by Turkmenistan on some Azeri oil fields in the sea illustrate the uncertainty and risks associated with the Caspian investment environment. The Caspian Sea is also very important for intra-regional transport. Without regional cooperation, a comprehensive transport strategy cannot be developed. This, in turn, demands a definite legal regime for the sea. The median-line delineation proposal for the Caspian is exclusively driven by commercial self-interest, one that will certainly inhibit regional cooperation.

Regional cooperation is further required to maintain security in the region. Current conflicts, although dispersed, have had a destabilizing influence on the region. Events in Chechnya, Karabagh, Turkey, and Afghanistan have impeded finalization of pipeline proposals and have delayed investment. These and other conflicts can also potentially spill over borders, thereby drawing in external actors and leading to a more complex development environment. More players will inevitably impede or complicate regional cooperation. Grouping selected regional countries into defense or military alliances is counterproductive. One such example is the pact being formed between the United States, Turkey, Azerbaijan, and Israel.

Entrenched Old Leadership Practices

Independence has not necessarily brought new and more appropriate practices of governance and commerce to the Caucasus and Central Asia. Heavy-handedness, autocracy, police power, nepotism, and corruption have increased with renewed interest in the area, albeit to different degrees in the different states. Almost all of the newly independent states are governed by former Soviet politicians and "strong men" socialized in a context of obtaining and maintaining centralized power. These former chief custodians of communism became, overnight, its primary opponents, adopting a new schema of belief and practice. In reality, they were people left without belief, deserted by their most treasured ideals. At the same time, their ambition and ego, and centralist and despotic culture, found a match in the drive of foreign companies, governments, and opportunists for the region's resources and markets. The result is a dangerous mix of reinforced egos, narrow-mindedness, short-term deal making, and favoritism.

Autocratic leadership is also prevalent elsewhere in the larger Middle East and Persian Gulf. It has not necessarily hindered social and economic development. There are, however, major differences between the oil-rich dictatorships of the Persian Gulf and those of the Caucasus and Central Asia. As noted above, even the most optimistic estimates of the newly independent states' hydrocarbon resources do not remotely compare with those of the Persian Gulf. Further, production and transit costs associated with Caspian oil and gas is much higher than in the Persian Gulf. Consequently, the hydrocarbon wealth of the Caspian region is not sufficient to allow for significant pilfering by the elite and their intermediaries as well as to provide broad socio-economic benefits to the larger populace.

Overcoming the power of this handful of strong individuals and their allies—and building robust and broad-based political, economic, and social institutions—represents a major challenge and a prerequisite for sustainable long-term development of the region as a whole and of individual states. The Caspian region requires development of a strong civil society to restrict abuses by the state. But a healthy civil society needs an open society of politics and markets. Unfettered private and public participation and competition is a prerequisite for such a development.

Imagining the Future

Given the issues identified above, what does the future hold? Can the Caspian region overcome its challenges and become a strategic frontier for energy and development in the twenty-first century? Focusing on certain themes that I believe form the basis for a better understanding of the Caspian region, this section explores future prospects.

The inhabitants and environment of the Caspian region have not always featured prominently in the renewed interest in the area. It is unlikely that this state of affairs will persist. The needs of people, including their engagement in political decision making, as well as the condition of the environment, will receive increasing attention in the future.

In recent decades, nationalist agendas have played a prominent role in the politics of the Caspian region, illustrated by the Iranian revolution and numerous conflicts, including those in Chechnya and the Karabagh. In most states of the region, ethnic diversity is the rule, and the populations for which particular states were "designed" do not necessarily constitute substantial majorities. A careful balance of the interests of predominant groups with those of the various minorities represents a major challenge to forging national integration of states, and, ultimately, development and growth. At the same time, inadequate progress in broad-based development would inevitably lead to a resurgence of as yet latent nationalist identities throughout the region.

People hold real power for potential change in the region. Social transformation is critical for development in the region. Poverty increases the vulnerability of states and provides impetus to fundamentalist movements. Often forgotten is the fact that the Iranian revolution had as much to do with the socio-economic expectations of people and their struggle for access to resources as it had with ideological considerations. The same applies to the various conflicts currently plaguing the region.

The emphasis of newly independent states on oil and gas exports is a risky strategy for overcoming structural underdevelopment. It could inhibit the development of local industry and job creation. Countries should strive for more diverse and robust economies, recognizing the impediments to oil and gas export in the Caspian Region. They need to devise options for more sustainable long-term growth and development, including investment of oil revenues in productive capacities, manufacturing in particular.

There are prospects for political transformation. It is unlikely, however, that newly independent states will easily give up their newfound freedoms. They have tried, with varying degrees of success, to liberate themselves from structural dependence on Russia. This effort is reflected in their search for alternative pipeline routes, Western investment partners, establishment of national armed forces, and following independent foreign policies. There is also a general reluctance to engage in new partnerships that will result in new hegemonic relationships.

Increasingly, the stability of governments will depend on the extent to which they can accommodate the different interests within their domains. As indicated by the 1997 elections in Iran, the youth and women have an increasingly decisive impact on politics. An important factor in political and social development is the extent to which the new states can integrate ethnic differences in their emerging nations. Prolonged conflicts have drained resources and seriously impeded development.

Iran plays an important role in the political transformation of the region. It is moving toward increasing political maturity, and as the country's commercial linkages with its northern neighbors strengthen, it will have a positive impact on political freedoms in the neighboring states. Rather than exporting rogue "fundamentalism," Iran can assist in the export of constructive values and practices to neighboring countries. It certainly appears to be in a better position to do so than Turkey, where the military continues to remain the real power behind the fragile democracy.

The Caspian Sea environment must be protected. The sea represents a highly significant resource base for the region, offering hydrocarbon, sturgeon, and other valuable resources. It also supports an important and growing tourism industry. Yet, as a closed lake, it is particularly vulnerable. Since the collapse of the Soviet Union, economic hardship and concomitant over-exploitation has caused a serious threat to the Caspian's sturgeon resource. Rapid population increase in the region, as well as increased development, will increase pressure on the environment. Sound management arrangements are required, because resource depletion will result not only in loss of an important biome, but also in increased competition for resources; this could lead to inter-state tension and political instability. Without ecological sustainability there can be no economic sustainability in the region.

A number of positive initiatives to address the Caspian's environmental pressures and manage it as a resource have commenced in recent years. This includes the formation of the Organization for Regional Cooperation of Caspian States (ORCCS), the center for Caspian Studies, and the Iranian fisheries company's sturgeon management plan. A key challenge is to strengthen these initiatives. International agencies should support these local initiatives rather than forming new organizations. The littoral states have made a great deal of effort to consolidate different ministries related to the energy industry. The same initiative must also be made in relation to coordinating the work of ministries impacting on the environment.

The Caspian region has the potential to develop a strong regional economy. Geographically, it enjoys an envious position in relation to established and emerging world markets. It enjoys maritime access to major shipping routes and is blessed with significant hydrocarbon resources. It also has the benefit of other resources, ranging from the Caspian Sea's maritime riches to agricultural and mineral resources. Further, it has a large, young population that can form the basis of a strong internal market. Iran, in particular, possesses technical know how in a range of fields that can benefit the region as a whole.

Regional alliances are the key to meeting the challenges and capitalizing on the opportunities of the region. In terms of geography, capital, existing infrastructure, and technical resources, most countries are impeded in meeting their challenges, albeit to varying degrees. Most of the countries also suffer from a legacy of underdevelopment and structural dependence on Russia. The potential of the region is great, but so are its challenges, and overcoming them would involve investment beyond the means of any one country. Regional cooperation, planning, management, and foreign investment are essential to maximize human, financial, and physical resources.

Examples of regional cooperation abound, including the integration and expansion of rail networks and Caspian shipping, the extension of oil and gas pipelines, swap arrangements for oil and gas, and peace keeping in the contested Karabagh territory. A clearer definition of the Caspian economic region will bolster regional cooperation and the region's ability to manage hegemonic and monopolistic interests. It could also assist in refocusing the newly independent states toward internal development, rather than on emphasizing the export of hydrocarbon resources. Ironically, some of the delays in decisions on pipelines and the legal regime of the Caspian Sea have probably had the effect of enhancing the prospects for regional cooperation. Current intra-regional initiatives at infrastructure improvements and trade may eventually lead toward a regional energy plan.

The United States has had no qualms about its strategic interest in the Caspian region. Yet as Ian Bremmer has noted, its policies toward the

Caspian do not reflect the realities of the Caucasus and Central Asia. Rather they are reactions to what Iran, Turkey, and Russia might or might not do.[5] The United States' attempt at containing Iran and integrating Russia into the West have fallen short of expectations. It has also overestimated Turkey's ability to expand Western interest in the region. U.S. Caspian policies are in need of serious overhauling.

For example, United States foreign policy toward Iran is based on a perception that the Persian Gulf is better secured through its existing alliances, and that American interests are protected as long as external threats to its allies can be contained. At one stage, it was the Shah of Iran, later Saddam Hussein, who served the United States' agenda; today, it is Saudi Arabia. What threatens these states now is their own internal tension; more specifically, their people's attempts to change the existing socio-economic order and power structures. How these governments will face this challenge will determine the security of the region and their ability to fund increases in oil production.

The United States' interest will be better served by less of a "single pillar" approach. It needs to focus more on the region as a whole and work with all relevant actors. This approach implies an understanding that the resources of the region should not be exploited by concentrating on a few countries alone. A "single pillar" approach will promote counter-alliances: for example, a Russia-Iran alliance. It will also provide a stronger foothold for Asian interests, a rather important market for the Caspian Region. To date, this approach in itself has played a part in inhibiting resource development. For example, United States' support for the Azeri position on the Caspian legal regime has prevented resolution of this very important issue, making many prospective investors hold back final decisions. A more inclusive approach to regional development is a better alternative.

Similarly, Russia's "near abroad" policy will not be a viable basis for future engagement with the area. The newly independent states do not favor prolonged hegemonic interest in the region, whether by Russia, the United States, Iran, or Turkey. Russia's current approach may strengthen the position of either the United States or Iran. Russia, however, faces a difficult task in overcoming authoritarian traditions and a collectivist culture. Given the country's institutional breakdown, widespread corruption, as well as antiquated infrastructure and investment bases, the potential for imperialist elite to attempt regaining lost "world power" status and regional influence through military means remains significant.

Iran has immense strategic significance in the Caspian region. Firstly, connecting the Persian Gulf to the Caspian Sea, Iran is the geographic and cultural nerve center of the two most energy-rich regions in the world. Iran itself possesses the world's fourth largest reservoir of oil and the second largest of natural gas. In the wake of the Soviet collapse, Iran now connects the Middle East, the Caucasus, and Central Asia. It possesses over 400 miles of the Caspian shore, 750 miles of the Persian Gulf's shore, and overlooks the Strait of Hormuz, through which over 11 million barrels of oil pass every day. For these reasons alone, Iran is a critical player in development of the region.

The second element in Iran's strategic significance is political Islam, which will remain a formidable force in Turkey, Egypt, Algeria, and the occupied territories. Political Islam cannot be overturned; it can only be accommodated. Iran's integration into the economic and strategic formations of the region provides it with the opportunity to use its experience to contain and moderate the potentially disruptive effects of militancy. Russia's uncertain future is another reason for Iran's increased strategic significance. Iran is well equipped to assist the newly independent states seeking structural independence from Russia.

Previous descriptions of Iran as a "rogue" or "outlaw" state are no longer valid, although periodic Iranian rhetoric may support this view. Iran is a state increasingly in pursuit of its own national interest, as illustrated by its policies in relation to Chechnya and other regional conflicts. What is important in Iran's stance is that it shows the way for engagement in the future, particularly in drawing the distinction between common and private issues, in mediating between independence and interdependence, and stimulating the latter without giving up the former.

China is likely to become a significant actor in future development of the Caspian Region. The economic, as opposed to political, orientation of the Caspian states invites greater involvement by China. China would not be challenged by Caspian states on issues of democratization, human rights, and international arms control, since the region itself encompasses states of great political diversity and minimal regard for the ideological or moral agenda of the West. Closer ties make economic sense to China. Xinjang, the resource-rich northwestern province, is too far from the Chinese ports for optimum development and could be integrated instead with the Caspian rail system and export outlets. China is already showing significant interest in the oil and gas resources of Central Asia and the Caucasus. The Chinese National Oil Company has arrived at an agreement with Kazakhstan to build a pipeline and engage in oil exploration.

Turkey is another important player in the Caspian theater, even though it is not a Caspian state. Russia's reduced control over the newly independent states and the expansion of United States commercial interest in the Caspian has provided Turkey with an opportunity to project its influence in the region. Pipeline proposals to Ceyhan and Supsa is the manifestation of this interest. Over the long term, these proposals should show their political bases, rather than economic and practical viability. This western-oriented approach to pipeline building underestimates both the significance of the East as a market for Caspian oil and gas, as well as the geographic and political impediments to the Turkish-based pipeline proposals. Turkey cannot base its engagement in the area on its Western alliances, as these may change. In particular, a change in United States-Iranian relations would diminish Turkey's influence in the region.

The stability and development prospects of the Caucasian and Central Asian states will be largely determined by how they manage relations between external investors and the growing expectations of their residents. Foreign investment is a prerequisite for development, but the benefits of development will have to be distributed widely. Also important is accommodation of the legitimate fears and expectations of ethnic groups for political and socio-economic participation. The states need to diversify their interest and focus from hydrocarbon, and work hard for regional cooperation.

Conclusion

Pressure for development in the Caspian region will inevitably mount. Should the governments of the newly independent states wish to remain in power, they will be compelled to democratize their societies and cooperate with each other. These countries also realize that nations do not have permanent allies, they have permanent interests. The newly independent states will exploit the opportunity for United States' investment, and support it to the full. At the same time, however, they will not be blind to the influence and opportunities for development that can be gained by establishing good relations with Iran and Russia. It is therefore likely that the United States will in the future have to engage in the region more in relation to local conditions rather than in terms of its own agenda.

A major challenge for future regional alliances is to disassociate different nationalist sentiments and tensions from economic interests. Exploiting ethnic and religious differences for self-interest or allowing internal ethnic strife to spill over international boundaries can seriously inhibit fragile economic alliances. Equally unproductive is competing on ideological or nationalistic ground. A case in point is the post-Soviet relations between Iran and Turkey. Partly by their own design and partly because of the way outside forces projected their interests in the region, these states were made to represent a "reference society" for other states. This development has hardly been conducive to regional cooperation and development, as exemplified by the problems facing the Economic Cooperation Organization.

It is unlikely that the current impasse on the Caspian Sea legal regime will persist, simply because the subject is critical to regional development. In the case of Iran, some form of line to indicate an area of exclusive sovereignty would have had to be invented, had it not existed already as a legal necessity, enabling it to deal with issues such as navigation, resource exploitation, and immigration. This applies in principle to all countries. At the same time, what is important about the "donut" principle is that it implies regional cooperation in resource extraction. A commitment to joint resource development should facilitate holistic management of the Caspian Sea as a natural resource. The primary motivating factor for median-line delineation, on the other hand, is a "separatist" approach resource development. This delineation approach has far-reaching implications for the environment.

The Caspian region could once again become a strategic crossroads of the world. Its reopening as a frontier for energy and development is, however, associated with serious challenges, imposed by the varied interests of different players and their own internal transformations in a new world order. The potential for growth and development is rich, but so is the prospect for serious impediments of a political-economic and strategic nature. Past metaphors and understandings for engaging the area in world affairs will not realize its substantial potential. A fundamental prerequisite for development and stability is mutual respect by the different players for each other's inter-dependent interests and needs.

A new approach to Caspian studies is fundamental to establishing such respect and overcoming other developmental challenges in the region. This includes formulating a clear definition of the region, a clear understanding of what constitutes appropriate development in the area, and of the roles that different actors could and should play in the development process. Such an education should also overcome fictional and simplistic notions of the region's development potential and challenges. It must address the destructive pursuit of self-interest so evident in the region today. The new approach to Caspian study needs to underscore the heavy-handed leadership, corruption, and nepotism that prevails in the political sphere. It has to revisit the prevalence of hollow rhetoric, the focus on global as opposed to local needs, and the zero-sum attitudes toward engagement.

Finally, the new approach to the Caspian challenges should foster historic reflection, realizing that the Caspian region has been at historic crossroads before, and that decisions made at different stages have affected its development prospects for many years thereafter. An environment of mutual respect and understanding would facilitate the strengthening of regional institutions and is critical in stimulating regional cooperation and development. This broader regional framework would assist individual countries to follow their own policies in relation to social and political development, depending on their own history and developmental contexts.

Notes

1. Richard Bulliet, in an unpublished article entitled "NIRA: A World Region Emerges," excludes Russia as well.
2. Ibid.
3. See Ali A. Mazrui, "Islamic and Western Values," *Foreign Affairs,* September/October 1997, pp. 118-132.
4. See Hooshang Amirahmadi, "Adopt a Longer-Term Perspective on Iran," paper presented at *Iran in Transition: An Economic, Political and Energy Conference,* Institute for the Study of Earth and Man at Southern Methodist University (Dallas), May 1996.
5. See Ian Bremmer, "Oil Politics: America and the Riches of the Caspian Basin," *World Policy Journal,* Spring 1998, pp. 27-35.

Part I

Peoples and Resources

CHAPTER ONE

The Lands and Peoples of the Caspian Region

John Schoeberlein and Alisher Ilkhamov

Introduction

For most of the twentieth century, the picture of the Caspian region has been not as a single entity, but as a set of geographic fragments, considered most often in isolation from one another. Today, the development and transportation of the region's oil and gas resources, as well as environmental considerations, require that one view the region as a coherent whole. Seen historically as well, the region can only be understood if viewed as an integrated domain, because the changing configuration of "states" and "empires" as well as population flows and overland trade evolved around the Caspian Sea as a focus of activity and not as a barrier separating isolated domains.

The view of the Caspian region as an integrated whole should not obscure, however, the diversity and differences that exist in the region. The range of issues facing the region can be best understood also in a comparative perspective. To that end, this chapter will describe the region's geography and examine national identity of each of the newly independent states of Azerbaijan, Kazakhstan, and Turkmenistan, as well as the Russian Federation's autonomous republics of Daghestan and Kalmukia, which border the Caspian. The analysis will serve to explain in part the sources of current politics and tensions of the region.

This chapter treats neither Russia nor Iran in any appreciable degree. Suffice it to say that Russia borders the Caspian by its Astrakhan Province, which encompasses the mouth of the Volga. Iran's Caspian littoral is shared by the coasts of its northern provinces; from west to east, East Azerbaijan, Gilan, Mazandaran, and Golestan (previously known as Gorgan).

The Caspian Basin

The Caspian Sea is the world's largest inland body of water. The Caspian Basin constitutes a major subregion within the continent of Eurasia. On the very easternmost edge of what is conventionally considered "Europe," it forms part of the boundary between Europe and Asia. As all major bodies of water, the Caspian Sea serves both to link and to separate. It brings Russia and Iran in contact with one another, though their land is separated by hundreds of kilometers. It separates Azerbaijan from the other major former Soviet republics with predominantly Muslim populations. Yet its shores and the sea itself have for millennia constituted an important transportation corridor.

The Caucasus region is geographically divided between north and south by the Caucasus Mountains, extending from the near of the Black Sea toward the Caspian Sea. The Southern Caucasus, also sometimes Russo-centrically referred to as the Transcaucasus, is composed of today's independent states of Armenia, Georgia, and Azerbaijan. To the south lies Iran's Azerbaijan Provinces.

The Northern Caucusus, which has remained, with the important exception of Chechnya, firmly within Russia since the breakup of the Soviet Union, is made up primarily of a series of ethnically defined autonomous territories, including Daghestan, Chechnya, Ingushetia, Northern Ossetia, Kabardino-Balkaria, the Karachai-Cherkess Republic, and the Adygei Republic. Bordering the Caspian Sea, Daghestan and Kalmuk-Khalmg Tangch Republic (Kalmukia) constitute autonomous national republics within the Russian Federation. They have their own republican government and parliament, and they enjoy a greater degree of independence in relation to the Russian central government than do Russian provinces such as the neighboring Astrkhan Province on the Caspian.

The combined territory of the littoral entities bordering the Caspian Basin amount to more than 5 million square kilometers,[1] approximately equal to two-thirds of the continental United States, or one-half of the territory of China. At the heart of the basin is the Caspian Sea itself. It occupies a vast area of some 393,000 square kilometers. It stretches for 1,200 kilometers south to north and is 320 kilometers from east to west. The countries occupying the longest stretches of its shoreline are Russia and Kazakhstan on the northwest and northeast, while the southern littoral is occupied by Azerbaijan, Iran, and Turkmenistan.

The largest part of the Caspian's perimeter—from the northwest in Russia to the southeast in Turkmenistan—is bounded by open plains that are grassy steppes in the north and barren deserts in the south. From its southeastern shore on the Iran-Turkmenistan border to the Russia-Azerbaijan border on its western shore rise mountains of varying magnitude. A low range, the Kopet-Dagh in Turkmenistan, grades into the much more precipitous Elburz Mountains in north-central Iran, rising to 5,601 meters very near the Caspian shore. Lower ranges in Azerbaijan rise again to the Great Caucasian Range on the Russian border, where lies Daghestan.

Iran's portion, which is geographically oriented toward the Caspian Basin, is relatively limited. The Elburz Mountains, which run along the Caspian's southern shore, effectively separate the littoral zone from the rest of the country. Yet, in this narrow strip lies a corridor of habitation and activity. Some of the most important fishing and transportation ports on the Caspian occupy this zone, and though land is scarce, it is one of the most densely populated segments of the Caspian shore. Consequently, this region is in a position to be affected as much as any by the pollution of the Caspian and its rising water level. The best access to transportation routes from southwestern Turkmenistan and the major economic centers on Azerbaijan's Caspian coast is along this shoreline corridor, though the main transportation corridor across northern Iran passes to the south of the mountains and the newly completed rail link to Central Asia does not cross into Turkmenistan before reaching the northeasternmost point of Iran at Sarakhs.

The Caspian draws its water from three major rivers—together constituting 80 percent of the incoming water—the Terek and Ural Rivers, and above all the Volga, which drains a vast basin in Central Russia. The bottom of the Caspian varies from shallows in the north of four to six meters and gently sloping in the east, to much steeper slopes and greater depths in the west, reaching as deep as 1,025 meters. The climate also varies considerably from north to south. The north experiences a generally cool, continental climate with cold winters, while in the east, in the deserts of Turkmenistan and southwestern Kazakhstan, precipitation is very low and temperature variation is great in the course both of the day and the year. The south and southwest enjoy a much more moderate climate with relatively warm temperature all year round. The warm southern climate is responsible for the vast amount of evaporation from the sea that balances the inflow of such mighty rivers as the Volga.

Since the sea has no outlet, climatic changes have a significant impact on the water level. From the 1920s to the 1970s, the level was dropping due to the irrigation water being drawn especially from the Volga. From the late 1970s, however, the level began again to rise due to climatic change. The lack of an outlet also means that environmental threats to the entire drainage basin effect the sea and all the states that share its shoreline. One impact of the deteriorating environment with global implications is the severe decline of the Caspian sturgeon, which have provided one of the world's richest sources of caviar.

As a means of transportation, the Caspian is but one end of a navigation network that extends along much of the 3,500 kilometers of the

Volga, and, further, along a system of canals in Russia. There are major ports in each of the countries bordering the Caspian: Astrakhan and Makhachkala in Russia, Aqtau (formerly Shevchenko) in Kazakhstan, Turkmenbashi (formerly Krasnovodsk) in Turkmenistan, Bandar-e Anzali in Iran, and Baku in Azerbaijan.

The land can be generally characterized as belonging to either desert, steppe, or mountain zones. In much of the Caspian countries, agriculture is heavily dependent on irrigation. Only in very limited areas is there any forest cover. The high percentage of pasture land that predominate the lands of the Caspian Basin, particularly in Kazakhstan and Turkmenistan, represents semi-arid or arid zones where vegetation is quite sparse. Much of the land of Azerbaijan, Iran, and Daghestan is dominated by mountains, thus greatly limiting agricultural activity.

The People

The population of the Caspian Basin region is extremely varied in its composition, density of distribution, and other characteristics. The region lies in the historical contact zone between nomadic pastoral populations who inhabited large, sparsely populated territories of the steppe and desert, and settled agricultural populations who occupied those areas such as desert oases, river valleys, and shorelines that were more conducive to intensive habitation. In the past two to four centuries, various parts of the region have seen the expansion of Russian domination and with it an influx of European population and the displacement of pastoralism with agriculture and industry. Simultaneously, there has been a shift of nomadic populations to settled life and a great intensification of settlement in many parts. This has brought a substantial influx of Christian population into a region in which Islam long predominated. The shifts in the pattern of population intensified in the Soviet period, which was characterized by forced settlement of nomads, a vast expansion of lands under irrigation, and a systemic effort to build new settlements and to expand urbanism and industry, drawing new immigrants.

The total population of the territories surrounding the Caspian stands at about ninety-five million—Azerbaijan with 7.5 million, Kazakhstan with 16.7 million, Turkmenistan with 488,000, Russia's Astrakhan Province with 991,500, Daghestan with 1.8 million, Kalmukia with 322,000, and Iran with 66 million. However, the greatest concentrations of population in such countries as Kazakhstan and Iran are outside of the Caspian Basin per se. On the whole, the heart of the Caspian Basin is quite sparsely populated, except in such places as on the Volga Delta, Baku, and along the coastlines of Daghestan, Azerbaijan, and Iran.[2]

Throughout the territories surrounding the Caspian Sea, a very high percentage of the population is rural and is employed in various forms of agriculture, ranging from 32 percent in Astrakhan Province to 55 percent in Turkmenistan and 57 percent in Daghestan. For this reason, agricultural land is a crucial resource.[3] The breakup of the Soviet Union has important implications for the future of converting pasture land to crop land, a process that depends in many cases on the expansion of irrigation. Recent irrigation accounts for the existence of much of the small amount of arable land in Turkmenistan, for example. While in Turkmenistan there are large territories of rich soil that, if irrigated, could become highly productive crop land, such an expansion would require drawing more water from the Amu-Darya River. This would have international due to the severe water shortage problems down river in the Khorezm Oasis and the desiccation of the Aral Sea.

Extremely high population growth rates are being experienced by some of the countries in the Caspian Basin, while declining populations are a problem in other areas. The highest growth rates are found in Iran (2.21 percent), Turkmenistan (1.82), and Azerbaijan (.78). On the other hand, Kazakhstan and the Russian Federation as a whole have negative growth rates (-0.15 and -0.07, respectively).[4] In the first years of independence there was a massive out-migration of Russians and other non-Central Asians from Kazakhstan. Russians and other European nationalities felt themselves vulnerable in independent Kazakhstan, Turkmenistan, Azerbaijan, and other states of the region. The out-migration, meanwhile, has now considerably diminished and to some extent reversed itself as the migrants find limited opportunities for settlement and employment in Russia.

With respect to infant mortality, the highest rates occur in Turkmenistan (81.6 deaths per 1,000 live births) and Azerbaijan (74.5), compared with 63.2 in Kazakhstan, 24.7 in the Russian Federation, and 52.7 in Iran. The relatively low rates in Azerbaijan, Kazakhstan, and especially Russia reflect the more urbanized populations as well as what remains of a relatively well developed health services sector from Soviet times. In countries such as Kazakhstan and Russia, the conditions in the far-flung rural areas that characterize the Caspian Basin are considerably worse than for these countries as a whole. Iran's relatively good standing in this regard, despite its predominantly rural population, is undoubtedly connected with its oil wealth and the state's commitment to social services.

The highest life expectancy rates are in Iran (67.4 years overall, male and female average), with the lowest in Russia (63.2) and Turkmenistan (61.5). The general health of the population in all of the former Soviet states has also been greatly diminished by widespread ecological problems, often brought on by severe negligence by the Soviet state and its successors. By contrast, the environment is much healthier in Iran, partly due to its lower level of industrialization. The general health and well-being of the population is also much better, due to community support, state social services made possible by oil exports, and the absence of alcoholism and other social problems characteristic of a society in turmoil.

While over 50 percent of the population is currently of working age in Turkmenistan, Azerbaijan, and Kazakhstan, between one-third and two-fifths of the population is now below working age, representing a very large population that will be entering the workforce or the ranks of the unemployed in the next ten to fifteen years. Between 1991 and 1995, in Azerbaijan, Kazakhstan, and Turkmenistan, the proportion of the employed among those of employable age dropped between 5 percent and 10 percent, increasing the unemployed to nearly 30 percent in Azerbaijan and nearly as high in other Caspian Basin states.[5] As there is a preponderance of young people in the working age population as well, these will also be both in need of employment and producing their own offspring in the coming years. Given the limited potential for the agricultural sector to employ this population, it is of crucial importance that other sectors of the economy expand. To this end, the governments face a tremendous challenge, not only in promoting economic conditions that allow for expanded employment, but also in providing for the necessary educated workforce.

The combination of a well-educated and readily mobilized young population, together with rising expectations and severely limited opportunities, may provide the organizational force for opposition to governments that they see as failing to provide economic opportunities and meet social needs.

National Identity and Integration

The past decade or so has been a period of tremendous turmoil in the politics of nationality and religion in the Caspian Basin. There are three important circumstances that form the background to nationalist politics in the former Soviet states of the Caspian Basin. First, the existence of the Soviet Union, in which Russian culture and language, and persons of Russian nationality, held a dominant position, meant that nationalist politics would be framed to one degree or another in opposition to Russia. Second, within the Soviet context, a certain sphere was officially allocated to the nation, as reflected in the creation of national republics, which meant that nationalist ideas had a congenial administrative and social context in which to develop. Third, while certain groups were officially recognized and allocated territories, other groups were given limited recognition or were not recognized, and this set them against those who dominated over them in a given territory—e.g., the various non-Azerbaijani groups inhabiting Azerbaijan.

Furthermore, the boundaries between national territories did not and could not reflect the distribution of national groups on the ground, due to the intermingling of populations, and this meant that national groups were located in administrative units that neighbored "their own" national territory and in which they were a potentially disaffected minority. Because in some cases national autonomous entities were created for groups that were not populous and that became very much intermingled with Russians and other groups migrating in—e.g., in Kalmukia and Kazakhstan—the national group constituted a minority in its territory, which subjected them to especially strong assimilative forces in the face of Russian domination.

The general balance of nationalities in the region may be understood as consisting of a group called the "titular nationality," for which the territory is designated. Only in Azerbaijan, Turkmenistan, and Astrakhan do the titular nationalities constitute substantial majorities. Elsewhere, diversity is the rule. In Azerbaijan, the titular nationality constitutes 90 percent of the population, with Russians being 2.5 percent and other nationalities making up the remaining 7.5 percent. In Kazakhstan, the titular nationality is an absolute minority; 37 percent of the population consists of Russians, while the other non-Kazakh nationalities make up 21.1 percent of the population. In Russia's Astrakhan, 72 percent of the population is Russian, with the remainder being other nationalities. In Daghestan, 9.2 percent are Russian, while the remainder is other nationalites. In Kalmukis, 45.5 percent of the people are Kalmuk, 37.7 percent are Russian, and 16.9 percent are other nationalities. In Turkmenistan, 73.3 percent are Turkmen, 9.8 percent are Russian, and 18.5 percent are other nationalities. In Iran, the national group—albeit difficult to define—consists of 51 percent of the population, while the remainder is made up of a variety of ethnic groups including Azerbaijani Turks, Kurds, and Baluch.[6]

Most of the Russians inhabiting Kazakhstan live in the extensive northern border region with Russia, where they are an overwhelming majority; during Soviet times they scarcely thought they were not in Russia. Russian nationalists on both sides of the border have called for the "return" of these territories to Russia. In Kalmukia, too, Kalmuk nationalism is doubly vulnerable, since Kalmuks barely outnumber Russians and their republic is subordinate to the Russian state.

Daghestan represents a relatively unusual case in this regard, in that while Russians constitute merely 9.2 percent of the population, there is no predominant national group. Though the great majority of the population belongs to an encompassing category that might be termed "Daghestani," meaning people who inhabit or come from Daghestan, who share cultural and religious commonalty. In Daghestan, the problems of nationalism are connected with the vulnerability that Russians feel in this part of the Russian Federation, where some Daghestanis showed substantial sympathy with the separatist cause of the Chechens, and also in the relative position of dominance of the various Daghestani nationalities over others.

Azerbaijan

Azerbaijan consists of 40 percent plains while the remainder is occupied by the Caucasus Mountains and their foothills in the west. The arid subtropical climate of the west and central portions of the country is moderated in the east by the proximity of the Caspian Sea. The lowlands in the east on the Kura and Aras Rivers offer the country's best agricultural lands, and this area with its oil resources supported some of the earliest and most extensive industrial development in the Caspian Basin as a whole. Thus, geographically, Azerbaijan is very much oriented toward the Caspian Sea, with its greatest concentration of population, its transportation corridors, and its richest economic zones in the area nearest the sea.

Due to the heavy concentrations of population and industrial activity in this region, land is an extremely scarce resource. In addition, the east is particularly effected by serious environmental problems due to the following factors: (1) nearly a century of oil production and refining, (2) use of chemical fertilizers, pesticides (including DDT), and defoliants in cotton cultivation, and (3) the rising level of the Caspian, which threatens inundation, especially of the Apsheron Peninsula, where the country's largest cities, Baku and Sumgait, are located.

The Azerbaijani identity came into being on territory where Turkic and Iranian cultures have met and mixed over many centuries. A major component of the Azerbaijani population inhabiting this region prior to the eleventh century spoke Persian. Subsequently, here were successive waves of conquest by Turkic-speaking armies. These brought in new population components and culminated eventually in the domination of Turkic language among the population. Meanwhile, the influence of Persian culture and Persia remained strong, and between the sixteenth and eighteenth centuries Azerbaijan was incorporated into Safavid Iran. As a consequence, Shi'ite Islam predominates among the Azerbaijanis. In the early nineteenth century, expansion of the Russian Empire extended south of the Caucasus Mountains and by the third decade of the nineteenth century, the Persian Empire ceded to Russia the lands on the northern side of the Aras River, which came to form eventually the Russian Azerbaijan; the lands to the south of the frontier constituted the Persian Azerbaijan.

Azerbaijan as a national category is of relatively recent origin, arising out of a nascent nationalist movement beginning around the turn of the twentieth century, reinforced by ideas then current in the Turkish cultural sphere, in opposition to domination by Russia and Iran. Also current at that time were the Russian categorization of this part of the

Caucasian population as "Tatars" together with many other Turkic-speaking groups in the Russian Empire, and the pan-Turkist view that Turkic speakers of both Persian and Russian Azerbaijan were simply a subcategory of the Turks.

In World War I, the Bolsheviks took power in Russia, and withdrew their country from the war. As a result, the British Army entered Baku in August 1918, but within a month they yielded the city to the Turks. Under the control of the Turkish Army, the Russian Azerbaijan was given the name "Transcaucasian Azerbaijan."

In May 1918, following the collapse of Russian central control, an "independent" republic of Azerbaijan was declared under the leadership of communists. However, when the Turkish Army took Baku, an Azerbaijani nationalist party came to power. When British troops again entered Baku in December 1918, in an effort to win the support of the local elite, they allowed the nationalist government to remain. In April 1920, power again changed hands when the Bolsheviks eliminated the government and reincorporated the territory under the name "Soviet Socialist Republic of Azerbaijan." In 1922, Azerbaijan was incorporated into the "Transcaucasian Soviet Socialist Federation." In 1936, once Moscow's central control had been consolidated on its southern periphery, the Transcaucasian Federation was dissolved, and Azerbaijan was given the status of a republic within the Soviet Union.

In the contemporary Republic of Azerbaijan, 6.9 million Azerbaijanis make up the majority, but these do not constitute the majority of Azerbaijanis in the region. Iranian Azerbaijanis number 10.3 million, while 1.1 million more live in other former Soviet states.[7]

The Azerbaijani state did not become an independent entity—either in 1918 or with the demise of the USSR in 1991—as the consequence of a nationalist struggle, as in the case of many European and post-colonial nation-states. Both the first brief independence and the more recent and enduring one resulted from struggles for power at the center of the empire, which rather unexpectedly and willy nilly left the local elite to its own devices. The realization of Azerbaijan as a national concept was as much the result of outside forces as internal demands. Yet in the course of the twentieth century it has become a fact of great consequence for the region. This fact gained force, first with the historical precedent of independence in 1918-1920, and then especially with the institutionalization of the nation as an administrative entity within the Soviet Union. With this grew a somewhat autonomous bureaucracy, a national elite, a national consciousness among the population, and ideological institutions corresponding to the nation including those charged with the production of literature, art, history, and education—in short, all that goes into building the contemporary concept of the nation.

It is not surprising, then, that in the wake of the Soviet collapse, politics has been dominated by expressions of nationalist aspirations, and all the major parties are keen to demonstrate their commitment to the nation. Suprisingly, Islam has had a relatively limited presence in the post-independence politics of Azerbaijan. Though a number of Islamic parties have formed, none of them has substantial support among the population. Rather, Islam has served primarily as a cultural and politically symbolic element in the broader nationalist ideology.

Relations with Armenia and Armenians have served as one of the major catalysts for the development of Azerbaijani national consciousness. Armenians and Azerbaijanis have lived closely intertwined for centuries, with Armenians for example forming a major component of the urban population in cities such as Baku. As Baku became a rich oil production center at the beginning of the twentieth century, Armenians were successful in reaping the benefits of this boom and they occupied prominent roles in the local administration. For this, they were much resented by the less well-to-do Azerbaijanis. Due to their prominent roles in many spheres of the political spectrum, Armenians were identified with both the Tsarist government's suppression of early nationalist uprisings in the lead-up to independence in 1918, and with the Bolshevik movement that eventually led to the reincorporation of Azerbaijan in 1920. When the division of territory between Azerbaijan and Armenia was conducted during Soviet times, inevitably substantial portions of each population remained on opposite sides of the border, and this evoked recriminations from both sides that the other was unfairly favored.

Another important thread of the enmity between Armenians and Azerbaijanis traces back to Armenia's separatist struggle against Ottoman domination and the widespread ensuing slaughter and expulsion of the Armenian population from what is now eastern Turkey. The Musavat Party, which was formed by Azerbaijani nationalists in 1911, had a strongly anti-Armenian strain and sympathized with Turkey in the Turkish-Armenian conflict. Armenians since that time have held all of their Turkic neighbors responsible for the "Armenian Genocide" and the recent war with Azerbaijan has been waged in the spirit of retaliation.

The Lezghis represent 2.4 percent of the population of Azerbaijan. They straddle the border with Daghestan, and speak a language in the Caucasian family related to Georgian and other languages of the Daghestan and the North Caucasus. There is a considerable degree of dissatisfaction among the Lezghis with their position within Azerbaijan, where they lack possibilities for education, press, and other cultural institutions in their own language. On such grounds, a separatist movement has grown, with its major impetus coming from the Lezghis in Russian Daghestan, demanding unification of an autonomous Lezghi territory under Russia. The Russian parliament has declared its support for this movement, seeking in this way to put pressure on the Azerbaijani government. This, combined with the Lezghis' refusal to fight in the war with Armenia exacerbated the contempt and discrimination toward them on the part of Azerbaijanis. Whether the Azerbaijani government chooses to grant them greater rights and autonomy or to continue past policies of discrimination and assimilation will hinge on the way that the Azerbaijani national concept is articulated in coming years. To date, there is little sign of Azerbaijan being more accommodating of diversity than it had been under the Soviet government.

Daghestan

The republic of Daghestan is attenuated from north to south along the Caspian's western shore. It rises from a low plain near Kalmukia in the north to the Great Caucasian Range on the border with Azerbaijan and Georgia in the south. Daghestan is composed of four main geographic zones: (1) the Noghay Steppe in the north; (2) a narrow coastal plain along the republic's Caspian shore that widens to the north of the republic's capital, Makhachkala, on the central coastline; (3) a foothill zone occupying the center of the republic; and (4) the southern portion characterized by precipitous mountains and narrow valleys watered by fast-flowing rivers. A series of rivers traverse the country from west to east, including the Kuma, Sulak, Terek, and Samur, which provide amply for agriculture in the valleys, though suitable land is scarce.

In Daghestan, the climate is generally warm and dry. Cattle-herding is one of the crucial activities both in the low plains and in the foothill and mountain zones. The key environmental problems in Daghestan include deforestation, caused partly by industrial air pollution, and, in the north, desertification of the steppes caused by salinization of the soil.

Unlike the other territories surrounding the Caspian Sea, Daghestan has no "titular" nationality—no particular group for whom the territory is named and whose culture has a special status. No single group makes up more than about a quarter of the population. The diverse population corresponds to a fragmented geography of predominantly mountainous domain. Through much of history, Daghestan constituted a remote territory on the rather distant periphery of larger states, and was divided among an array of small, semi-autonomous principalities. As such, Daghestan had no history as a distinct territorial entity before its establishment as an autonomous territory within the Russian Federation and the Soviet Union.

In recent centuries, this land was periodically contested over by the two major states to the south—Persia and Ottoman Turkey. At times it was incorporated into broad imperial conquests—those of the Arabs, the Mongols, Timur, and the Ottomans—but during most of history, there was little direct control from outside. The Islamization of Daghestan began with the early Arab conquests in the seventh century, but proceeded often very slowly to remote areas, and it was only in the sixteenth century that this process was more or less complete.

In 1806, the Russians conquered Derbend on Daghestan's southern Caspian shore, and in 1813 Russia established its domination of the territory with full "pacification" not achieved until 1864, at the expense of about a million Daghestanis and others from neighboring areas making a mass exodus to Turkey.[8]

Under the Bolsheviks, the national autonomous unit of Daghestan, with its capital at Makhachkala, was created in 1920. Daghestan has over 30 officially recognized native nationalities. Six groups constitute the majority of the population: the Avars (27.5 percent, according to the 1989 census), Darghis (15.6), Qumuqs (12.9), Lezghis (11.3), Laqs (5.1), and Tabarsarans (4.3). In addition, immigrant Russians number at 9.2, Azerbaijanis at 4.2, Chechens at 3.2, and other small groups contributing the remaining 6.7. The republican government of Daghestan is controlled chiefly by Avars, who during Soviet times came to predominate in the intellectual and administrative elites, and subsequently in the various political and nationalist movements that arose in the perestroika era.

There are three linguistic families represented in the native population—Caucasic, Turkic, and Iranian—and a great many languages, including the Iranian language family subsuming a small number of Tats who speak a language related to Persian and live mainly in the Derbend area. There are ten official languages in Daghestan and thirty-nine administrative districts, which are predominantly inhabited by the various groups.

A number of very serious points of tension exist among various groups in Daghestan, many of them connected with the problem of competition over resources. While the majority of the population of many of the groups is rural, land suitable for agriculture is extremely limited, accounting for only 15 percent of the republic's territory and actually shrinking due to the desertification of the Noghay Steppe. Furthermore, the republic has a relatively high natural population growth rate, which is augmented by an influx of refugees from conflict zones in Chechnya and Azerbaijan as well as discriminatory treatment in other parts of Russia.[9] Another point of tension is connected with the deportation by Stalin in 1943 of the Chechens from the North Caucasus, including the Aukha region of Daghestan. The Chechen lands were then reallocated to 1,200 Laq and 500 Avar families.

When the Chechens were allowed to return in 1957, the local authorities were not prepared to return lands to them. The matter came up again in 1991 during perestroika, when Chechens demanded their rights, and this led to serious tensions between the Chechens and Avars. The Laqs, in turn, insisted on resettlement to lands that had traditionally been inhabited by Qumuqs.[10] A movement called Tenglik formed, demanding autonomy for the Qumuq areas, and as these constitute some of the best agricultural lands in the republic, this is perceived as a serious affront by both the government and other ethnic groups. A similar separatist movement of Lezghis has demanded unification with Lezghi lands in Azerbaijan. The resettlement of the mountain peoples has resulted in the Birlik ("Unity") movement's call for a reunification of the lands of the Noghay Steppe, which were divided among three neighboring territories in 1957.

The appearance of Islamic movements of varying degrees is another source of radicalism. The official Islamic hierarchy assumes a relatively moderate position, while such groups as the Islamic Renaissance Party, made up of various groups, including some that are reminiscent of "Muslim Brotherhoods" elsewhere in the Islamic world, advocate a more fundamental Islamization of the society. The activity of these various groups was evident in 1991-92 when, in a period of a year and a half, 526 mosques and 203 Islamic schools were built.[11] The various Islamic parties together have approximately 20,000 active members, and the portion of the population that would support them is estimated at 25-30 percent.[12] The significance of these groups is evident in the election of the candidate of the Russian Muslim Union to the State Duma in Moscow, who received 30 percent of the vote, beating the Communist Party candidate.[13] The influence of Islamic organizations, meanwhile, is limited by the disunity among them, as they are each dominated by different elements of the population.

Among the general population, the influence of Islamic revival is substantial, and there is widespread discussion of the establishment of an "Islamic Republic of Daghestan."[14] The example of neighboring Chechnya has played a significant role in this, where Islam and nationalism have combined to serve effectively in mobilization against Russian domination. Meanwhile, in Daghestan, there is no significant movement for separation from Russia. Several factors explain this: (1) the Daghestan government derives 85 percent of its revenues from the Russian federal budget, and the economy of Daghestan is simply not capable of sustaining independence in its current condition; (2) many Daghestanis are dependent on the opportunity to migrate to neighboring regions of Russia to find work; (3) being within Russia has a moderating effect on the internal balance of power, as minorities feel that the Russian central government can help to guarantee their interests against the more powerful local groups; and (4) separation from Russia could give impetus to separatist movements such as those of the Qumuqs and Lezghis.

Of all regions of territories surrounding the Caspian Sea, Daghestan is perhaps the one with the most indefinite future, contingent on a broad array of unpredictable variables. Stability will require a careful balance of the interests of the predominant group and the multitude of minorities.

It will also depend on the policies of neighboring states as well as Russian central and neighboring local governments, which could effect both the position and separatist aspirations of minorities and the economic well-being of the republic. The Russian central government may have to learn how to accommodate a constituent republic with a strongly Islamic orientation in its administration and public life.

Kalmukia

To the north of Daghestan and south of Russia's Astrakhan Province is Kalmukia, on a large, sparsely populated plain to the southwest of the Volga. The whole of Kalmukia's territory is extremely low, some areas being as low as 28 meters below sea level. The greater portion of the republic occupies the Caspian Depression, much of which is below sea level. The highest area, lying in the west of the territory—the Yergeni Hills and the Sal'lk-Manych Ridge—does not exceed 221 meters above sea level.

Heavy rains frequently lead to the flooding of Kalmukia's many depressions, and the warm summer climate and abundance of precipitation provides for rich grassy vegetation, and, in the Yergeni Hills, woodlands. This climate is conducive to the traditional Kalmuk activity of pastoralism, as well as to agriculture. Nevertheless, the republic is very sparsely populated with few significant populated centers aside from some small coastal towns in the east and the republic's capital, Elista, in the west.

In the Caspian Basin, the Kalmuks are an anomaly, not only in being a Mongol people, speaking a Mongol language called Oyrat, but also in their traditional practice of Tibetan Buddhism (or "Lamaism"), which makes them distinct from the mainly Turkic and Muslim nomads who have cohabited this area of the western steppes.

In their early occupation of their current territory, the Kalmuks were much in favor with the Russian state. Originating from the area straddling the border between contemporary Kazakhstan and Xinjiang in China, the Kalmuk migrated to the Ural region in the sixteenth century and continued on to the lower Volga Basin in 1632. They came to occupy their current territory, between the Volga and the Don, in 1661, when they concluded a treaty with Russia, whereby they would have use of these lands in exchange for accepting Russian suzerainty and defending the southeastern boundary of the empire. In general, the Kalmuks were allowed considerable autonomy, but the relationship with Russia was mutually suspicious, which led in 1771 to about two-thirds of the Kalmuks abandoning the Volga Basin and returning eastward.[15] Due to a harsh winter and attacks from the Kazaks in Jungaria, this migration had disastrous consequences, with about one half of this population per-

ishing.[16] In the Volga Basin, too, those who remained fared poorly as the
Russian government liquidated the Kalmuk Khanate and its territory
was divided between the adjacent administrative units of Russia, with a
special administrative entity for the Kalmuks appearing again only with
the creation of the Kalmuk Autonomous Province in 1920.[17] In 1925, the
Cyrillic script was introduced, which was in turn replaced in 1930 with
a Latin script, and then again in 1938 with a new Cyrillic one.[18]

In 1943, the Kalmuks were declared a traitorous people and, together
with other people of the region, including the Chechens, Ingushes, and
Qarachays, were accused of collaboration with fascism. As collective pun-
ishment, the Kalmuk Autonomous Republic was liquidated, and the
Kalmuks were deported in the dead of winter to various parts of Siberia—
the Altay, Krasnoyarsk, Novosibirsk, and Omsk regions, even though
some 30,000 Kalmuk men were at this same time serving on the war
front, and 14,000 were given war medals.[19] The Kalmuk lands were reallocated to neighboring provinces and settled by others; Kalmuk place-
names were erased from the map.

The plight of the Kalmuks took a severe toll on the population count,
from 190,000 in 1897, to 130,000 in 1926, to 106,000 in 1959.[20] In
1957, the Kalmuk Autonomous Region was reestablished and the Kalmuks were permitted to return. Even with some recovery in subsequent
years, taken as a whole, the period from 1897 to 1989 saw a drop in the
Kalmuk population from 190,600 to 173,000, while the overall population of the region had increased by 227 percent since 1926.[21]

In the post-Soviet period, there has been an enthusiastic revival of
Kalmuk religion and culture, as the republic ratified the "Steppe Code"
as its local law, established cultural relations with the related peoples in
Mongolia and China, and promoted national symbols.

Many problems face Kalmukia at present, both in terms of ethnic relations and economic conditions. Kalmuks now constitute only 43 percent
of the population, 1 percent more than the number of Russians. There
have been tensions between these groups, as Kalmuks have sought to
assert a predominant role in the republic and Russians have felt their
position threatened. Other groups that have played a role in open, though
limited, conflicts include the Cossacks and immigrants from the Northern Caucasus.[22] In continuity with their pastoral past, the economic activity of the Kalmuks is heavily concentrated in stock-breeding, which
constitutes 73 percent of the agricultural production of the republic.[23]
Kalmukia is one of the poorest regions of the Russian Federation. For
example, it occupies the 56th place out of 72 regions in regard to the level
of development of the transport infrastructure.

Despite these problems, there have been only limited manifestations of
inter-ethnic tensions. Indeed, the process of Russification has had a substantial impact, especially on the urban population, which constitutes

about 50 percent of Kalmuks according to the 1989 census. The cultural commonalties between not only the Kalmuk elite and the Russians of Kalmukia, but also a broad segment of ordinary urbanized Kalmuks, allow for a broad social consensus. Furthermore, the republic leadership has shown determination to take steps to dissipate any tensions.[24] Kalmukia's unquestionable future of remaining within Russian, the presence of a large Russian minority, the small overall number of Kalmuks, and the specter of the Chechen conflict all serve to moderate Kalmuk nationalist ambitions.

Kazakhstan

Kazakhstan is composed of three geographic zones. A mountainous zone, the Altay and Tien-Shan Ranges on the border with China, occupies a small percentage of the country at a great distance from the Caspian Basin. About one half of the country is steppe and low rolling hills that extend across the north from the Caspian to the Altay Mountains. The remainder—about 35 percent of the territory—is arid lowlands, including part of the Qizil-Qum Desert, the desert plains surrounding the Aral Sea, and the Mangyshlak Peninsula extending into the Caspian in the west.

Three major river basins orient segments of the country toward neighboring geographic zones: the Irtysh, flowing from the Altay region, toward western Siberia; the Sir-Darya, in the south, toward the heart of Central Asia; and the Ural River, toward the northern Caspian Basin and southern Russia. None of these rivers flow into the central or southwestern regions of Kazakhstan, limiting the possibilities there for agriculture. These arid territories support primarily animal-herding now, as they have throughout the history of the Kazak people. Irrigation agriculture in the vicinity of the Sir-Darya and the southeastern mountains supports cotton and horticulture, while a population primarily of immigrants from European Russia practices predominantly dry-land wheat farming in the steppe zone along the long northern border with Russia. Throughout most of Kazakhstan, a continental climatic zone closely connected with western Siberia provides the region with cold, snowy winters and hot, dry summers.

The portion of Kazakhstan most closely integrated into the Caspian Basin is itself relatively poorly integrated into Kazakhstan as a whole. This area is not especially conducive to habitation, with only a very sparse rural population and a handful of relatively minor urban centers that were developed in Soviet times. Only the recent development of oil resources has brought this region into the national focus and promises to result in some greater investment in the region.

Kazakhstan is a vast territory of semi-arid steppes and deserts that, until recent times, was inhabited almost exclusively by nomads. In the early eighteenth century, the Russians played the different Kazakh juzes ("hordes") against one another in order to gain dominance. In the mid- to late eighteenth century, Russian expansion into Kazakh territories continued with the construction of a series of forts occupied by Cossack troops. Already by the nineteenth century, approximately 400,000 Russians, as well as Cossacks and Ukrainians, were settled in the Kazakh steppes. With the subjugation of the Great Juz in 1848, Russia consolidated its control. The colonization process continued into Soviet times, with new heights of in-migration achieved from 1953 to 1965 under the so-called Virgin Lands program, which brought most of the Northern Kazakhstan steppes under grain cultivation and overwhelming settlement by European nationalities.

During the colonial period, the territory of Kazakhstan was governed by the military administration of the "Kirghiz Steppe Territory." When the Bolsheviks seized control in Petrograd in late 1917, the political movement Alash Orda was formed in January 1918 to promote Kazakh self-determination but, by the end of 1919, it surrendered to the Reds. By 1920, the Red Army gained control of the territory and formed the "Kirghiz Autonomous Republic" as an administrative unit within the Russian Federation.[25] In 1921, the Soviet government introduced the New Economic Policy, but it was abandoned in Kazakhstan by 1925. The determination to impose settlement upon the Kazakh herdsmen resulted in imposition of land redistribution and confiscation of cattle from the wealthier Kazakhs. With the intensification of collectivization since 1929, an estimated 1.5 million Kazakhs died, while 80 percent of Kazakh herds were destroyed between 1928 and 1932, and hundreds of thousands of Kazakhs fled to China and neighboring Soviet republics.[26] In 1924, the administrative division of all Central Asia was established in approximately its current form, but the Kazakh Autonomous Soviet Socialist Republic remained subordinate to Russia, and only achieved the status of a separate republic in 1936.

In the first years of independence, the Kazakhstan government sought to encourage Kazakhs living abroad to "return" to Kazakhstan. While in 1997 the percentage of Kazakhs in Kazakhstan was put at 51 percent,[27] the CIA *World Factbook* put the percentage of Kazakhs in 1996 at 41.9 percent. Regardless of whether Kazakhs form a majority in the republic as a whole, Russians constitute a very sizable minority, and they significantly outnumber the Kazakhs in more than one third of the republic's provinces.

The government of Kazakhstan sees a serious threat to its security and stability in the fact that along most of its long northern border with Russia, the majority of the population is Russian and might prove more loyal to Russia than to Kazakhstan. There has recently been a revival of an assertive Cossack identity on both sides of the border, some supporting the view that parts or all of Kazakhstan should be annexed to Russia. Russian nationalism of this type is represented by such organizations as "Lad," "Russian Community," and the "Society for Slavonic Culture."

In the south of the country, there is a substantial population of Kazakhs and other Central Asians—especially Uzbeks in such cities as Turkistan and Shimkent—among whom Islam plays a particularly important role. Likewise, the Kazakh population in this region is much more inclined to mobilize behind Islamic organizations, whether of a social or a political nature. The end of the Soviet era saw in Kazakhstan, as elsewhere in the region, a tremendous revival of popular interest and participation in Islamic practices and organizations, especially in the southern regions. In 1990 alone, 100 new mosques were opened.[28] The nationalist party Alash adopted an Islamic as well as a nationalist agenda and found support particularly in the southern regions, and among rural Kazakhs and recent immigrants to urban areas.[29]

Because Kazakhstan's Qaziyat was subordinate to the official Muslim body in Uzbekistan's capital of Tashkent,[30] the influence of the Uzbek clergy has been strong in Kazakhstan, especially in the southern provinces of Shimkent and Jambil (recently renamed Taras). The government's efforts to promote Kazakh Imams, however, provoked opposition in the southern provinces.[31] However, due in part to the lack of unity, meanwhile, following some initial enthusiasm for Islamic revival, the Islamic movement has failed to attract broad support beyond the southern regions.

Alongside the concepts of nationalism and Islam as defining factors in the national politics, the notion of "Eurasianism,"[32] introduced by President Nursultan Nazarbaev, too has failed to provide a substantial mission or identity for the Kazakh nation. Faced with confusion about Kazakhstan's relation to Russia, to Islam, to Central Asia, to Western culture, and to its internal divisions, the new state is in a particularly ambiguous political position, and is confronted with significant challenges to its national integration. Kazakhstan's presidential political system has essentially defeated the notion of an independent judiciary or legislature, and President Nazarbaev has positioned himself in an authoritarian role that is reminiscent of Turkmenistan's president Niyazov, the Turkmenbashi.[33] Perhaps the greatest challenge to Kazakhstan's national integration, meanwhile, is the intensified stratification of the society that feeds on the country's natural resource exports. Already there is a severe alienation between the general population, whose lot is worsening, and the narrow national elite, whose interests seem divorced from the country as a whole.

Turkmenistan

Turkmenistan is dominated by the desert. The Qara-Qum desert occupies roughly 90 percent of the country, making its central region usable for little other than the pasturing of herds of sheep, which has been at the core of the traditional Turkmen economy. More hospitable lands are found only around the southern and eastern edges of the country—in the vicinity of the Amu-Darya River in the east, and in the oases that line the foothills of the Kopet-Dagh Mountains in the south. In addition, the Qara-Qum Canal, a massive irrigation channel drawn from the Amu-Darya, was completed in 1967, creating a band of irrigated land running across the region to the north of Turkmenistan's border with Iran.

Turkmenistan's climate is the most severe in the region. In the summer, desert temperatures not infrequently exceed 50 degrees Celsius (122 degrees Fahrenheit), while in winter they drop to -30C (-22F). The moderating effect of the Caspian Sea diminishes the severity of the climate in the southwest of the country, and this, combined with the abundance of oases along the Kopet-Dagh, makes the southern and western rims of the country the most suitable for agriculture and human habitation in general. These regions, meanwhile, are also the most severely effected by the impacts of human activity, most notably the destruction of soil and groundwater resources by heavy use of agricultural chemicals, salinization exacerbated by irrational irrigation practices, and the rising of the water table in the vicinity of poorly constructed canals.

The core of the Turkmen population stems from the Oghuz tribes of Turkic pastoral nomads who previously inhabited the areas of Mongolia and southern Siberia, whence they migrated to the Caspian Basin in the tenth century. As they entered Central Asia, they adopted Islam, and one of their branches eventually formed the basis of the Ottoman dynasty in Anatolia, while another remained on the territory of present-day Turkmenistan. The Turkmen language differs markedly from other Central Asian Turkic languages, while bearing strong similarities to Turkish.

The Turkmens overwhelmingly remained nomadic pastoralists until the twentieth century. They occupied the desert lands between the Central Asian oases, the Caspian Sea, and the Iranian Plateau, and often lived in some degree of tension with their settled neighbors. As Shi'ism came to predominate in Iran in the sixteenth century, the Turkmens remained Sunni Muslims, and the tension took on a sectarian dimension. On the grounds the Persians were not truly Muslims, the Turkmens conducted raids on the Shi'ite communities and captured slaves whom they sold on Central Asian markets. Despite their habitation of remote regions, the Turkmens were continuously dependent on Central Asian markets for grain, metal, and other essentials that their nomadic lifestyle did not provide. Due to this continuous interaction, the results of mixing populations may be seen, for example, in the fact that Uzbek dialects in western Uzbekistan bear marked similarities to Turkmen.

As an important consequence of the retention of pastoralism, Turkmen social organization has been dominated by the principle of lineage descent, and among the Turkmens a number of major divisions, or "tribes," have maintained their significance to the present day.

When Russia's conquest of this part of Central Asia began in the mid-nineteenth century, the Russian Army was able to take advantage of tribal rivalries, making successive alliances with various groups to defeat the others. The only group to actually put up substantial resistance to the Russian conquest was the Tekes, and the Russians defeated them with the assistance of others. The last major defeat of the Turkmens was in the early 1880s at Gok Depe, but complete control proved difficult to achieve. Throughout the colonial period, there were periodic outbreaks of resistance, the most important of which was the uprising of 1916, provoked by the Tsarist government imposing military conscription on the Central Asians for battle in World War I.

Many of the Central Asian states had no real historical precedent in analogous territorial-political entities prior to the establishment of the current configuration in the early Soviet period. By contrast, while Turkmenistan was not a state in pre-conquest times, it was a more or less autonomous territory between states, and when it was incorporated into the Russian Empire, the territory of today's Turkmenistan was essentially kept as one administrative unit in the "Trans-Caspian Province." After the Bolshevik Revolution of 1917, which was accompanied by the promise of national autonomy to the various nationalities, there was much struggle over territory. Initially, the Bukharan Emirate and the Khiva Khanate were retained and renamed as the Bukharan and Khorezm People's Soviet Republics. These included some of the oases where Turkmens had begun to settle.

In 1924, there was a general territorial reorganization, Bukhara and Khorezm were liquidated, and the Turkmen Soviet Socialist Republic became one of the first fully fledged republics in the Union of Soviet Socialist Republics. At this time, the expanded administrative entity included a number of the oases around the Turkmen traditional tribal territory, including, in the northeast, Tashauz (now officially Dashhowuz)—which had been in the Khorezm Republic—and in the southeast, Charjou (now Charjev)—in the Bukharan Republic. These oases had substantial non-Turkmen populations, but their inclusion in Turkmenistan was justified in part on the grounds that it was necessary for the economic viability of the new republic that was otherwise dominated by desert.

The creation of Turkmenistan was in no way the result of a Turkmen nationalist movement calling for recognition of a national territory. This decision was made in Moscow by people who undoubtedly felt little need to accommodate any Turkmen nationalist leaders, who were not a significant force at this time. Much more influential among the Bolsheviks were leaders from the Bukharan and Uzbek territories, though the ceding of the mentioned oases to Turkmenistan rather than Uzbekistan—or indeed, instead of keeping Central Asia as a single Turkistan Republic, as some of these Central Asian Bolsheviks had advocated—was undoubtedly an effort to rein in the Uzbek leaders and give more strength to the otherwise quite weak Bolshevik Turkmen leadership.

If the Turkmens lacked a substantial nationalist movement, there was nevertheless a relatively coherent national identity as compared with some of the other Central Asian groups that were allocated "national republics" in the early Soviet years. Linguistically and culturally they are quite homogeneous, as compared, for example, with the Uzbeks. They are also relatively distinct in these regards from their neighbors. Thanks to intermarriage and cultural influences in the border areas—especially in the oases where the Turkmens had begun to settle—the cultural and linguistic lines between groups were somewhat blurred, but the core population was nomadic and this formed a sharply defined concept of the group.

The most important factor leading to the formation of the contemporary Turkmen national identity was the experience of the Soviet period, during which a Turkmen bureaucratic administration and cultural institutions were created. While the Turkmens remained predominantly rural and urban life was dominated by Russian culture more than in most other countries of the region, nevertheless, the administrative sphere was one in which, on most levels, Turkmens predominated over other groups. Here, all Turkmens came together, albeit in a struggle for power which was more often than not articulated in terms of sub-national identities.

Among the elements of the common Turkmen national culture is Islam. During the Soviet period, Islam suffered in Turkmenistan even more than in other parts of the Soviet Islamic world. Repressive measures against Islam were implemented more strongly in Turkmenistan. The population was more systematically prevented from religious practices than in neighboring republics, with the militia even physically preventing entry to mosques on religious holidays. At the present time, however, Islam is a weakened force, both in the political life of the country and in the cultural sphere.[34] The current policy toward Islam is more permissive than during the Soviet period, as current leadership has sought to demonstrate its Islamic credentials. Former communists are conducting the Hajj to Mecca. Meanwhile, the official clergy represents no opposition to the government. At the same time, all Islamic parties are banned, as indeed all parties except the president's "Democratic Party," which is the heir to the Communist Party.

The new national ideology of Turkmenistan is predominantly Western in its orientation. This is reflected in the fact that Turkmenistan has been the first of the Central Asian states—after Azerbaijan in the Caucasus—to actively pursue a shift from Cyrillic script to a version of the Latin alphabet. It is characteristic, meanwhile, that the new Turkmen Latin alphabet strongly differs from all other Turkic Latin alphabets. At the core of Turkmenistan's new official national identity is the notion that it is a country unto itself, pursuing an independent line, with little interest in the involvement of international organizations or in multi-lateral treaties with its formerly Soviet partners.

There is relatively little diversity of nationalities within Turkmenistan, with 72 percent of the population being Turkmen and less than 10 percent each of Russians and Uzbeks. Yet this diversity does present some issues for the country. Russians have generally been migrating out of Central Asia, and given their key role in the industrial and urban sectors of Turkmenistan's economy, it is important that they be encouraged not to leave. One indication of Turkmenistan's sensitivity to this issue is that in contrast to its neighbors, it has granted to Russians the right to dual citizenship, diminishing the imperative that they make their choice in the short term.

The most populous and influential tribes today include the Tekes, Goklens, Yomuts, Ersaris, and Sariqs.[35] The preeminent group is the Teke, to which Turkmenistan's President Niyazov belongs. The second most important position is occupied by the Yomuts, who are also relatively well represented in the central administrative apparatus, though not in the most influential positions.

The most significant circumstance in the politics of post-Soviet Turkmenistan, meanwhile, is the widespread acceptance of the complete authority of the former Communist Party First Secretary, Saparmurad Niyazov. The unassailability of his position has been guaranteed by skillful use of the bureaucratic patronage system, cooptation of the religious elite, Soviet-era methods of political control, a balance between opposing tribal interests, the support of his own Teke tribe, and fears of social instability. As a result, there has only been relatively limited scope for the development of any kind of nationalist movement or even of national consciousness. Politics in Turkmenistan is dominated by the authority of one individual and an ideology of the state taking its own path autonomous of all neighboring states, in place of any kind of more developed national idea.

Conclusion

While the expansion of oil and gas production in the Caspian Basin has greatly increased this region's importance in the world arena, there are undoubtedly other issues—just as in the Middle East—that dominate political developments in the region. While oil wealth could provide a significant boost to the economies of several countries and of the region as a whole, if this wealth is not used wisely, there may not be conditions for enjoying the benefits of it. Such issues as the rapid growth of the population and of unemployment; the problems of providing education, social security, and the infrastructure for social development; and the emergence of severe tensions between various national and religious movements in these countries, present serious challenges. More than half a decade after independence, there are few indications that these problems are diminishing.

Notes

1. The total land area in square kilometers of the littoral entities bordering the Caspian Sea are as follows: Azerbaijan, 86,100; Kazakhstan, 2,669,800; Turkmenistan, 488,100; Russia's Astrakhan Province, 44,100; Daghestan, 50,300; Kalmukia, 76,100; and Iran, 1,636,000. The land area suitable for agriculture, including arable land, meadows, and pastures, make up 43 percent of Azerbaijan's total land area; 72 percent of Kazakhstan; 71 percent of Trukmenistan; and 35 percent of Iran. Four percent of Kazakhstan and 11 percent of Iran's total land area consists of forests and woodland. See generally *The World Factbook* (Washington, D.C.: Central Intelligence Agency, 1996); *Russian Regions Today: Atlas of the New Federation* (Washington, D.C.: The International Center, 1994).

2. Based on the land area for each country (see note 1 above), the pouplation density per square kilometer is about 87 for Azerbaijan; 6.2 for Kazakhstan; 1.0 for Turkmenistan; 22.5 for Astrakhan Province; 35.8 for Daghestan; 4.2 for Kalmukia; and 40.4 for Iran. See generally, CIS Interstate Committee for Statistics (1996); *Itogi Vsesoiuznoi perepisi naseleniia 1989 goda* ("Results of the All-Union Census of the Population of 1989") (Minneapolis: East View, 1992-1993), 12 vols. [hereinafter referred to as the "1989 Census"]; "Iran," in *The World Factbook* (1996).

3. For details, see the data on the rural population in *The World Factbook* (1996); State Committee for Statistics of the Russian Federation (1990); CIS Interstate Committee for Statistics (1996); All-Union Census of 1989; data on agricultural lands in *The World Factbook* (1996); CIS Interstate Committee for Statistics (1996); *Social Indicators of Development, 1996* (Baltimore: World Bank/Johns Hopkins University Press, 1996).

4. See generally under each country's entry in *The World Factbook* (1996). For Russian Federation statistics, see generally Michael Ryan, ed., *Social Trends in Contemporary Russia: A Statistical Source-Book* (New York: St. Martin's Press, 1993).

5. For details, see *Demograficheskii ezhegodnik, 1994* ("Demographic Yearbook, 1994") (Moscow: 1995); *Rynok truda v stranakh Sodruzhestva Nezavisimykh Gosudarstv: V tsifrakh i diagrammakh: Statisticheskii sbornik* ("The Labor Market in the Countries of the Commonwealth of Independent States") (Moscow: Mezhgos. statisticheskii komitet sodruzhestva nezavisimykh gosudarstv, 1996); "Russia," in *The World Factbook* (1997).

6. See generally, "Iran," in *The World Factbook* (1997).

7. "Iran," in *The World Factbook* (1996).

8. Alexandre Bennigsen, "North Caucasians," in Stephen Thernstrom, ed., *Harvard Encyclopedia of American Ethnic Groups* (Cambridge, Mass.: Harvard University Press, 1980), p. 750.

9. See generally *Budushchee Dagestana* ("The Future of Daghestan") (Makhachkala: Izd. DNTS RAN, 1994).

10. *Dagestan: Etnopoliticheskii portret: Ocherki, dokumenty, khronika* ("Daghestan: An Ethno-political Portrait") (Moscow: TsIMO, 1993-1995), vol.3, p.25.

11. Rafis Abazov, *Islam I politicheskaia bor'ba* ("Islam and Political Struggle") (Moscow: Informatsionno–ekspertnaia gruppa "Panorama," 1992) p. 34.

12. *Dagestan: Etnopoliticheskii portret,* vol.2, p.41.

13. *OMRI Russian Regional Report* (Prague: Open Media Research Institute), December 11, 1996.

14. *Dagestan: Etnopoliticheskii portret,* vol.2, p.742.

15. N. N. Pal'mov, *Etiudy po istorii privolzhskikh kalmykov, XVII i XVIII veka* ("Essays on the History of the Kalmuks of the Volga Basin, 17th and 18th centuries") (Astrakhan: Izd. Kalmytskogo obl. ispolnitel'nogo komiteta, 1926), p. vi.

16. George V. Vernadsky, "Istoricheskaia osnova russko-kamytskikh otnoshenii ("The historical basis of Russo-Kalmuk relations"), in Arash Bormanshinov and John R. Krueger, eds. *Kalmyk-Oirat Symposium* (Philadelphia: Society for the Promotion of Kalmyk Culture), pp. 34-35.

17. P. D. Bakaev, *Razmyshleniia o genotside* ("Reflections on Genocide") (Elista: 1992), p.4.

18. *Ocherki istorii Kalmytsskoi ASSR: Epokha sotsializma* ("History of the Kalmuk ASSR: The Period of Socialism") (Moscow: Nauka, 1970), pp. 423 and 425.

19. Ibid., p. 11.

20. Ibid., pp. 5-6.

21. Ibid., p. 5. See also T. S. Guzenkova, *Mezhetnicheskaia situatsiia v Kalmykii* ("The inter-ethnic situation in Kalmukia"), in series *Issledovaniia po prikladnoi i neotlozhnoi etnologii* ("Research in Applied and Urgent Ethnology"), no. 34 (Moscow: Rossiiskaia Akademiia Nauk, Institut Etnologii i Antropologii, 1992), pp. 4-5.

22. Guzenkova, p. 20.

23. *Russian Regions Today: Atlas of the New Federation* (Washington, D.C.: The International Center, 1994), p. 274.

24. Guzenkova, pp. 23-24.

25. "Kirghiz" was the term used by the Russians in pre-Soviet times to indicate the Kazakhs, and the Bolshevik retained this usage until 1925.

26. Martha Brill Olcott, *The Kazakhs* (Stanford, Calif.: Hoover Institution Press, 1987), pp. 184-185.

27. *RFE/RL Newsline* (Washington, D.C.), vol. 1, no. 42, pt. I, May 30, 1997 (quoting Kazakhstan's National Agency of Statistics).

28. Abazov, p. 15.

29. Mehrdad Haghayeghi, *Islam and Politics in Central Asia* (New York: St. Martin's Press, 1995), p. 86.

30. This body was known as the Sredneaziatskoe dukhovnoe upravlenie musul'man ("Central Asian Spiritual Administration of Muslims"), or SADUM.

31. Reef Altoma, "The Influence of Islam in Post-Soviet Kazakhstan," in Beatrice F. Manz, ed., *Central Asia in Historical Perspective* (Boulder, Colo.: Westview Press, 1994), pp. 171-173.

32. The concept originated with and is explained in Nikolai Sergeevich Trubetskoi (pseudnym I. R.) *Nasledie Chingiskhana: Vzliad na russkuiu istoriiu ne s Zapada, a s Vostoka* (Berlin: Evraziiskoe izdatel'stvo, 1925).
33. Martha Brill Olcott, "Nursultan Nazarbaev and the Balancing Act of State Building in Kazakhstan," in Timothy Colton and Robert Tucker, eds., *Patterns of Post-Soviet Leadership* (Boulder, Colo.: Westview, 1995), pp. 175-180, 185-187.
34. For details, see generally Rafis Abazov, note 11.
35. See generally, E. J. Brill's edition of *Encyclopedia of Islam (1913-1936)* (New York: E. J. Brill, 1987), vol. 8, p. 897.

CHAPTER TWO

Survey of Caspian's Oil and Gas Resources

Ottar Skagen

Introduction

This chapter examines the extent of Caspian's oil and gas resources, as presently estimated, and discusses the potential for some Caspian countries to become significant producers. A general description of the petroleum industries in these countries will be followed by an analysis of recent developments in oil and gas production in the Caspian area. Regional oil and gas consumption will be discussed in order to arrive at an estimate of potential volumes that may be exported from producers. The brief treatment of current and possible future export markets for Caspian oil and gas will be followed by a survey on the status of key export pipeline projects.[1]

In this chapter, the term "Caspian area" will refer to the Caucasian and Transcaucasian countries of Armenia, Azerbaijan, and Georgia, and the Central Asian countries of Kazakhstan, Kyrgyzstan, Tajikistan, Turkmenistan, and Uzbekistan, even though only Azerbaijan, Kazakhstan, and Turkmenistan border the Caspian Sea. The term "Caspian countries" will refer to former Soviet republics on the Caspian Sea. For the purposes of this chapter, Iran and Russia are not included in the terms "Caspian area" and "Caspian countries."

Background

Since the break-up of the Soviet Union, the Caspian area's petroleum resources have attracted strong and continual interest from at least five different quarters. First, the governments of the republics where the resources are located consider them crucial for economic recovery and future prosperity. Second, the politicians from the Organization for European Cooperation and Development (OECD) countries view the resources as important from an energy security perspective, and would like to see them utilized for geo-politically constructive purposes. Third, the Russian oil and gas industry thinks it should recoup the expenses and be compensated for the expertise that it poured into the development of the republics' energy sectors as part of the overall Soviet Union's programs. Fourth, the Russian political leaders are worried about the geo-political implications of the southern republics of the former Soviet Union forging petroleum-based links with Western interests. Fifth, Western oil, gas, and engineering companies are seeking and securing major contracts.

The Caspian area would not have attracted the bulk of international oil companies if it had not offered exciting possibilities, and petroleum industry developments in the area are well worth monitoring. However, that said, following the oil and gas affairs of the Caspian countries from a distance may leave the impression that contracts are being signed, fields are being developed, and pipelines are being built in all directions, and that the sky is the limit for Caspian oil and gas production. The first of these notions is wrong, the second questionable.

For each deal that is made, apparently half a dozen memoranda of understanding and letters of intent are signed, with reporters often failing to note the differences in character between contracts and agreements with political rather than practical economic purposes. Trumpeted projects are again and again being delayed, with declines rather than increases in output as the result. As for the future, the levels to which oil and gas production may rise remain very much to be seen. Reserves are considerable, but perhaps not as vast as Caspian countries would like them to be; gas production could quickly run up against market constraints.

Oil and Gas Reserves

The estimates of Caspian area oil and gas reserves vary. This is because of a lack of reliable information and different interpretations of the existing data. Estimates also show changes from one year to the next not wholly explained by new discoveries and production; the pool of geological and reservoir information keeps growing, warranting sometimes major revisions of prior conclusions.

The estimates of Caspian area oil and gas reserves by the local governments tend to be rosier than that given by foreign companies or independent experts. This is understandable; playing up one's assets and long-term potential is part and parcel of the competition for foreign investment. Moreover, governments striving to build national identity and pride, and trying to contain unrest in the face of economic and social hardships, cannot be blamed for using exaggeration or over-estimation in order to raise hopes for a brighter future.

According to one estimate, as of January 1, 1996, among the four important oil and gas producing countries—Azerbaijan, Kazakhstan, Turkmenistan, and Uzbekistan—the remaining recoverable oil and gas reserves read as 26,000 million barrels (mbls) and 5,200 billion cubic meters (bcm), respectively. The proved and probable oil and gas reserves for Azerbaijan stands at 7,008 mbls of oil and 880 bcm of gas; for Kazakhstan at 15,800 mbls of oil and 690 bcm of gas; for Turkmenistan at 978 mbls of oil and 2,740 bcm of gas; and for Uzbekistan at 2,154 mbls of oil and 884 bcm of gas.[2]

Azerbaijan owes its oil and gas reserves to the south Caspian basin, which extends from Georgia across Azerbaijan territory and the Caspian Sea to western Turkmenistan. The bulk of Kazakhstan's oil and gas reserves are thought to be located in the north Caspian basin including the portion of the Caspian Sea north of the 45th parallel, northwest Kazakhstan, and Astrakhan and much of Kalmukia in Russia. Discoveries could also be made in the north Ustyurt Basin including the Buzachi peninsula, the middle Caspian Basin extending across the sea from the 45th parallel in the north to the Apsheron Sill in the south and including the Mangyshlak peninsula, and the southern Turgay and Chu-Sarysu Basins in central Kazakhstan.

Turkmenistan's petroleum reserves are located in the Turkmen portion of the south Caspian Basin and in the Amu Darya Basin. A string of fields running eastward from Cheleken on the Turkmen side of the Caspian Sea and then southward along the coast tracks the eastern limits of the south Caspian Basin. These fields hold the bulk of Turkmenistan's oil reserves. Most of the country's gas reserves are located in the Amu-Darya Basin in the central and eastern parts of the country.

Uzbekistan shares the Amu-Darya Basin with Turkmenistan, and most of the former's oil and gas reserves are located in a strip of territory some 100 kilometers wide on the Uzbek side of the border. Discoveries have also been made in the north Ustyurt Basin extending from western Uzbekistan into Kazakhstan, the Fergana Basin extending from eastern Uzbekistan into Tajikistan and Kyrgyzstan, and the south Tajik Basin extending from southern Uzbekistan into Tajikistan.

National leaders generally take a more optimistic view of reserves than foreign observers. Thus, the president of the State Oil Company of Azerbaijan Republic (SOCAR) holds that oil reserves under Azeri jurisdiction amount to some 17.5 billion barrels (bbls). The Kazakh government estimates Kazakh reserves at 24 bbls of oil and 1,700 bcm of gas. The Turkmen government in 1995 put reserves at more than 50 bbls of oil and 21,000 bcm of gas. The Uzbek government estimates of proved and potential liquids and natural gas reserves are given at 40.5 bbls and 2,000 bcm respectively, with potential reserves estimated at 277 bbls and 7,400 bcm.

Oil and gas discoveries have been made in the other Central Asian and Transcaucasian republics, too. Observers see neither Kyrgyzstan nor Tajikistan, Armenia, and Georgia developing into significant producers, let alone exporters. However, if foreign investors were given the right incentives, fuel self-sufficiency ratios could increase to an extent that would matter locally, lessening fuel import payment burdens and reducing the impact of fuel import irregularities.

Azerbaijan

The most influential player in the Azeri oil and gas sector is SOCAR. Established in 1992 by the merger of a number of enterprises inherited from the Soviet era, this huge organization with 80,000 or so employees is operating Azerbaijan's oil and gas fields, refineries and crude oil pipelines; it is also negotiating production sharing agreements (PSAs) for new fields with foreign oil companies, preparing tenders, evaluating bids, selecting winners, and negotiating contracts with equipment supply and service companies. Various international firms/groups have signed agreements with SOCAR to develop the Guneshli, Azeri, and Chirag fields and the Shah Deniz, Karabakh, Lankaran-Talysh, Dan Ulduzu, and Ashrafi prospects, with more agreements to come.

Azerbaijan has been an oil producer and exporter since the 1870s. Production peaked at 440,000 barrels a day (bls/d) in 1940, dipped to about 300,000 bls/d in 1950, but climbed back to about 400,000 bls/d by 1970. However, the Soviet Union's efforts to develop Western Siberia led to the marginalization of the Azeri oil industry, with output falling to 250,000 bls/d in 1990. With the break-up of the Soviet Union and the onset of the Azerbaijan/ Armenia war over Nagorno-Karabakh, funding for exploration and development became even more scarce, resulting in the 1996 Azeri production of only 182,000 bls/d.

Azerbaijan's onshore fields make up the majority of its producing fields, but they are typically small, old, and characterized by low well productivity; currently over 80 percent of oil and condensate production and 96 percent of gas production is offshore. These rates will increase as the present fields and those slated for development by international consortia come on stream.

Since 1994, another influential player in the Azeri oil and gas sector has been the Azerbaijan International Operating Company (AIOC), consisting of some 11 foreign oil companies, and SOCAR. The AIOC-SOCAR deal, which provides for the development of the Guneshli, Azeri, and Chirag fields, has been serving as a model for Azerbaijan's more recent PSA agreements. Each of these PSAs is legally self-contained; it lays down the parties' rights and obligations to the last detail and it must be ratified by the Azeri parliament before entry into force. The AIOC faces capital expenditures amounting to $7.4 billion and operating expenditures of almost $6 billion during the project's 30-year lifetime. Production is to start at some 100,000 bls/d in late 1997 and increase to 0.7 mbls/d.

Traditionally, Azerbaijan's crude-oil production has been refined at the country's two refineries with a combined distillation capacity of about 20 million (metric) tons (mmt) a year. As domestic crude production dropped below the capacity of the refineries, Russian and Kazakh crude was imported to keep them busy. In recent years, however, imports have dropped significantly, driving the refinery capacity utilization rate below 50 percent. Both refineries are old and primitive by Western standards.

Gas production has plummeted from about 13 bcm per year in the mid-1980s to 6.1 bcm in 1996, for the same reasons that oil production has dropped and, additionally, because there are few facilities for collecting associated gas produced offshore. Azerbaijan, which in 1980 was a net gas exporter to the tune of 20 percent of its output, in 1990 had to import about 40 percent of its needed supplies. The period between 1991 and 1995 saw declines in imports due to economic contraction, rising import prices, and supply cut-offs resulting from gas users' inability to pay Azerigas and Azerigas' inability to pay Turkmenistan, its main supplier of imported gas. Consequently, in 1995, President Heidar Aliyev ordered a complete halt in gas imports. Since then much of the country has been without supply.

Azerbaijan's oil production may increase to between 0.8 and 1 mbls/d by the year 2010. Forecasting Caspian oil and gas output is, however, extremely difficult. Output could be less if the building of export pipelines proceeds at a slower pace than planned, but it could also be greater if prospects slated for development live up to—or exceed—expectations. Although gas production is not forecast to increase that much, an optimistic Azerbaijan, too, could see itself as a future gas exporter.

Kazakhstan

Prompted by low levels of efficiency and pockets of opposition to reform in the energy sector, the Kazakh government in early 1997 ordered a major reorganization of the energy sector. Three ministries dealing with mineral resource management and licensing, oil and gas, and coal and electricity were merged into one ministry of energy and natural resources. A separate committee for investment would handle negotiations with foreign oil companies. Finally, a national oil company, Kazakh Oil, would manage the government's oil and gas business interests.

The government welcomes investments by foreign oil companies. The best known, attractive fields not already developed have been or are in the process of being awarded to joint ventures (JVs) between Kazakh and foreign firms. The JVs in 1996 accounted for 23 percent of total Kazakh oil production and about 50 percent of its gas production, and rising.

Oil was discovered in Kazakhstan in 1911 by the Nobel brothers. The subsequent decades saw the discovery and development of a number of smaller fields. In 1979, the north Caspian basin became the talk of the Soviet oil industry as exploration at greater depths along the coast resulted in the discovery of the huge Tengiz field. Crude oil and condensate production—all of which is onshore—peaked in 1991 at 533,000 bls/d. By 1996, output had slowed down to 433,000 bls/d, however.

Because of the enormous distance between western Kazakhstan, where the bulk of oil production takes place, and the northeastern and southeastern parts of the country, where the main consumption centers are located, Kazakhstan exports a high share of output to Russia and, via Russia, to other former Soviet republics and the world markets; in return, it imports Siberian crude to its eastern refineries. This arrangement with Russia is cumbersome, however, leaving Kazakhstan vulnerable to Russian refineries to allocate orders to Kazakh producers, to Transneft to allocate pipeline space to Kazakh producers, and to Russian producers to sell crude to Kazakh refineries. The Kazakh output dropped between 1991 and 1995, mainly because Russian refineries and export pipelines were swamped with Russian crude.

The Kazakh ambition of joining the ranks of world-class oil exporters centers on the Tengiz field and the Kazakh sector of the Caspian Sea.[3] As for Tengiz, a production association that was established in Soviet times to develop the field ran into technical problems and production did not commence until 1991, and at a low level, at that. The Gorbachev government left the development of the Tengiz field up to a Soviet-American joint venture with Chevron, an arrangement to which Kazakhstan has continued to adhere to, counting on Chevron to take production to some 0.7 mbls/d. Progress has been slow; Transneft's restrictions on the venture's access to pipelines is one reason, and the problems encountered by the project to build a new pipeline from Tengiz to Novorossiysk is another reason for the slow pace of progress.

Fields as large or even larger than Tengiz could be discovered in the Kazakh sector of the Caspian Sea. Between 1993 and 1996, a Kazakh and six foreign oil companies carried out a major seismic survey of the 103,000 square kilometer sector, with some encouraging results. The foreign companies have established a consortium to further explore and possibly develop fields in a 6,000 square km area of their own choice. Other companies will be invited to tender for development rights in the remaining sectors.

The Pavlodar and Shymkent refineries in eastern Kazakhstan have distillation capacities of 8 mmt/y and 6.5 mmt/y, respectively, while a third refinery located at Atyrau on the north side of the Caspian Sea has a capacity of about 5 mmt/y. By standards of the former Soviet republics, the Kazakh refining industry is relatively sophisticated. However, the output mix remains dominated by heavy products.

Gas production reached 6.2 bcm in 1992. Low levels of activity at the key Karachaganak field, and declines in associated gas production reflective of declines in oil production, resulted in a drop in output to 4.2 bcm in 1996. Located across from the Russian Orenburg field, across from the Russia-Khazakstan border, Karachaganak was discovered in 1979 and put on stream in 1984. Like Tengiz, Karachaganak proved difficult to develop, and after the breakup of the Soviet Union, Kazakh authorities decided to bring in foreign capital and know how to complete the job. Two Western companies, British Gas and Agip, were awarded the right to negotiate a field development plan and a PSA. More recently, Gazprom and Texaco have become partners, with the former reportedly considering to transfer its stake to LUKoil. However, disagreement on how to dispose of the output from the field and—again—a lack of willingness on the part of Gazprom to grant access to its export pipelines, have led to declines rather than increases in investment.

Kazakhstan's gas balance for 1995 shows a deficit of about 4.5 bcm, corresponding to roughly half of domestic use. The republic exports Karachaganak gas to Russia while the main consumption centers in the east are supplied exclusively with gas imported from Turkmenistan, Uzbekistan, and Russia.

Assuming the building of an oil export pipeline from Tengiz to the Black Sea, the Kazakh government until recently projected oil production to increase to about 0.9 mbls/d by the year 2000 and 1.6 mbls/d by the year 2010 and gas production to reach 27 bcm by the year 2000. During the spring of 1997, however, the oil output target was apparently revised to 3.4 mbls/d. Kazakh officials seem to think that offshore reserves could support much higher production levels than previously assumed. However, they have not articulated publicly a sound plan aimed at getting the additional oil to world markets.

Turkmenistan

Until 1993, the Turkmen petroleum industry comprised a number of independent "concerns," such as Turkmenneft in charge of oil production and Turkmengaz responsible for gas production and transport, both being the successors of the Turkmen divisions of the Soviet Union's petroleum industry. In 1993, President Saparmurat Niyazov abolished all of them and put their component enterprises under the supervision of the ministry of oil and gas. In 1996, however, he issued a new decree whereby the ministry of oil and gas was replaced by an integrated ministry for the oil and gas industry and mineral resources. Apparently the reorganization implies a centralization of petroleum-related decision-making powers.

Western oil companies do not play too important a role in the Turkmen oil and gas sector. Those few that have entered into JVs with the government to produce and export oil have had a strained relationship with their partner. One, Argentinean Bridas Corporation, considered taking the government to court for revoking its export permit. However, in 1996, an Asian company signed a PSA with the Turkmen government, and an American major oil concern has signed a protocol covering a wide range of upstream activities. Besides, Russia is eyeing Turkmenistan's resources; in 1995, Gazprom and the Turkmen government formed the TurkemenRosGaz JV, with a mandate extending from exploring for oil and gas on Turkmen territory to exporting Turkmen gas to, and possibly beyond, the former Soviet Union.

Turkmenistan has been an oil producer for 85 years. Output peaked in 1970 at about 320,000 bls/d. By 1990, production was about 114,000 bls/d, and during the first half of the 1990s it fell by another 40 percent. The oil industry is characterized by extremely low well productivity. Most of the crude, produced domestically and/or imported from Uzbekistan, is processed at Turkmenistan's two refineries, which have a combined distillation capacity of about 8.5 mmt/y; both are primitive by Western standards.

Gas was discovered in Turkmenistan in 1951. Production peaked in 1989 at almost 90 bcm; it dropped to 35.2 bcm in 1996. Output has fallen mainly because of: (1) decline in gross national product; (2) decline in energy consumption in the former Soviet republics; (3) cuts in deliveries to the former Soviet republics because of disagreement over price and non-payment by importers; and (4) decision by Russia to terminate an arrangement whereby Turkmenistan had delivered gas to Russia and had received credit in return with exports of similar amounts of Russian gas to Europe.

A long-term program for the oil and gas sector adopted in late 1993 envisages increases in oil production to reach 1.6 mbls/d and in gas output to 220 bcm/y by the year 2020. Evidently, the program assumes the construction of several new export pipelines. Although much can happen over more than 20 years, such optimism appears unfounded. Russian researchers have revisited the results of the last official evaluation of Turkmen gas reserves, carried out in 1988, and noted that, while hundreds of billion cubic meters of gas have been produced, few discoveries have been made since then. This leads to the conclusion that reserves can support a production of only 65 to 75 bcm/y between now and 2005. Moreover, it may prove highly unlikely that markets can be found for up to 200 bcm/y of Turkmen gas.

Uzbekistan

In 1992, a national oil and gas company, Uzbekneftegaz, was established on the basis of the existing enterprises in charge of oil and gas upstream activities, oil refining and product distribution, gas transmission and distribution, and petroleum facility construction. More recently, also, the enterprises responsible for petroleum exploration have been placed under the Uzbekneftegaz umbrella.

Not many foreign oil companies are active in Uzbekistan. In 1993, various blocks were put out to international tender, but apparently no bids were submitted. More recently, however, several major foreign companies have decided to carry out joint exploration projects with Uzbekneftegaz, and others are involved in the planning and implementation of downstream projects.

Uzbekistan has been an oil and gas producer for more than 100 years, but field development on a large scale commenced only in the 1950s. In 1996, crude oil and gas condensate output amounted to about 150,000 bls/d. Unlike the trends in the rest of the former Soviet republics, Uzbek production continued to increase after the breakup of the Soviet Union. There were three reasons for this: (1) the bulk of output was consumed domestically, (2) the Uzbek economy had contracted less than the average for the former Soviet republics, resulting in only a minor decline in domestic oil and gas use, and (3) on the supply side, there had been a timely phasing in of new fields as production from old ones were declining.

Uzbekistan has in recent years exported some crude oil and condensate to Turkmenistan, but the bulk of output is refined domestically at two refineries located in the Fergana valley, in the eastern part of the republic. A third refinery is under construction in the Bukhara region. Uzbekistan is more or less self-sufficient in refined products.

Gas production in 1996 amounted to 48.1 bcm. The bulk of output is processed at the Mubarak gas processing plant and the Shurtan gas chemical complex. Uzbekistan is a relatively minor gas exporter; since 1985 sales to Kazakhstan, Kyrgyzstan, Tajikistan, and Turkmenistan have fluctuated between three and five bcm/y. Several times, deliveries have been reduced below contracted levels or halted, as buyers have been unable to pay. Uzbek authorities hope to increase oil production to 180,000 bls/d shortly after the turn of the twenty-first century, and increase gas production to 60 bcm/y by the year 2000. Unlike some other Central Asian republics' output targets, the Uzbek estimates seem realistic.

Others

Producing a total of 5,500 bls/d of oil and 65 million cubic meters (mcm) of gas in 1996, Kyrgyzstan, Tajikistan, Armenia, and Georgia are almost invisible in the context of global hydrocarbons production; they are also tiny by regional standards, more so in the last 15 years, during which their combined oil and gas production plummeted by 95 percent and 85 percent, respectively.

Presently, Kyrgyzstan's oil production meets only some 6 percent of its domestic oil production needs; the production is sent to Uzbekistan for refining at the Fergana refinery. Kyrgyzstan itself has no refinery; a small-scale refinery has been under construction, however. Gas production covers only about 5 percent of the electricity industry's and residential consumers' needs, with Kyrgyzneft—the state enterprise responsible for oil and gas exploration and production—starving for funds.

The Tajik oil and gas production stems from a handful of fields in the Fergana and south Tajik basin. Like Kyrgyzstan, Tajikistan has to send its crude to Uzbekistan for refining. Currently, crude output corresponds to about 1 percent of refined product consumption while gas output meets some 5 percent of domestic needs.

In Georgia, oil production peaked at about 60,000 bls/d in the early 1980s, declined to 3,000 b/d by the time of the breakup of the Soviet Union and amounted to 2,500 bls/d in 1996. A refinery at Batumi, in 1996, reportedly operated at 0.2 percent of its original nameplate capacity. Gas production is close to nil.

Armenia does not produce any oil or gas. It has seen quite a bit of exploration, however. Many of the exploration wells drilled in Soviet times did not reach planned depths, and so Armenia is trying to entice foreign companies to come and find out for themselves what the republic may have to offer.

Existing Oil Pipelines

Presently, there are two operable oil export pipeline systems out of Central Asia. One line with a capacity of about 240,000 bls/d runs from Atyrau on the north coast of the Caspian Sea to Samara in Russia. Oil from the Tengiz field is fed into this pipeline. Another line with a capacity of 140,000 to 160,000 bls/d runs from the fields northeast of the Caspian Sea to Orsk in Russia. A third pipeline running from Atyrau along the north coast of the Caspian Sea to Astrakhan, Komsomolskiy to Grozny in Chechnya, is apparently not operable but, as discussed below, it will be integrated into the Caspian Pipeline Consortium (CPC) project.

Kazakhstan's crude oil exports to areas beyond the former Soviet Union are routed through the Russian port of Novorossiysk on the Black

Sea. The combined throughput capacity of the pipelines serving this terminal is reported at 640,000 bls/d, and the capacity of the terminal itself is of the same order of magnitude. While the pipelines carrying Russian and Kazakh oil to Tikhoretsk in the Krasnodar territory have spare capacity, the pipelines running from Tikhoretsk to Novorossiysk are frequently operated at above full capacity and constitute a bottleneck for Russian as well as Caspian oil exports. Also at the Novorossiysk terminal itself there have been congestion problems.

The Central Asian republics import Siberian crude oil via a pipeline that crosses the Russian-Kazakh border south of Omsk and runs to the Pavlodar refinery in northeastern Kazakhstan and Shymkent refinery in southern Kazakhstan and then enters Uzbekistan and finally ends at the Chardzhou refinery in Turkmenistan. Its capacity is reported at 540,000 bls/d to Pavlodar and 340,000 bls/d for the rest of the way. For intra-Central Asian oil trade there is a small pipeline with a capacity of some 40,000 bls/d linking Kyrgyzstan and Tajikistan with Uzbekistan.

As for the area lying west of the Caspian Sea, a crude oil pipeline between Grozny and Baku with a nameplate capacity of some 180,000 bls/d was built in the early 1980s to transport Chechen crude to Azeri refineries. This line has recently been idle, but is being rehabilitated and modified to move some of the oil produced at the first stage of the Azeri-Chirag-Guneshli project, in the opposite direction. There is also an old and partly inoperable pipeline connecting Batumi and Supsa, on the coast of Georgia, with Tbilisi; the line will be rehabilitated and upgraded to carry the rest of the AIOC's "early oil" out of the region.

Existing Gas Pipelines

The bulk of Central Asia's gas exports are transported out of the region via a pipeline system running northwestward along the Turkmenistan-Uzbekistan border and on through western Uzbekistan and southwestern Kazakhstan. South of the Aral Sea, a line branches off in a northbound direction, skirts the sea and runs through Kazakhstan to Chelyabinsk in Russia. At Makat, in Kazakhstan, the main system forks into one line running westward along the north coast of the Caspian Sea and turning southward to link up with the grid in the northern Caucasian and Transcaucasian republics, and one line continues northwestward to Alexandrov Gay in Russia. All these lines are supplied from fields in eastern Turkmenistan and Uzbekistan. Gas from western Turkmenistan is sent northward on a pipeline joining the main system at Beyneau in southern Kazakhstan.

About 63 bcm of gas a year can be moved from Gazli in Uzbekistan to Makat, while the line from Makat to Alexandrov Gay has a capacity of 35 bcm/y and the one from Makat to the Caucasian republics has a capacity of 28 bcm/y. The line to Chelyabinsk at present can handle about 3.2 bcm/y, while the line from Okarem to Beyneau has a capacity of some 1.6 bcm/y. The Soyuz export pipeline system originates at the Russian Orenburg gas field across the border from the Kazakh Karachaganak field, traverses northwestern Kazakhstan and picks up Central Asian gas at Alexandrov Gay; it has a design capacity of about 75 bcm/y.

Within Central Asia there is a pipeline system allowing for exports of Turkmen and Uzbek gas to Tajikistan, Kyrgyzstan, and southern Kazakhstan. The Transcaucasian republics receive Russian and Turkmen gas via a pipeline system that enters Georgia from Ingushetia and crosses Georgia from north to south and Azerbaijan from west to east. Armenia is served by various offshoots of this system, of which only one is currently operable. Azerbaijan is also at the receiving end of the Iranian Igat-1 export pipeline.

Regional Oil and Gas Consumption

Energy consumption data for the Central Asian and Transcaucasian republics are scant and unreliable. The observations and estimates that exist leave no doubt, however, as to the following: (1) in most of the republics, total primary energy supply (TPES) per unit of gross domestic product (GDP)—that is, the energy intensity of the economy—is very high by international standards; (2) generally, TPES has declined sharply since the breakup of the Soviet Union; (3) GDP has dropped even more, however, implying that energy intensities have further decreased; and (4) decline rates have varied significantly across countries, reflecting differences in resource endowment, political circumstances, and economic policies.

The Caspian countries differ strongly from each other also in terms of the fuel structure of their energy consumption. Whereas Kazakhstan and Kyrgyzstan depend on coal for about 60 percent and 30 percent of TPES, respectively, in the other republics solid fuel's share of primary energy supply ranges from 0 percent to 4 percent. Apart from Kazakhstan, the petroleum producing countries consume mainly oil and gas, with Uzbekistan and Turkmenistan depending on gas for three-fourths and two-thirds of TPES, respectively. The smaller countries in the area rely to a higher extent on primary electricity.

Forecasting the Caspian countries' energy consumption in general and oil and gas use in particular is extremely difficult. Economies in transition move in unpredictable ways, and even more so in an area politically as volatile as the Caspian. Moreover, little can be inferred from the countries' own history, and even less from other countries' experiences, about how economic growth, energy price changes, and other factors will affect energy use and fuel shares. Suffice it to say, the lifting of supply constraints on fuel use, outlook for economic growth, structural changes, energy price increases, plans for the electricity sector, and similar matters may cause increases in the 40 percent to 50 percent range in both oil and gas use between 1995 and 2010.

Gas Export Markets

Oil companies investing in oil fields do not need to worry about exactly where and to whom to sell the oil. They only need to know that the crude can be moved to world markets, i.e., piped or trucked or moved in any other way to an export terminal on a world ocean or into a trunk pipeline system leading directly to refineries. Above all, they do not need to worry about selling the oil before it is produced.

A company investing in a gas field, however, needs to locate buyers and sign contracts even before starting to develop the field. Gas is a network-bound product and cannot easily be re-routed following market signals. Only in the United States are there reliable gas spot markets. Producers in the Caspian area may only gain access to European and Asian markets that by and large are governed by long-term contracts and they, therefore, must sell their gas in addition to providing for transportation in advance of starting field development projects.

Caspian gas could in principle be exported to the following destinations: (1) other former Soviet republics, including in particular Ukraine and perhaps Russia; (2) Turkey and Iran; (3) eastern, central, and western Europe; (4) Pakistan, India, and other points in southwest Asia; and (5) China, Japan, and other points in east Asia.

For years, Ukraine had been the main market for Turkmen gas, and since 1993 it has been the only market outside the Caspian region for Turkmen gas. Ukrainian fuel debts have mounted and Turkmenistan has repeatedly stopped deliveries to force payment. However, Ukrainian gas importers have their own problems collecting money from industrial and municipal gas users. For the time being, Ukraine may be perceived as a market of last resort, and exports may be held up mainly because the alternative, which is to leave it in the ground, is even less attractive than continuing the struggle over prices and payments. However, Ukraine's problems are transition problems and, though may drag on, they are not expected to become permanent features of the Ukrainian economy. When the recovery kicks in and the country regains its ability to pay, the Ukrainian gas market, which is forecast to increase from about 75 bcm to 110 bcm over the next 15 years, could become a prime market for Caspian as well as other gas producers.

The Turkish gas market is small but dynamic; Turkish authorities anticipate growth in consumption from 6.4 bcm in 1995 to 30, 40, or even 60 bcm/y by the year 2010. Turkey imports gas from Russia and will continue doing so. However, the state gas importing company, Botas, is negotiating in all directions for further supply. By spring of 1997, contracts, memoranda of understanding, and letters of intent had been signed with some 10 different suppliers, including Turkmenistan. Independent observers generally find Turkish gas demand forecasts excessive and suspect that Botas is over-contracting. Thus, how much Caspian gas will actually find its way to Turkey remains to be seen.

In addition to the gas produced domestically, the central and eastern European (CEE) countries currently consume only Russian gas. This will change. The Czech Republic has contracted for delivery of 53 bcm of Norwegian gas over a 20-year period and Poland has signed a letter of intent for the importation of 2 bcm/y of Dutch gas. The other CEE countries also talk about diversifying imports, and Caspian gas—initially imported via Russia, later perhaps via Turkey—is one option. However, both the Czech Republic and Poland have emphasized that Russian gas will remain the backbone of their supply. Other countries will have even greater difficulty breaking away from Russia's Gazprom. Few of them can afford to ignore that Russian gas comes cheaper than most other gas and that Gazprom is more flexible on payment terms than most other suppliers, accepting goods as well as transit services and project work in lieu of cash. Central Asia could get to supply a share, perhaps not a large one, of the CEE gas market, which is expected to total 70 to 80 bcm/y by the year 2010.

The Western European gas market holds a special attraction for the Caspian gas producers/exporters. It is big, has recently been quite dynamic and, unlike some other actual or potential markets, it is not ridden by payment and currency instability problems. Western European gas use is projected to increase by some 50 percent, from 350 to about 525 bcm between 1995 and 2010, mainly because of increased use of gas in the power sector. However, already-contracted supply corresponds to a large share of forecasted demand, and competition among suppliers over that part of the market that is not already provided for, could be fierce. Turkmenistan is mentioned among those distant suppliers western Europe might have to call upon around the year 2010 but, again, volumes would not be large.

To the south, Pakistan and India are as yet minor gas users, but consumption could increase from about 37 bcm in 1995 to perhaps 120 bcm/y by the year 2010. Currently, demand is supplied entirely from indigenous production, but it will not be for much longer. Neither country has proved reserves to match the forecasted demand; both are drawing up import schemes. The Caspian countries would be well-placed to supply Pakistan and India, if India could bear the thought of being at the end of a gas pipeline passing through Pakistan. A pipeline bound for southwest Asia could cross Afghanistan, but that country for years has been without a functioning central government and it is at the bottom of most if not all lists of countries eligible for project financing. For the moment, Iran, Qatar, and other Middle Eastern countries could be ahead of the Caspian countries in the race to gain footholds in the Pakistani and Indian gas markets, but the outcome is not predetermined. Aside from the former Soviet republics, Pakistan remains a strong candidate to become the biggest taker of Caspian gas.

China and Japan are sometimes mentioned as possible future markets for Caspian gas. Indeed, both of these countries face increasing import needs and will have to sign up new suppliers. The building of pipelines from Turkmenistan and Kazakhstan to the Yellow Sea have been under discussion for a while. Distances are daunting and costs would be very high, however. Russian gas and south Asian liquefied natural gas (LNG) seem more competitive alternatives. Doors to the Chinese gas market will hardly be flung open for Central Asian gas producers any time before the year 2010, possibly not even before 2020.

Possibly, Iran and Russia each could become takers of Caspian gas, even though both have enormous gas reserves of their own, with Russia being the world's biggest gas producer and exporter. Iran is actually preparing for imports of Turkmen gas; in northern Iran this gas will come cheaper than Iranian gas produced in the south of the country. For Russia, it could be cheaper to supply incremental European demand for Russian gas by importing Caspian gas to southern Russia, thereby freeing up more Siberian gas for exports, instead of developing the giant but high-cost fields on the Yamal peninsula. Whether Gazprom is thinking along these lines remains to be seen, but if the Russian company decides to optimize its sourcing of gas in this fashion, then the volumes of import of Caspian gas could be significant.

Pipeline Projects

The potential Caspian oil production by the year 2010 is estimated to be in the range of 3 to 5 mbls/d. Meanwhile, according to estimates, the local oil consumption by then hardly would reach 1 mbls/d. The capacity of existing export pipelines is about 400,000 bls/d. Therefore, a rapid build-up of export pipeline capacity must take place to ensure the flow outward. If such a build-up does not happen, output cannot be increased to its potential. The same holds for gas. While production could be raised to perhaps 200 bcm/y by the year 2010, local gas use may not grow beyond 100 bcm/y, implying a need for export pipeline capacity of up to 100 bcm/y, which far exceeds current capacity. Some pipeline projects are at an advanced stage of preparation, others remain on the drawing board. It remains very much to be seen whether pipeline capacity will materialize to the required extent and on time.

The AIOC Oil Pipeline Project

In early 1993, Azerbaijan signed an agreement with Turkey for construction of an oil pipeline system with a capacity of 800,000 b/d, to extend from Baku via Armenia or Georgia to eastern Turkey and onward to an export terminal at Ceyhan on the Mediterranean. Russia quickly started to promote the Baku-Novorossiysk option for Azerbaijan's oil, arguing that investment costs would be lower and that Russia as a major oil producer and exporter itself would be able to guarantee deliveries. During 1995, another tug-of-war turned on whether to export "early oil" produced by the AIOC via Novorossiysk or via a terminal on Georgia's Black Sea coast. Discussions became heated as both Turkey and Russia assumed that the choice of a route for "early oil" would tip the balance in favor of that route for "main oil"—that is, the volume that will be produced when all fields come on stream. In October, 1995, a decision to split "early oil" exports among the two routes was announced, allowing both sides to declare victory.

In late 1995 and early 1996, Azerbaijan, Russia, and Georgia signed framework agreements, while AIOC, SOCAR, Transneft, and the newly established Georgian International Oil Corporation entered into commercial agreements. The implementation of plans to rehabilitate existing pipelines and build new segments along the two routes has taken longer than expected. No work on the Russian segment of the Baku-Novorossiysk line was done due to the conflict in Chechnya, and as of spring 1997 it remained unclear whether Transneft would meet the August 1997 deadline for getting the line operable. As for the Georgian route, rebuilding contracts have been awarded and the line could be ready on schedule by the end of 1998. Meanwhile, by mid 1997, the AIOC was to decide on one or more export routes for "main oil."

The CPC Oil Pipeline Project

In 1992, Kazakhstan and Oman agreed to set up the Caspian pipeline consortium with the purpose of moving Kazakh crude from western Kazakhstan to international markets. Later, Russia decided to join the CPC. CPC opted for constructing a pipeline from Tengiz around the north coast of the Caspian Sea to Novorossiysk, making use of existing but inoperable pipelines and other infrastructure along the way wherever possible.

Financing for CPC remained a problem. The consortium partners wanted Chevron—the Kazakh government's partner in developing Tengiz—to put up most of the money, but Chevron refused because of dissatisfaction with the proposed distribution of shares and voting rights in the pipeline. The Oman Oil Company, representing the Omani government, undertook to raise money for a first stage of the project, but failed to meet deadlines.

In April 1996, a new version of the CPC, called "CPC-II," was launched. The governments had their shares reduced to a total of 50 percent with eight oil companies taking up the other 50 percent. The government shareholders would provide rights of way and existing facilities while the burden of arranging for finance would fall on the companies. Working out the final distribution of shares and detailed pipeline management and tariff arrangements took much longer than expected, however. In May 1997, the parties managed finally to agree on the building of the line, with work scheduled to start in 1998. The capacity of the line will be increased gradually from 560,000 bls/d to 1.34 mbls/d.

The Central Asian Oil Pipeline

Unocal, an American company, and Delta Oil Company of Saudi Arabia have proposed to construct a crude oil export pipeline with a capacity of about 1 mbls/d from Turkmenistan via Afghanistan to a terminal on the Pakistani coast on the Arabian Sea. The companies argue that this is the most logical route for supplying Central Asian, Russian, and Azeri crude to the rapidly growing Asian oil markets. They also plan to build a gas export pipeline from southwestern Turkmenistan to Sui in Pakistan and they see a potential for synergism between the two projects.

The pipeline would originate at Chardzou in eastern Turkmenistan, the site of one of the country's two refineries and the terminus of a pipeline carrying mainly Russian crude oil to refineries in Kazakhstan and Turkmenistan. It will run southward through the area where the bulk of Turkmenistan's gas reserves are located, enter Afghanistan north of Herat, continue southward parallel to the Afghan-Iranian border, enter Pakistan's Baluchistan province, and extend southwestward onto an as-yet-unspecified point on the coast. With a total length of about 1,670 km, the line would have to transit some 700 km of Afghan territory.

While the project may make economic sense, the parties have to deal with the fact that a pipeline from Turkmenistan to Pakistan would have to transit Afghanistan from north to south, and that the latter country seems completely ungovernable. The emergence of the Taliban was greeted as a promising development, but once in control of Kabul, the Taliban have acted in ways that lead Western companies to doubt their suitability as partners in major projects. Besides, their future control of all of Afghanistan seems far from assured.

Turkmenistan-Turkey-Europe Gas Pipeline

Since 1993, Turkmenistan has promoted the idea of building a gas export pipeline from Turkmenistan via Iran to Turkey. In 1993 and 1994 the government tried to get an ambitious two-stage project off the ground. Initially, a system with a capacity of 15 bcm/y would be built with a view to supplying the Turkish market. Exports would start at 2 bcm in 1998 and increase gradually to 15 bcm/y year from the year 2010 onward. In time, the capacity of the system would be increased to 28 bcm/y with a view to supplying markets further west. However, funding remained an insurmountable problem. International financial institutions (IFIs), foreign private lenders, and foreign companies alike refused to participate in a scheme involving Iran, and neither Turkmenistan nor Iran could afford to foot the bill. Currently, the basic idea is being pursued through a series of less grandiose schemes, namely the construction of a pipeline from Turkmenistan to northeast Iran feeding gas into the Iranian transmission pipeline network, continuing from Tabriz in northwest Iran to the Iran-Turkey border, onward to Erzurum in eastern Turkey, ending in Ankara.

Turkmenistan-Pakistan Gas Pipeline

In 1993, Turkmenistan's president, Islam Niyazov, and Pakistan's prime minister, Benazir Bhutto, signed a memorandum of understanding on the building of a gas export pipeline from Turkmenistan through Afghanistan to Pakistan, allowing for exports of around 20 bcm/y. Early in 1995, it appeared that the Bridas Corporation, an Argentinean company, would be in charge of the project. Later in 1995, however, a consortium made up of Unocal and Delta Oil of Saudi Arabia launched a competing project, and quickly gained the support of the Turkmen government, which revoked Bridas's export license. This fall from grace has caused Bridas to seek legal remedies against Unocal for having stolen its plans, while mobilizing support and financing to go ahead with its project. Again, however, exporting Caspian gas to Pakistan may be a very good idea from an economic point of view, but the Afghanistan factor could impede the project for the foreseeable future.

Turkmenistan-China-Japan Gas Pipeline

In 1994, Turkmenistan and China signed a protocol of intent to build (1) a gas pipeline with a capacity of 28 bcm/y from Central Asia to the east coast of China, and (2) facilities for either a sub-sea pipeline or an LNG system to transport part of the gas onward to Japan. The Mitsubishi Corporation, Exxon, and other companies have looked into the project and apparently have not dismissed it. However, building a 6,000 to 7,000-

km long pipeline across several countries would be a colossal undertaking; costs are tentatively put at $12 billion. Given supply costs and the outlook for gas prices in China and Japan, the competitiveness of Turkmen gas in eastern China and Japan appears highly uncertain. Exports of Central Asian gas eastward appears to be a better option for now and for the foreseeable future.

Kazakhstan-China Pipeline

In September 1997, China and Kazakhstan signed a series of oil agreements valued at some $9.5 billion. One agreement awarded to China the rights to develop some of Kazakhstan's largest oilfields. Another agreement provided for the building of a 1,860-mile oil pipeline from Kazakhstan to Xinjiang province in western China. More significant than delivering Kazakh oil to China, the China-Kazakh deal will further weaken Russia's hold on Kazakh crude exports through its territory, while at the same time balance the Western, primarily American, role in the development of Kazakh hydrocarbon resources.

Conclusion

Currently, the lack of export pipeline capacity is often cited as a constraint on Caspian petroleum exports and therefore on production. This may be true for oil, but it is hardly true for gas. Oil exports require only access to an export terminal on a world ocean. Gas exports, however, require contracts with buyers, and Caspian gas may not be as easy to place as some Caspian countries seem to believe.

In the longer run, reserves and production costs could emerge as constraints on oil production. If especially the Kazakh sector of the Caspian Sea lives up to expectations, oil output growth rates could be formidable. If not, output could level off earlier and without making a tremendous difference to oil importers' security of supply, let alone world oil prices. Gas output growth rates will probably remain constrained by market rather than reserves conditions.

Regardless of how big the Caspian area will become as a petroleum producing area, clearly Azerbaijan and Kazakhstan and probably also Turkmenistan and Uzbekistan will receive major growth impulses from their petroleum industries. Their longer-term prospects would be bright provided they do not continue supporting inefficient energy consumption, delay overdue economic restructuring, allow oil and gas revenues to be snapped up by swollen bureaucracies, provide subsidies, or spend on white elephants.

As a transit country for a share of Azerbaijan's oil exports and, possibly, other Caspian petroleum exporters, Georgia will earn transit fees, gain access to comparatively cheap oil, and benefit in other ways; the decision to route some "early oil" via Georgia has triggered considerable foreign interest in the country, and a favorable decision on "main oil" could make foreign investors flock to Georgia. It is harder to see how the other smaller countries in the Caspian area could benefit from petroleum-driven growth in Azerbaijan, Kazakhstan, and possibly Turkmenistan and Uzbekistan. Consequently, the future could well bring a widening of income differences in the area.

Notes

1. The data reported in this chapter are based on the information found in the following sources: Sabbath Bagirov, "Azerbaijani Oil: Glimpses of a Long History," in *Perceptions* (June-August, 1996); *East Bloc Energy* (various editions); *East European Energy Report* (various editions); Economist Intelligence Unit, *Country Profiles* and *Quarterly Reports* (various reports on Central Asian and other republics); European Bank for Reconstruction and Development, *Transition Report* (London, 1996); European Bank for Reconstruction and Development, *Transition Report Update* (London, 1997); *Gas Matters* (various editions); International Energy Agency, *The IEA Natural Gas Security Study* (Paris, 1995); International Energy Agency, *World Energy Outlook* (Paris, 1996); International Energy Agency, *Energy Policies of Ukraine* (Paris, 1996); International Energy Agency, *Energy Statistics and Balances of Non-OECD Countries, 1993-94* (Paris, 1996); Keun-Wook Paik, *Gas and Oil in Northeast Asia* (London, 1985); PlanEcon, Inc., *Energy Outlook for the Former Soviet Republics* (Washington, D.C., June 1996); Petroconsultants (Geneva), *Foreign Scouting Service: Commonwealth of Independent States* (various editions); *Petroleum Economist* (various editions); PetroStudies Company, *Gas Business in Russia and other FSU Countries,* Vol. 7: Central Asia (Malmö, Sweden, 1995); Richard Pomfret, *The Economies of Central Asia* (Princeton, 1995); *Russian Petroleum Investor* (various editions); Jonathan P. Stern, *The Russian Natural Gas "Bubble"* (London, 1995); The World Bank, *Kazakhstan Energy Sector Review* (Washington, D.C., 1993); The World Bank, *Azerbaijan Energy Sector Review* (Washington, D.C., 1993); The World Bank, *Uzbekistan: An Agenda for Economic Reform* (Washington, D.C., 1993); The World Bank, *Turkmenistan* (Washington, D.C., 1993); The World Bank, *World Development Report 1996* (Washington, D.C., 1996); The World Bank, *Natural Gas Trade in Asia and the Middle East* (Washington, D.C., 1996).
2. Petroconsultants, *Foreign Scouting Service: Commonwealth of Independent States* (Geneva: January 1997) and other editions.
3. The notion of national sectors of the Caspian Sea is disputed. Azerbaijan and Kazakhstan hold that the entire sea should be partitioned among the littoral states; they act as if such a regime for the sea has already been adopted. Russia and Iran propose that only a coastal zone should be reserved for littoral state management while the area in the middle should be jointly exploited. Turkmenistan has been vacillating on the issue.

CHAPTER THREE

Survey of Iran's Economic Interests in the Caspian

Bijan Khajehpour-Khouei

Introduction

Iran is known predominantly as an oil economy; oil exports constitute the country's major source of foreign exchange. Regardless, Iran's agriculture, at 20.9 percent of the gross domestic product (1996/97), leads all the other productive sectors; it is followed by oil, gas, and energy (17.6 percent), trade and commercial services (15.7), industry (14.5), transportation and communications (5.8), construction (4.4), and mining (0.5), while non-commercial services account for 21.1 percent of the gross domestic product (GDP).[1] By the same token, oil and gas dominate every international discussion about the Caspian; very little is said or analyzed in terms of the region's demography, non-petroleum resources, and the other forms of economic activity. The Iranian coastal plain wedged between the Alborz Mountain chain and the Caspian Sea is a thriving, robust part of Iran's overall economy. This chapter examines Iran's diverse economic interests in the Caspian.

Geography and Demography

Iran's Caspian region is divided into the provinces of Gilan in the west and Mazandaran in the east, together having a total area of 61,465 square kilometers (approximately 24,000 square miles), which represents about 4 percent of Iran's total area.[2] The coastline stretches for some 657 kilometers (410 miles). The region has some of Iran's best agricultural soil and arable lands, resulting in this tiny part of Iran being responsible for some 40 percent of the country's agricultural activity. In view of the government's emphasis on self-sufficiency in the food sector, the Caspian region's role in this regard may be even greater in the future.[3]

The combined population of Gilan and Mazandaran in 1996 stood at some 6.27 million, giving the region Iran's highest population density.[4] The region's labor force stands at about 1.4 million. Given the agricultural economy of the region, some 51.2 percent and 42.9 percent of the labor force of Gilan and Mazandaran, respectively, work in the agriculture sector, which includes also fisheries and forestry.[5]

Agriculture

Iran's Caspian region produces rice, wheat, tea, barley, citrus fruits, vegetables, and tobacco. This is made possible by excellent soil conditions, an abundance of low salinity water reserves, an annual average rainfall of 200 millimeters, the Alborz range which traps the region's humidity, moderate temperatures in part due to the Caspian, and availability of labor.[6]

Some 479,000 farming units covering an area of 1.14 million hectares are registered in Iran's Caspian provinces, employing some 629,000 people. Underlying the region's agriculture is also animal husbandry, which occurs at levels well above the national or other provincial averages. In regard to forestry, some 25 percent of Iran's forests are located in the Caspian provinces, where a large number of forestry projects are being carried out, contributing to the growth of the wood industry.

Fisheries

Iran's fisheries company, Shilat, estimates that the Caspian Sea's fauna consists of some 114 different species and 63 subspecies of fish, chief among them being different species of sturgeon, herring, pike, carp, and, occasionally, seals.[7] Historically, however, the most important fish type, from an economic point of view, has been the sturgeon, among whose offerings, besides the flesh, is the roe, caviar.

The export value of caviar exceeds $50 million annually, making caviar one of the most important non-oil export items for the Iranian economy. It is estimated that 90 percent of all sturgeon production and consequently 90 percent of the world's caviar production originates in the Caspian Sea. After Russia, Iran is the next largest producer of caviar and supplies 20 percent of the world market.[8]

Due to the high value of the sturgeon fish, the entire management of this specific fishery lies exclusively with Shilat, a government-owned organization. This centralized state management offers two advantages: first, potential for malpractice in areas of capture and resource management are minimized because of tight government control of catch activities; second, government resources are brought to bear in fields of research and restocking. The relatively positive aspects of the sturgeon management in Iran notwithstanding, over the past decade the level of catch has decreased significantly, which indicates a depletion in the resource.[9]

In addition to sturgeon, other main species include the bony fish, kilka, as well as aquaculture production of different species. With the decline in the catch of sturgeon in recent years, the importance of kilka has risen. Shilat has encouraged the private sector to exploit kilka on a large scale, and the performance has exceeded expectations of the Second Five Year Plan (1995-2000). Currently, the bulk of the kilka catch is used for the production of fish powder; projects are under consideration to produce canned kilka meals for human consumption, which would have a domestic and possibly regional market.[10] Due to the favorable climate and availability of low salinity sea water, aquaculture, with a specific emphasis on carp and other species, is becoming an important area of fisheries in Iran's Caspian Region.

Currently, some 13,000 fishermen are registered in Iran's Caspian regions. It is believed, however, that more people derive their living from fishing in the Caspian than the figures indicate, some perhaps involved in artisanal or unauthorized activities such as catching and distributing sturgeon.

Iran's Shilat has engaged the other littoral states since 1993 in a dialogue aimed at (1) establishing a fisheries quota system, and (2) promoting a joint effort to protect the endangered sturgeon stock. A breakthrough was announced recently when the five littoral states of the Caspian Sea reportedly adopted a $150-million, 10-year conservation program to reverse the depletion of sturgeon. At current levels of fishing, the Caspian's sturgeon catch would fall to some 1,000 tons by the year 2000. With the implementation of the conservation program, special fish farms will introduce several million young sturgeon to the sea each year, hopefully enabling future annual hauls of up to 12,000 metric tons by the year 2010. While this objective seems too ambitious, at least the clear need for collective action on the part of the littoral countries has been recognized.

Tourism

Iran's Caspian shores are a major attraction for Iranian tourists, especially in the summer months, giving this sector a significant role in the regional economy, even though in recent years the rising Caspian waters has laid waste to many tourist facilities and beaches, resulting in a decline in the number of domestic tourists. Regardless, the opening of the borders with neighboring Azerbaijan and Turkmenistan has encouraged a new flow of tourism from Central Asia, Caucasia, and Russia to Iran's Caspian provinces and vice versa. For example, the "Mirza Kouchak Khan" cruise offers tours to Azerbaijan, Russia, and other destinations in the Caspian Region, resulting also in an increase in border trade, which is of increasing importance to the development of Iran's Caspian littoral. In particular, Iran's ministry of roads has initiated ferry services to Baku in Azerbaijan and Astarakhan in Russia, thereby contributing to an increase in tourism among the three countries. The introduction of border markets has developed an interest on both sides for products of the other side, increasing at the same time tourism and small-scale trade.

Industry and Mining

In addition to agriculture and fisheries, Iran's Caspian region is also host to a variety of industries that, in order of importance, include food, textiles, wood, paper, chemicals, non-metallic minerals, metals, machinery, and mining. Industrial activity of the Caspian provinces accounts for just 5 percent of the total industrial output of the economy. Iran's trade with Central Asian and Caucasian countries is expected to result in further industrial activity in Iran's Caspian region. The mining sector there represents some 10 percent of the country's overall mining activity; it employs 9,000 people, producing mainly marble, bauxite, lime stone, lead, zinc, and coal.

Oil and Gas

The Caspian basin as a whole sits atop of the world's third largest hydrocarbon reserves after the Persian Gulf and Siberia. The natural gas and oil reserves of the Caspian are estimated at 10 billion tons (73 billion barrels), exclusive of the onshore reserves.[11] In February 1996, Iran commenced its very first exploration project in the Caspian; the oil platform "Iran Khazar" deployed for the purpose has the capability of drilling in waters up to 6,000 meters deep. The platform itself was built at a cost of $68 million. Iran is well positioned not only to exploit the Caspian's oil and gas resources on and for its own, it also can assist the other littoral countries in their efforts by offering technological services.

Transportation and Ports

The Caspian Sea is an important transportation medium for the littoral states. There are ferry services connecting the ports of Baku in Azerbaijan, Mahach Galeh in Dagestan, Astarakhan in Russia, Aktao and Guriev in Kazakhstan, Turkmenbashi and Aktash in Turkmenistan, and Nowshahr and Anzali in Iran. After the collapse of the Soviet Union, the cargo link between Anzali and Baku was expanded to include regular passenger services. The connections between Nowshahr and Mahach Galeh in

Dagestan and between Anzali and Astarakhan account for a significant share of the seaborne traffic on the Caspian.

The Caspian Sea itself is linked to the Black Sea by way of Russia's Volga-Don Canal. This connection is important because it is the most economical connection between Iran and Europe, it cuts the transportation time by ten times, and offers a climate more favorable for transporting certain types of goods and products.

Under its Second Five Year Plan, by the year 2000, Iran would have spent some Rls. 94 billion (app. $30 million) on improving and expanding its Caspian ports. All Central Asian and Caucasian republics are dependent on their neighboring countries for transit of their goods to terminals on the open seas. Iran offers a relatively developed infrastructure for linking the Caspian region to the rest of the world.

Iran has two main commercial ports on the Caspian, Anzali and Nowshahr. Nowshahr is a purely commercial port, whereas Anzali handles also passenger traffic. The recent designation of Anzali as a "special economic zone," which involves free trade and liberal foreign investment and monetary regulations, will help increase the role of this port city in the regional development.

It is clear that following the collapse of the Soviet Union, Iran has put more emphasis on its northern facilities. The increased attention to the Caspian ports is reflected clearly in the rising volume of goods handled by Anzali and Nowshahr. For example, in 1976, the weight of goods unloaded at Anzali stood at 338,000 tons, while Nowshahr unloaded 118,000 tons; in 1991, Anzali unloaded 417,000 tons and Nowshahr 174,000 tons; in 1993, the volume unloaded at Anzali and Nowshahr rose to 1.3 million tons and 388,000 tons, respectively; in 1995, Anzali unloaded 2.2 million and Nowshahr unloaded 1.6 million tons.[12]

Pipelines

In the period between 1992 and 1995, Iran's trade relations with its northern neighbors increased multifold, accounting in 1995 for 10 percent of Iran's overall foreign trade. Iran has targeted its northern neighbors for its non-oil exports. At the same time, Iran has become an intermediary in the trade between different former Soviet republics and/or the outside world. There are in place a number of bilateral and multi-lateral agreements for the swap and/or transit of goods through Iran, including oil, gas, and raw materials.

The centerpiece of international economic relations among the Caspian countries is Iran-Russia relations. In the early 1990s Iran and Russia found themselves in agreement opposing Azerbaijan's rush to stake out an Azeri "sector" in the Caspian Sea and to begin tendering the same to Western oil concerns, all without consultation with Russia or Iran. This coincidence of interests in the Caspian was further strengthened when the United States began to apply pressure on Iran, a result of which was to question Russia's involvement with certain projects in Iran, including the Iran-Russia project to complete the nuclear power generation plant in Iran's port of Bushehr on the Persian Gulf. The improved and deepened relations between Iran and Russia provide a counterweight to plans by Azerbaijan and Kazakhstan, who insist on the division of the entire Caspian Sea into exclusive areas of sovereign national jurisdiction. For now, at least, Russia has selected Iran as its strategic partner in the Middle East and especially with regard to the Caspian region. There is no doubt that if the present anti-Iranian climate in Western countries persists, Iran will have little choice than to cozy up to Russia even more.

It is a given that pipeline is the best medium for exporting natural gas from the Middle East and the Caspian basin. The countries involved in the pipeline projects also realize that the routing of a pipeline is as important as the production of oil and gas, because the location of the pipeline results in geo-political dependency on the host country through which the pipeline passes. Many pipeline proposals are currently under consideration or development/construction; chief among them are five lines, which all together link Iran to Azerbaijan and Georgia, Turkmenistan, Armenia, Nakhjavan, and Turkey and Europe.[13]

In addition to the foregoing, there are two proposed loop systems. One system, named PEACE (Pipeline Extending from Asian Countries to Europe), would create a gas loop system within Iran to serve the interests of Iran as well as other regional gas exporting countries. First published in May 1993 by the Institute for Political and International Studies in Tehran, this proposal seeks to connect Iran's gas fields with those in Central Asia, Azerbaijan, and, potentially, Iraq.[14] In terms of cost, PEACE compares cheaper than the other loop project entitled UNIDO/Chiyoda/ENI, which covers most of the Middle East, extending to Europe by way of Turkey, and which would include two liquefied natural gas terminals, one in Iran's southeastern port of Chah-Bahar and another in Oman.[15]

Iran's existing IGAT-I and the IGAT-II gas pipeline under construction provide the foundation for the development of a loop/network, which can: (1) unitize and integrate the gas resources of Central Asia, Caucasia, and the Persian Gulf, thus avoiding exorbitant investments by individual exporters in creating multiple lines; and (2) minimize the effects of interruption from individual exporters, as the total volume of gas present in the loop will be fungible. The loop/network would also prompt the participating countries to switch from oil to gas as the cleaner of the two sources of energy. The main gas producers/exporters participating in the loop/network would be Turkmenistan, Uzbekistan, Kazakhstan, Iran, Iraq, Kuwait, Qatar, Saudi Arabia, United Arab Emirates, and Oman. In addition to the exporters themselves, the consumers would include Azerbaijan, Armenia, Georgia, Turkey, Pakistan, India, and Afghanistan.

To actualize any of these plans, the regional actors would have to give themselves over to confidence-building measures. Bilateral agreements may be a suitable starting point. For example, the signing of a 22-year, $20-billion agreement between Iran and Turkey for exports of Iranian and Turkmen gas is one such step in developing mutualities. The agreement, signed in August 1996, provides for the construction of a pipeline to carry annually up to 350 billion cubic feet (app. 13 billion cubic meters) of gas to Turkey, worth some $1 billion.

In addition to gas pipelines, Iran is also actively pursuing the policy of becoming the main export route for the Caspian oil. Debate over the routing of oil pipelines to world markets dominates the developments in the Caspian basin. Friction between the five littoral states on the lake's legal regime and tension between the United States and Iran have been two key factors impeding the emergence of a unanimously acceptable solution for the transport of Caspian resources.

For various reasons, the United States has worked against what seems to be the most economical and reasonable route, that is, a pipeline route through Iran, which would transport the Caspian production to Iran's ports of the Persian Gulf. Even though presently the Caspian oil is delivered to Iran in the north as swap for the oil delivered by Iran at its Persian Gulf outlets, the plan of a viable trans-Iranian pipeline is yet to be tabled by Iranian officials.

In May 1998, Iran announced plans for the construction of a pipeline connecting Neka on the Caspian Sea to Tehran, followed in June by announcing a $400 million tender for the project. The planned 392 km Neka-Tehran pipeline would be the first leg of a project aiming eventually to pipe the oil all the way down to the Kharg terminal in the Persian Gulf.

Through the new Neka-Tehran link and the existing Tehran-Tabriz pipeline, the initial flow of Caspian oil would be pumped to the two large refineries in Tehran and Tabriz with a combined capacity of 350,000 barrels per day. In return for this volume, Iran would provide crude to international customers in the Persian Gulf. In the meantime, the already existing network of pipelines inside Iran, which serve the pumping of crude oil from the country's southern oilfields to the Isfahan, Arak, and other northern refineries, would be reversed to pump oil from north to south.

Once Caspian's crude production outstrips the volume (350,000 bpd) needed by the refineries in Tehran and Tabriz, the surplus could meet the demand for crude at Arak and Isfahan (400,000 bpd). This would free up Iran's crude production in the south for delivery to customers in the Persian Gulf. In the meantime, Iran would expand the existing pipeline to connect Neka directly to the Persian Gulf terminal on Kharg Island.

The trans-Iranian pipeline from the Caspian to the Persian Gulf is often compared, with futility, to the feasibility of alternatives such as an underwater pipeline through the Caspian and a pipeline traversing the war-stricken Afghanistan. Turkmenistan, a country most dependent on Iran's petroleum infrastructure, has voiced support for the Iranian option. A decision by Kazakhstan to export its Caspian oil by way of Iran would generate sufficient volumes to make the Iranian option economical. It is highly likely that Kazakhstan and customers of Kazakh oil would eventually settle on Iran as the outlet for the bulk of the Kazakh production. For now, at least, there is no sign of an economically viable, or even competitive, alternative to the Iranian option.

Caspian Sea Cooperation Council

In the post-Soviet era, Iran proposed the creation of the Caspian Sea Cooperation Council (CSCC). Though the CSCC has not been established formally, its objectives are nevertheless discussed regularly among the Caspian littoral states meeting more or less with regularity. As envisaged by Iran, the CSCC's agenda includes the promotion of: (1) cooperation between the Caspian Sea littoral states in exploitation of the sea's resources, agriculture, energy, and industry; (2) peace in the region through economic cooperation; (3) cooperation in intra-regional and extra-regional trade, especially in the areas of oil and gas; (4) agricultural production with the view to regional self-sufficiency; (5) sea, rail, and road transportation services and industries, shipping, and railway links among the member countries; (6) tourism; and (7) cooperation in fisheries and seafood production, with an emphasis on caviar production.

The creation of free-trade zones is expected to generate new volumes of trade, resulting ultimately in economic development in the Caspian region. However, due to the tensions among some of the Caspian countries, progress has been relatively slow. For the time being, Iran, for one, has pursued the CSCC's objectives in bilateral relations. These have included: (1) linking Iran's railway to Turkmenistan's in May 1996, a clear step toward connecting Iran's transportation infrastructure to the rest of Central Asia, one which could in time facilitate regional trade and development of a common market; (2) establishment of the "Caviar Cartel" in 1994 by the Caspian Sea littoral states in order to regulate the caviar-related economic activities; and (3) creation of a special economic zone in the Iranian port of Anzali, as another move toward promoting regional trade. Furthermore, the Iran-Turkmenistan gas pipeline project is in a way also the realization of Iran's desire to bring about a regional pipeline network for gas exports.

Problem Areas

From Iran's perspective, there are a number of regional issues that require discussion and resolution. These include the legal regime of the Caspian Sea and the state of the environment. Iran insists that the Caspian Sea is a "lake" and that the Caspian resources should be exploited and managed by all littoral countries on an equal basis. Iran argues that due to the Caspian's nature as a closed sea, the "law of the sea" concepts such as territorial waters, continental shelf, and exclusive economic and other zones do not apply. This view is supported by Russia. Azerbaijan and Kazakhstan, on the other hand, insist that the Caspian is a "sea" and it ought to be divided up into territorial waters for individual exploitation by littoral countries. These two opposing views have been the main source of tension in the Caspian region, creating uncertainties with respect to investment in and/or development of offshore resources.

Related to the foregoing, Iran is concerned also about the strategic implications of the Caspian being considered subject to the "law of the sea" regime. In that case, the presence of the major maritime powers, especially the United States, in the region would be unavoidable. That possibility will challenge directly Iran's geo-political position, just as has occurred in the Persian Gulf, where foreign powers are said to be "guarding" the flow of oil. To avoid such a scenario in the Caspian, Iran insists that the Caspian countries alone should be present in the Caspian.

Presently, the Caspian basin is deluged with a whole host of actual and potential environmental challenges. Iran, for one, is concerned about industrial pollution flowing into the Caspian by different rivers, especially the pollution disgorged by the Volga. Another area of concern is the effect of the Caspian's hydrography on the dispersion of polluted waters; the main current in the Caspian flows counterclockwise, which sweeps the pollution from the north and northwest down toward and into Iranian coastal waters.[16] The seaborne pollution generated by urban and industrial centers, coupled with oilfield leaks and seepage will, as it has already, threaten the delicate ecosystem of Iran's wetlands, resulting also in economic loss to Iran's fisheries and agriculture.

Conclusion

In terms of its share of Iran's gross domestic product, the Caspian may not ever rise to the same stature as the Persian Gulf. Regardless, its contribution to Iran's economy will not be insignificant, particularly in the area of agriculture. It is estimated that some 55 percent of the Iranian population is under 20 years of age. Within the next few years, an explosive number of young Iranians will enter the job and housing markets. The

Caspian region would have to provide its share of jobs and growth industries.

In terms of regional geo-politics, Iran's position is an enviable one: cultural and ethnic connections with almost all the other littoral countries, a vast frontage on the Persian Gulf and Gulf of Oman, a longstanding superior technical skill and technology in oil and gas development, a large population in terms of human resources and potential market for others, and a strategically significant entente with Russia. For the foregoing, Iran will continue to play no doubt an important role in Caspian's regional economic and political developments.

The individual Caspian countries' vociferous posturing notwithstanding, it is likely that the legal regime of the Caspian will eventually be resolved along the lines of the compromise reached in the Ashgabat (Turkmenistan) Agreement by Iran, Kazakhstan, Russia, and Turkmenistan, on November 12, 1996. This provides for: (1) each littoral country to receive an exclusive offshore economic zone of 45 miles (approximately 75 kilometers) in width: and (2) there be set up a joint management and exploitation of the remainder of the sea not within areas of national jurisdiction.[17] Needless to say, this agreement may have little value without Azerbaijan's endorsement. The oilfields claimed by Azerbaijan are located some 100 to 120 miles offshore, which under the Ashgabat Agreement would be shared by all the littoral states. The United States' support for the Azerbaijan position notwithstanding, it may prove costly in political and economic terms for Azerbaijan to hold out from the Ashgabat Agreement, simply because with the Nagorno-Karabakh situation unresolved, Azerbaijan may well not be able to afford alienating Iran and Russia.

Russia's constant efforts to become the only major conduit of the Caspian's oil and gas exports will not be completely successful. Foremost, pipeline economics would argue much more favorably for extensions through Iran because of the shorter distances involved, as well as direct links to the Persian Gulf and the Indian Ocean. Further, risk management concerns will dictate that oil and gas exporters and importers in Caucasia, the Caspian region, and Central Asia diversify their exit/export and import routes so as to minimize or offset Russia's otherwise dominant control over the flow of their energy resources. That leaves Iran, Turkey, and Georgia as the only other options.

An Iranian route for Caspian's gas exports provides the most economic outlet, yet Iran on its own cannot afford or finance the investments needed for the construction of the required pipelines and infrastructure. In November 1995, Iran's ministry of oil introduced the so-called buy back arrangement whereby the original investment by the foreign participant and an agreed upon return on the investment may be repatriated through in-kind gas and oil exports. Experience has shown, however, that this new policy alone would achieve only limited success, unless the country reforms its overall economic and investment policies toward mobilizing domestic and foreign investment.

Iran's location between its oil and gas exporting neighbors to the north and the Persian Gulf to the south has resulted already in the development of swaps. Thereby, Azeri, Turkmen, and Kazakh oil or gas is piped to Iran for consumption in Iran, meanwhile an agreed upon volume of Iran's oil or gas is shipped/piped out to Azeri, Turkmen, or Kazakh customers overseas. While the swap agreement with Kazakhstan has run into problems because of the quality of Kazakh oil, it is expected that the idea of swaps continues as a functional alternative to direct exports.

Reminiscent of the same ambitious human enterprise that produced the Suez and Panama canals, Iran has had dreams of connecting the Caspian Sea with the Persian Gulf by means of a 1,500-kilometer long canal. The idea is almost 30 years old; however, due to the enormous cost involved, the project never materialized. Not so long ago, Iran's former president, Hashemi-Rafsanjani, asked the ministry of construction *crusade* (*jihad sazindegy*) for a feasibility study on such a canal. The ministry put the initial investment at some $15 billion, with an annual maintenance/operating cost of some $6-7 billion, much of it needed for maintaining the canal's huge elevators and locks. On the plus side, however, the canal would cut the transportation cost of oil from the Caspian Sea to southeast Asia by some $15 per ton. This cost/benefit analysis may furnish the reason for international oil companies to support such a project; the actualization of the project on an international level, however, would require considerable political capital, a commodity presently not in abundant supply.

Notes

1. *Iran Statistical Yearbook* (Tehran: Iran Statistical Center, February 1998), p. 693.
2. Bahram Amirahmadian, "Geography of the Caspian Sea," *Central Asia and the Caucasus Review*, 2:14, p. 17. Iran's total area is 1,648,000 square kilometers.
3. The second Five Year Economic Plan, introduced on March 21, 1995, sets the framework for the country's economic development, including agriculture, into the year 2000.
4. Iran's population is estimated at 60 million, with 51.4 percent of it under the age of 19. On the basis of the official population growth rate of 1.7 percent per year, the populations of Gilan and Mazandaran may be estimated at about 2.31 million and 4.16 million, respectively. Compared to the country's average population density of 33.8 per square kilometer, Gilan has a density of about 149 and Mazandaran 80 per square kilometer. *Iran Statistical Yearbook*, p. 43; Hassan Motiei Langroodi, *The Eco-*

nomic Geography of Iran (Mashad, Iran: Ferdowsi University Press, 1996), p. 35.

5. Motiei Langroodi, p. 35. Gilan's labor force is estimated at 618,761 and Mazandaran's at 798,236. In Gilan, 13.8 percent of the working population works in industry, 32.4 percent in services, with 2.6 percent as unclassified; in Mazandaran, 17.5 percent works in industry, 38.9 percent in services, with 0.7 percent as unclassified. Ibid., p. 72.

6. For details, see Parviz Kardavani, *Iran Aqua Ecosystems: Caspian Sea* (Tehran, Iran: Qomes Publishing, 1995), pp. 177-181.

7. *Report on the Mazandaran Province* (Tehran: Jihad Sazindegy Ministry, July 1994), p. 7.

8. See, *Iran Fisheries Sector Study* (Tehran: Shilat Compnay, 1997). See also, Charles Gurdon & Sarah Lloyd, eds., *Oil & Caviar in the Caspian* (London: MENAS Associates, September 1995).

9. For example, in 1976, Iran produced 221 tons of caviar and 1,821 tons of sturgeon flesh; in 1986 production of caviar rose to 303 tons but sturgeon flesh dipped to 1,687 tons; in 1992, production of caviar dropped to 262 tons and sturgeon flesh came in at 1,583 tons; in 1996, caviar production hit an all-time low of 195 tons, while sturgeon flesh production further declined to 1,295 tons. *Iran Statistical Yearbook,* p. 149. The figures for caviar and sturgeon include catch of all sturgeon species (beluga, acetra, and sevruga).

10. In 1976, from a total of 8,428 tons of catch, 4,929 was bony fish, 2,368 was sturgeon, and 1,131 was kilka. In 1993, Iran fished from the Caspian some 22,328 tons of bony fish, 1,710 tons of sturgeon, and 28,730 tons of kilka. In 1996, the catch of bony fish and sturgeon dipped slightly to 15,500 and 1,600 tons respectively, while kilka declined to 41,000 tons from the previous year's 51,000 tons. *Iran Statistical Yearbook,* p. 148.

11. Leonid E. Sklyarov, "Caspian Sea Basin Countries in the 21st Century," *Central Asia and the Caucasus Review,* 5:14, p. 79.

12. *Iran Statistical Yearbook,* p. 380.

13. Narsi Ghorban, "Middle East Natural Gas Pipeline Projects: Myth and Reality," *The Iranian Journal of International Affairs,* vol. 8, no. 3 (Fall 1996), p. 645.

14. Narsi Ghorban and M. Samir, "Political Developments in the Persian Gulf and Their Impact on the International Oil Market," in *Iran Focus,* vol. 6, no. 7 (July-August, 1993), p. 3.

15. Ghorban, "Middle East," p. 646.

16. *Report of the International Maritime Organisation Mission to the Islamic Republic of Iran* (1994), p. 3.

17. *Iran Focus,* vol. 9, no. 11 (December 1996), p. 15.

Part II

Development and Environment

CHAPTER FOUR

Development of Caspian Oil in Historical Perspective

Firouzeh Mostashari

Introduction

This chapter examines the role of the Russian state in the development and growth of the Caspian oil industry in the nineteenth century. The chapter contends that while, initially, tsarist economic policy, including attraction of domestic and foreign investment, provided the main impetus for the Caspian oil boom, the subsequent policies resulted in the demise of the industry. The parallels between that experience and the beginnings of the new era in oil and gas development by the littoral countries of the Caspian Sea may portend a repetition of the harsh economic lessons of the past along with its adverse social and political consequences.

Historical Background

The connection between oil and the Caspian region is longstanding. Knowledge about the existence of oil—white and black—and of natural gas dates to antiquity. In the period from 600 B.C. to the twelfth century, Zoroastrians traveled to Baku on the Apsheron Peninsula to worship at the temple of Surakhani, where burned eternal fires fueled by natural gas.[1] There is evidence that crude oil was an article of commerce as early as the tenth century and was being exported extensively from Baku by the thirteenth century.[2] In his description of Greater Armenia, Marco Polo, the thirteenth-century Venetian traveler, referred to the remarkable "fountain of oil which discharges so great a quantity as to furnish loading for many camels."[3]

In 1723, the tsar, Peter the Great, annexed parts of the Persian-controlled eastern Transcaucasia, including Baku. Subsequently plans were drawn to extract and export the area's white oil to Russia. On Peter's death, in

1725, the plans fell into abeyance and shortly thereafter Nader Shah of Persia recaptured the erstwhile Persian possessions on the Caspian littoral. The development of the Baku/Apsheron oil reserves was reignited shortly after the fall of the Baku Khanate to Russia and conclusion of the ensuing 1813 Gulistan Treaty between Persia and Russia.[4] This marked the turning point in the history of the Baku oil industry; the khan's monopoly rights to the oil and oil-bearing lands of the Khanate passed to the Russian state.

In 1821, the state began to farm out its monopoly on the oil fields for four-year periods (the otkup system). Oil pits were leased to individual contractors for 13,000 pounds, by the Crown Department of the Georgian administration, which was the highest governmental institution in the Caucasus at the time. Due to the limited term of the lease, contractors had little incentive for making capital investments; they extracted the most out of the ground, without making improvements. In 1834-49, leases to private entrepreneurs were suspended and the government itself assumed the extraction of oil. Due to the continued use of primitive methods, production remained low and so, in 1850, the government reinstated the contract system, with all of its shortcomings.[5]

Due to technological, labor, and market constraints, the Baku oil production in the 1850s stayed at 250-300 thousand puds a year.[6] The method of extraction remained manual; and by 1871 the number of oil wells rose to slightly over twice the 116 in 1816.[7] In 1859 oil was discovered in the United States and the superior quality of American oil began to challenge Baku oil, even in the Caucasus. In 1863, with the introduction of a new process of chemical refining, Baku's products could compete only with American oil in regions where distances and transportation costs favored Baku oil, such as in and around the Volga region. The quality of Baku oil stayed inferior, because the Baku refiners continued to mix kerosine with heavier residuals.[8]

The labor employed in Baku's oil fields and factories added to production inefficiency. Until 1864 oil was being extracted by Azerbaijani debtor-peasants, and child laborers. The Russian administrators, however, justified child labor as "an introductory school for further more significant activity."[9]

Another source of the Russian oil industry's inefficiency was its internal organization. Government monopolies at such an early period of development hampered the growth of the industry. To illustrate: Even though production under a monopoly held by the Armenian merchant, I. M. Mirzoev, in the period 1863-72 increased from 340 to 1,536 thousand puds, it still compared unfavorably with the development of the industry after 1873, when the state monopoly was abolished.[10]

Abolition of State Monopolies

Production had increased by 10 percent between 1871 and 1872; it doubled during the first year of the new system. By the turn of the twentieth century, the extraction of oil increased 165-fold.[11] The number of workers in the oil industry also increased significantly after 1872. Numbering 680 in 1873, it had increased to 27,673 individuals by 1901.[12] Similarly, the number of derricks also grew impressively during this period, increasing from 2 in 1872 to 1,740 by 1901, with output increasing from 1,536 to 670,900 puds.[13]

The decision to free the oil industry from monopolistic control had not been a hasty one. The Russian ministry of finance had been contemplating the move since 1867, when a special commission was set up in Tbilisi to review the government's policy toward the oil industry. The government was concerned with increasing the profitability of the industry, since it had hitherto yielded a paltry yearly sum ranging between 111 and 162 thousand roubles.[14] The rescinding of government monopolies was part of the general plan of the minister of finance, M. Kh. Reutern, drawn up in 1866, in order to decrease the government deficit and encourage the development of industry.[15] Among issues debated at the time were the size of the land plots to be rented out, as well as the advisability of taxing the industry. On February 1, 1872, Tsar Alexander II approved a plan for oil lands to be auctioned off to the highest bidders, who were to pay an initial sum (the bid) plus 10 roubles per desiatina per year.[16] The duration of the contract was to be 24 years, after which the yearly rent increased to 100 roubles. The contracts stipulated that the land had to be developed within two years of concluding the contract.[17]

Following the abolition of state monopoly of oil lands, initially, 54 plots of oil-bearing land were traded in 1872. Of these, 40 plots were located in Baku province, primarily around the Balakhani Plateau, known for its large oil and natural gas reserves. The government, however, kept a substantial amount of the land off the market, renting out only 169 desiatinas from the 345 available in the Balakhani fields. In addition, the distribution of the auctioned lands favored the Russian and Armenian merchants. Ten of the fifteen most promising plots were bought by the influential Armenian merchant I. M. Mirzoev, and his Russian competitors, V. A. Kokorev and P. I. Gubonin, who had founded the Baku Oil Company in 1874 and owned the state-subsidized Caucasus & Mercury Shipping Company.[18]

For the time being, it seemed as if "the state monopoly had only been replaced by a private duopoly;" the transactions were ever more complex. The Russian government had its own agenda for the granting of oil lands and hoped to use land grants as an enticement to lure Russian officials to the Caucasus. Parts of the lands that had been set aside by the government had been given away between 1878 and 1881 "forever and in heredity" to elite civil and military administrators.[19]

The discovery of oil reserves in agricultural lands other than Balakhani prompted a wild rush for the lands of the villages surrounding Baku. In the early years after the abolition of the state monopoly's contract system, those investing in the oil industry came from various social groups: merchants, entrepreneurs, government officials, and peasants. Among these the most powerful two groups of local entrepreneurs were the merchant (commercial) capitalists and captains of industrial capital.[20]

The opening of the exploitation of the region's oil supplies to a large number of candidates quickly resulted in an increase in government revenues. Within a year of de-monopolization, the Caucasian administration was singing the praises of the program. In 1873, the governor of Baku reported, "The recent cancellation of the oil otkup showed how the freeing of industry is more profitable for the treasury,"[21] suggesting that state monopoly of salt and fishing too be abolished, as "the otkup has a harmful influence on the development of trade and the well-being of the region."[22] This shift from state tutelage to capitalism marked the general movement toward rapid industrialization in Imperial Russia.

The Excise Period (1872-1877)

The state monopoly of oil lands had been abolished, but the oil industry was not left to develop freely; the treasury's interest in exacting taxes soon began to interfere in the natural growth processes of the oil industry. Effective from February 1, 1872, the treasury imposed an excise duty on all petroleum products. Emulating the model for the taxation of vodka, the refiners were charged based on the capacities of their stills and the average time of operation. The estimated tax averaged between 25 and 40 kopecks per pud,[23] resulting in 1.2 million roubles in revenue for the government in the 5-year period.[24] Consequently, the excise caused producers of kerosine to refrain from refining significant portions of their crude oil; the ratio of crude oil to refined oil soon shifted in favor of crude oil.[25] Moreover, the quality of the refined goods became inferior to that of the American products, as the Russian refining was carried out hastily in order to reduce the tax, which was calculated based on refining time. True, the excise provided the state with additional revenues, but retarded the growth of the oil industry and discouraged refining activities. Consequently, between 1873 and 1877, the volume of American kerosine exported to Russia exceeded the volume of Baku kerosine sent to inner Russia by roughly 300,000 puds.[26]

During the excise period, the Baku oil industry was characterized by a multitude of small and primitive production units, worked by native Azerbaijanis and Armenians. Most of the well owners lacked the financial resources required for innovation; nor did they have any incentive to do so. Most of the drilling was conducted manually, even though engines were introduced beginning in 1873. Oil was transported using horse-drawn carts (arabas), barrels, and leather sacks.[27] The low technical level of the industry proved inadequate to meet the challenges of oil fountains appearing in 1873, producing disastrous results.

When the Khalify Company, one of the smaller oil companies, struck the first oil fountain or gusher in Baku, millions of barrels went to waste in the four months that the gusher was out of control. However, enough was salvaged to drive down the price of crude oil from 45 to 5 kopecks per pud.[28] In some cases the appearance of gushers led to the bankruptcy of their owners, since the technology to cap the gusher and harness it was lacking. The oil fountain would topple the derrick and create an oil lake. Yet, due to evaporation, these lakes could not be used as reservoir for long, since they turned into tar pits. Sometimes small amounts of the oil could be recovered. Yet in cases when the oil would flow onto other people's property, the claims for compensation would far surpass income derived from the sale of the oil.[29]

By 1876, the Baku oil industry was in a state of crisis. The price of crude oil had dropped to 2-3 kopecks per pud.[30] Consequently, a majority of smaller producers with inadequate storage and transportation facilities went bankrupt or were on the verge of bankruptcy. Finally responding to the plight of the local producers, the Russian Technological Society set up a special commission to study the state of the oil industry. The renowned chemist D.I. Mendeleev, an active participant in the proceedings, visited Pennsylvania to observe the American oil industry. Upon his return, he stressed the necessity of government support for the industry as well as the need to free the industry from the burden of taxation.[31]

Based on Mendeleev's recommendation, the state council abolished the excise tax the following year. Appearances of laissez-faire aside,[32] the oil industry was not entirely free from state interference. Companies, whether Russian or foreign, were still required to obtain permission from the government in order to start their activities in any industry. Joint-stock companies wishing to increase their capital, float bonds, change their structure, or make other major changes needed the permission of the central administration.[33] In sum, all major innovations had to be confirmed by the government. Thus the construction of pipelines, railroads, and shipping, all depended on governmental approval. In addition, foreign capital could not enter into the Baku oil industry without being sanctioned by the government. Therefore, the Russian government continued to strongly influence the development of this industry, albeit in a less direct manner.

The Nobel Brothers

The freeing of the Baku oil industry from the excise tax on kerosine and other refined products made this industry attractive to prospective Russian and foreign investors. Nobel Brothers Corporation (Nobel Corporation) was the first foreign investment group to participate in the Azerbaijan oil industry. It inadvertently contributed to the technical development of the Baku oil industry. Having laid down its initial capital in the industry during the 1877-78 period, Nobel Corporation soon consolidated its hold on the means of transportation, replicating the experience of the Rockefellers, who had overcome their competitors by establishing a transportation monopoly, owning nine-tenths of the pipeline networks in America.[34]

The transportation and sale of oil products had been the weak link in the Baku oil industry. Inspired by the example of Pennsylvania oil, Nobel Corporation sought to construct an eight-mile pipeline joining the Balakhani oil fields to the refineries in the so-called Black town. The capacity of the pipeline was estimated at 15,000 puds of oil annually.[35] The project threatened the livelihood of thousands of araba drivers and barrelmakers, who were about to lose their monopoly on the transportation of oil. The local government, possibly fearing the consequence of wide spread unemployment, also resisted the construction of the pipeline and did not permit it to run on treasury lands, much of which lay between the oil fields and the refineries. However, in 1877, Nobel Corporation's influence among official circles enabled it to convince the government to approve the project.[36] The pipeline put a virtual end to the livelihood of those, primarily villagers from the Balakhani area, who had engaged in carrying oil from the source to the factories.[37] Whereas previously, between 15 and 20 thousand araba drivers had worked there, now these numbers had been reduced to 200 at most.[38]

The displacement of unskilled local laborers exacerbated ethnic and class tensions in the region. By April of 1879, the chancery of the viceroy of the Caucasus was troubled about other implications of the Nobel pipeline. In particular, the local administration feared that other producers were excluded from using the pipeline and sought to prevent the monopolization of the means of transportation, which could lead to the monopolization of the oil industry as a whole.[39] The pipeline made Nobel Corporation highly competitive; it cut the company's transportation costs drastically. By 1883, the company furnished over 50 percent of all the kerosene shipped to Russia.[40] To cut shipping costs drastically, bulk transportation had to be adopted, and the shipping vessels themselves could be used as containers. Hence the idea of an oil tanker came into being. In January of 1878, the company introduced the first sea-going oil tanker, called the *Zoroaster*. This was soon followed by a number of other ships. The company successfully maintained its transportation monopoly until the completion of the Transcaucasian Railroad in 1883.[41]

The Transcaucasian Railroad

The oil industry had been freed, albeit arguably, from direct state control and taxation. The government, however, had not devised any concrete plan to organize the transportation of oil products to internal and international markets. This had damaged the interests of smaller producers whose market shares were squeezed by the likes of the Nobel Corporation, who owned their own means of transportation. More than a decade after the freeing of the oil industry, the government saw to the completion of the railroad connecting Transcaucasia, Russia, and the Black Sea. This added yet another dimension to the structure of the Russian oil industry in general.

The Russian administrators of the Caucasus were among the promoters of a railway across the region. The Grand Duke Mikhail Nikolaevich, viceroy of the Caucasus (1863-1881), saw railroads as a means of solidifying Russian control over the Caucasus, as Russian hold over the region was not yet secure, even though Russia maintained there a 200,000-man army at an annual cost of 19 million rubles. The region was drawn more to its Persian and Turkish neighbors than to Russia, and in the event of a war, like the Crimean War, the absence of railroads to provide troop transportation would have proven disastrous for Russia. So, Nikolaevich argued, connecting the region to Russia via rail was not only in the interest of the Caucasian economy, but also served Russia's strategic interests.[42] On June 16, 1878, a plan was drawn up to join the oil districts around Baku to the Transcaucasian Railroad. In May 1883, the railroad was inaugurated and the 25-year-old plan to connect the rails around Baku to those of Tbilisi and Potti, on the Black Sea, were finally realized.[43]

The construction of the Transcaucasian Railroad fundamentally altered the balance of power within the Baku oil industry. It broke Nobel Corporation's monopoly over the means of transportation. Moreover, it introduced the state as a major player again; the railroad was state owned and hence the differential railroad fares greatly influenced the oil producers' choices in the refining process. Furthermore, it facilitated Baku's access to the world market.

The Transcaucasian Pipeline

The Transcaucasian Railroad was not able to satisfy the ever-growing demands of the Baku oil producers. In 1883, Herbert Twedle, an American, along with his Russian business partner, Lieutenant Khanykov, peti-

tioned the ministry of finances for a concession to build a pipeline joining the Caspian and Black seas to the Rostov-Vladikavkaz railroad
branch.[44] The Transcaucasian pipeline would join the ports of Baku and
Batum, turning Batum into a terminal for exporting oil products from
the Caucasus. While the Russian state agencies were convinced of the
need to build the pipeline, there was internal disagreement on the source
of its financing. The indecisiveness of the government with relation to the
oil industry was partly due to the competing and overlapping jurisdic-
tions of the ministries of finances and state domains over the oil industry
and the oil lands; both ministries repeatedly thwarted each other's efforts.

In January 1884, the minister of state domains, M. N. Ostrovskii,
wrote to the minister of finances stating that "one of the major reasons for
the current crisis in the oil industry must be recognized as the lack of
cheap means of transportation for the sale of oil products in trade markets
abroad."[45] In his opinion, the Transcaucasian Railroad was inadequate;
the high tariffs placed on oil and its derivatives further caused significant
increases to "the price of these products and hence our oil factory owners
cannot compete with the Americans outside of the Russian market."[46]
The minister suggested that the Russians follow the example set by the
Americans and replace railroads with pipelines in order to substantially
cut costs.

In April 1887, the committee of ministers approved the construction
and exploitation of the Transcaucasian pipeline by a private company.[47] In
May 1887, the concession for construction of the pipeline was granted to
I. P. Ilimov, a member of the Russian Physics-Chemistry Society, who was
to create for the purpose a joint-stock company called the Caspian-Black
Sea Pipeline Company. Financing the pipeline ran into bureaucratic difficulties almost at once; its construction had to be postponed until 1907.[48]

Throughout the 1890s, the Russian government continued its deliberations on the question of how to best finance the Transcaucasian
pipeline. First, the government denied a request from the British to build
the pipeline in return for 3 percent of its revenues, boasting that the
Russian treasury itself was going to finance the operation. In 1890, the
treasury admitted, however, that it could not supply the sums necessary
for its construction.[49] This lack of funds delayed the government's plan to
monopolize the transportation of oil products and, thereby, as one senior
official advised Tsar Alexander III, "[u]niting in its hands the exploitation
of the Transcaucasian railroad and the pipeline, the government will
attain the possibility of regulating the profitability of both enterprises."[50]
Meanwhile, the delay in constructing the Transcaucasian pipeline not
only resulted in the incurring of considerable opportunity cost with
respect to oil exportation, but it also meant the ruin of many smaller companies, which did not have access to adequate means of transportation.
Caught in the cross-fire between different ministries, urgent planning

and financing issues were postponed for years, a delay that the Baku oil industry could not afford; its share of the world market continued to decline well into the first decade of the twentieth century.

Foreign Capital's Monopoly of the Oil Industry

Beginning in the mid 1880s, smaller companies, most of which were of local origin, began to experience severe difficulties in transporting their products out of Baku. The Transcaucasian Railroad was congested, and most of the wagons were owned by the larger companies. Shipping too was concentrated in the hands of a few firms. With the re-imposition of the excise tax in 1881, many smaller companies could no longer remain competitive. By the end of the 1880s, lack of capital led many Baku firms to sign unprofitable contracts with larger firms. In addition to the Nobel Corporation, the Rothschilds, who had recently appeared on the scene, began to take advantage of the newfound opportunities.

The Rothschilds arrived in the Caucasus ten years after the Nobel brothers. They closely resembled the classic exporters of "foreign capital;" they used their financial wherewithal to take over enterprises and became leading exporters of oil. In 1883, they purchased a significant number of shares in the Baku Oil Producers Society, followed in 1885 by another major purchase, making them a partner in the Batum Petroleum and Trading Company. In consequence of their default on loans, these two firms fell under the complete control of the Rothschilds. In this manner, some 135 middle and small-size enterprises eventually came under the Rothschilds' influence.[51] By 1896, the transfer of the management of national companies to foreign hands had reached such alarming proportions that the local administration, while conceding that foreign capital "certainly enlivens and stimulates industry," questioned the future direction of the industry. The Baku governor expressed concern that the "epidemic character" of the take-overs would result in the complete ownership of the oil industry by foreigners, a result that "could hardly coincide with state interests."[52]

In 1889, the Rothschilds signed contracts with 50 oil producers to supply their firm with kerosine for trading abroad, in return for a 4 percent commission.[53] Having contributed to the financing of the Transcaucasian Railroad, they owned also over one half of its tank cars. Thus, by controlling the sale and transportation of kerosine, the Rothschilds were in a position to influence the price of oil. As reported by the director of a local bank, the Rothschilds were deliberately hoarding kerosine in their reservoirs in Batum, to the full extent of their capacity. Under these conditions, they artificially created an excess of kerosine and lowered prices, to the detriment of the producers.[54] Using such tactics, the Rothschilds became the Russian Empire's leading exporters of oil products. Between

1886 and 1888, they exported kerosine to London, Austria-Hungary, and Turkey. After 1888, their operations extended to the Far East.

By the turn of the twentieth century, the question of whether foreign capital should be allowed to expand its activities in the oil-rich areas of Baku province was once more on the forefront of inter-ministerial discussions. A special commission including the ministers of finances, agriculture and state domains, and interior, and the high commissioner of the Caucasus reviewed the laws governing the Baku oil industry. While certain members of the commission had suggested that only Russians be permitted to rent oil lands, the final resolution was in favor of the introduction of foreign capital to the industry. The commission concluded that international competition with the Americans required significant amounts of capital not available in Russia. Foreign participation having been deemed beneficial and desirable, the commission concluded that no change was needed in the existing rules governing the oil industry.[55] Meanwhile, the local firms were ever more unable to independently export kerosine due to the control of transportation, storage facilities, and sales and distribution channels by firms controlled by foreign capital.[56]

The Baku Oil Crisis (1901-1905)

The 1901-03 crisis in the international oil markets, including Baku, resulted in the fall of demand for oil. Surplus drove down the price and slowed down activity in all of the oil related fields, including drilling, extraction, and refining. During this period the number of refineries in the Caucasus decreased from 91 to 86, with only 57 of these remaining in operation.[57] Beginning in 1900, the price of oil had started to plummet and by the year's end it had decreased by over 48 percent. By 1901, the Caucasian administration was openly referring to an "oil crisis," which it attributed to the oil prices, the decreasing demand for oil products, and the concentration of oil export in the hands of foreign firms.[58]

As enterprises cut production and dismissed large numbers of workers, a wave of protests and strikes swept the region. By 1903, workers, particularly the Armenians, began to openly demonstrate in the streets and had to be restrained by the Cossacks. In the first half of 1903, incidents of disagreement between workers and management were rising sharply, leading the factory inspectorate and the police to regularly intervene. By mid July, in Baku and the outlying regions, over 20,000 workers participated in a "general strike," which embraced almost all of the industries, factories, and enterprises. The influence of revolutionary circles soon was discerned.[59] In Baku, strikes resumed in the following year, and on December 31, 1904, it led to the first successful case of collective bargaining in the empire, between the oil producers and the oil workers, resulting in nine-hour working days, one unpaid day off per week, a three-month period of paid sick leave, a minimum wage for each category of worker, dining halls, and other benefits.[60]

Throughout January 1905, Baku's police chief reported the resurgence of strikes in factories and other enterprises. By the end of the month in Baku and other Caucasian cities, large demonstrations and simultaneous strikes were held as a sign of solidarity with the St. Petersburg workers, massacred on "Bloody Sunday," January 9, 1905.[61]

The strikes continued into February and assumed an ethnic dimension as well. On February 14, 2,000 representatives of the Armenian and Jewish workers' intelligentsia gathered on the streets and called for an armed uprising against the autocracy, which they accused of stirring ethnic enmity.[62] The ethnic disturbance in Baku reached its height on February 6-10, when a Muslim Azerbaijani villager, Aga Reza Babaev, fired into a crowd of Armenians gathered at a cathedral in Baku. A member of the Armenian nationalist party, the Dashnaktsutiun, killed Babaev; the incident erupted into a murderous rampage of Azerbaijani Muslims and Christian Armenians, indiscriminately killing one another with daggers and revolvers. Individuals were attacked in their homes and businesses, which were set on fire. After February 8, the violence spread to Baku's oil-producing areas.[63]

The damage to the oil industry as a result of fires amounted to 19,500,000 rubles.[64] Enterprises suffered a loss of over 20 million rubles and many remained inoperative for over two months.[65] The minister of finances, V. Kokovtsev, reporting to the tsar on February 15, wrote: "the Baku disturbances, *smuty*, were of an extremely serious character and in the course of several days, complete anarchy had ruled over Baku and its industrial suburbs."[66] The minister identified the nature of the disturbances as "national-religious," mainly between the Christian Armenians and the Muslim Azerbaijanis. He reported that the February incidents would have serious repercussions on the local economy as well as on that of the empire as a whole, noting that masses of Russian workers and specialists were leaving the area, with the disturbances affecting also Russia's credit worthiness abroad.[67]

Curiously, the tsarist police had been conspicuously absent from the scene in Baku. Senator Kuzminskii, sent by the tsar to investigate the Baku events, concluded: "The Baku police displayed complete passivity and inactivity in the face of the killings and devastation taking place in Baku. All throughout the February events the police's lower cadres were almost invisible on the streets."[68] The police chief, although he had Cossacks under his command, did not intervene in the confrontations nor did he disarm the participants. Senator Kuzminskii openly expressed his doubts as to whether the indecisiveness of the local administration was in fact unintentional.[69] The events of 1903-1905 showed that in the Caucasian borderlands, the autocracy was ambivalent in its defense of the oil industry interests. When faced with a rising revolutionary movement directed against it, the autocracy diverted the revolutionary movement and turned the participants against one another, resulting in the loss of lives and millions of rubles of loss to the oil industrialists.

Conclusion

Initially, Russia's state policy provided the impetus for the development of the Caspian oil industry by freeing it from government monopoly. Subsequent questionable policies did not fair as well, because state participation proved "much less beneficial than is generally believed."[70] Conflicts between the ministries of finances and agriculture and state domains often resulted in the victory of the former, and consequently the interests of the local industrialists were sacrificed for the short-term raising of revenues and the support of the largest of the industrialists. Indecision and lengthy deliberations delayed for over two decades the construction of the necessary transportation infrastructure, particularly the Transcaucasian pipeline. This contributed to the annihilation of the smaller companies for want of outlet to markets and led to the monopolization of the industry by the larger companies with access to ships, railroad wagons, trucks, and other vehicles.

The levying taxes on refined products benefited the treasury, but it proved detrimental, often fatal, to the smaller companies and compromised their competitiveness by increasing the cost and lowering the quality of the refined products. The larger firms, with their connections to state personnel and the directors of state banks, were in a favorable position to benefit from credits and loans, in order to withstand financial troubles. The smaller, mostly local, firms had little choice but to sink.

In the three-way rivalry among the state, foreign, and local capital, the state threw in its lot with that of the foreign capitalists and larger Russian enterprises, which it praised for their superior work ethics and knowledge. Alienated, the elite members of the Azerbaijani and Armenian entrepreneurial classes lost confidence in the state and began to pin their hopes on their respective community's sense of economic self-determination and control, each ethnic group donating money as they did to their respective nationalist groups during the Baku crisis.

If there is a parallel in the Baku experience of the nineteenth century and today's efforts by Azerbaijan to develop its oil and gas resources, it must be in the volatile admixture of foreign presence justified by investment, disparity of wealth and income between the underclass, which does not participate in the petro-promise, much less see a penny from petrodollars, and the overclass, tied to foreign investors obsessed with luxury and conspicuous consumption. While the Baku crisis was the harbinger of revolutionary change with respect to the ownership of means of production, Iran's 1951 oil nationalization marked—albeit briefly and perhaps arguably—the beginning of the end to the domination of the local Middle East petroleum industries by foreigners. Yet, as the 1979 revolution in Iran pointed out, the challenge facing the ruling elite in Azerbaijan, as in other former Soviet republics, in the long run, will be less about how to temper the foreign investor, but rather how to guard against the government's own excess and shortcoming with respect to the general population and their expectations for development and prosperity.

Notes

1. Charles Marvin, *The Region of Eternal Fire: An Account of a Journey to the Petroleum Region of the Caspian in 1883*, (Westport, Connecticut: Hyperion Press, 1976), p. 2.
2. Ibid., p. 164.
3. *The Travels of Marco Polo,* Thomas Wright, ed. (London: Henry Bohn, 1854), Book First, Ch. IV, pp. 31-32. According to him, "The use made of it is not for the purpose of food, but as an unguent for the cure of cutaneous distempers in men and cattle, as well as other complaints; and it is also good for burning. In the neighbouring country no other is used in their lamps, and people come from distant parts to procure it" (p. 32.)
4. J. D. Henry, *Baku: An Eventful History* (London: A. Constable and Co., 1906), p. 29: and Marvin, p. 169.
5. Henry, p. 30
6. The pud, which was a unit of weight used in imperial Russia, was equivalent to 36 pounds.
7. V. A. Nardova, *Nachalo monopolizatsii neftianoi promyshlen-nosti Rossii, 1880-1890-e gody* [Early Monopolization of the Russian Oil Industry: 1880-1890s], (Leningrad: Nauka, 1974), p. 7.
8. Marvin, p. 239.
9. RGIA (The Russian State Historical Archives, St. Petersburg), f. 1268, op. 10, d. 134, p. 157.
10. William J. Kelly and Tsuneo Kano, "Crude Oil Production in the Russian Empire: 1818-1919," in *Journal of European Economic History* (1977): 313. Marvin also observed that "the protective system of the Russians, following upon centuries of free trade under the Persians, stunted the growth of the petroleum trade. The industry grew, but its development was nothing like what it would have been, had there been no Government restriction." Marvin, pp. 203-204.
11. M. A. Ismailov, *Kapitalizm v sel'skom khoziaistve Azerbaidzhana na iskhode xix-nachale xx v* [Capitalism in the Rural Economy of Azerbaijan at the End of the 19th and Beginning of the 20th Century], (Baku: Izd-vo Akademii Nauk Az SSR, 1964), p. 23.
12. B. I. Akhundov, "Razvitie Bakinskoi neftianoi promyshlennosti posle otmeny otkupnoi sistemy," [The Development of the Baku Oil Industry After the Abolition of the Otkup System], in *Izvestiia Akademii Nauk Azerbaidzhanskoi SSR* 10 (October 1949): 64.

13. Akhundov, p. 66.
14. I. A. D'iakonova, *Nobelevskaia korporatsiia v Rossii* [The Nobel Corporation in Russia], (Moscow: Mysl', 1980), p. 52.
15. L. E. Shepelev, *Tsarizm i burzhuaziia vo vtoroi polovinie XIX veka* [Tsarism and the Bourgeoisie in the Second Half of the 19th Century], (Leningrad: Izds-vo ANSSSR, 1981), pp. 100-108.
16. John P. McKay, "The Development of the Russian Petroleum Industry, 1872-1900," in *Research in Economic History* 8 (London, 1982): 53. Each desiatina is equal to 2.7 acres.
17. Nardova, p. 113.
18. Nardova, pp. 114-115.
19. ANSSSR, *Monopolisticheskii kapital v neftianoi promyshlennosti Rossii: 1883-1914* [Monopoly Capital in the Oil Industry of Russia: 1883-1914], L. E. Shepelev, ed. (Moscow: Akademiia Nauk, 1961), p. 672. The recipients of these land grants included P. S. Lazarov, Commander-in-Chief of the Caucasian Army, D. S. Starosel'skii, Caucasian High Commissioner, Prince Ferdinand Zeinwitgenstein-Berleburg, Commander of the Cossack Brigade, and General V. M. Pozen, the Governor of Baku Province.
20. Arastun B. Mekhtiev, *Rossiiskii kapital v stanovlenii neftianoi promyshlennosti Bakinskogo raiona v poslednei tretii xix v, Avtoreferat* [Russian Capital and the Formation of the Baku Oil Industry in the Last Quarter of the 19th Century—Autoreferat], (Moscow, 1990), p. 12.
21. RGIA, f. 1268, op. 18, d. 182, p. 498.
22. Ibid., 499.
23. McKay, p. 58.
24. Marvin, p. 208
25. Kelly and Kano, p. 331.
26. Nardova, p. 93.
27. RGIA, f. 1282, op. 3, d. 246, p. 4.
28. Marvin, p. 207. Marvin asked, rhetorically, "how a magnificent oil fountain of this description should be able to make its owner a millionaire in one hemisphere and a bankrupt in another" (p. 213.)
29. Ibid., pp. 213-216.
30. D'iakonova, p. 52.
31. Nardova, p. 94.
32. Marvin, p. 204.
33. See generally John P. McKay, *Pioneers for Profit: Foreign Entrepreneurship and Russian Industrialization 1885-1913*, (Chicago: University of Chicago Press, 1970). According to McKay, one contemporary of these developments had observed, "there is not an industry in Russia today where the laissez faire doctrine is carried to such lengths as in the Baku petroleum trade" (p. 276).
34. A. A. Fursenko, *Neftianye voiny: konets xix-nachale xx v* [Oil Wars: The End of the 19th and Beginning of the 20th Century], (Leningrad: Izd-vo Nauka, 1985), pp. 14-16.
35. RGIA, f. 1268, op. 24, d. 231, p. 92.
36. Robert W. Tolf, *The Russian Rockefellers*, (Stanford, Conn.: Hoover Institution Press, 1976), p. 52.
37. RGIA, f. 1268, op. 25, d. 265, p. 117B (report by the governor of Baku).

38. Ibid.
39. Nobel Corporation's ambitions to monopolize the Russian oil market were clearly spelled out unambiguously in a confidential report delivered to its shareholders in April of 1883: "First and foremost, the task of this enterprise is to oust American kerosine from Russia, and then to proceed to export kerosine abroad. Our entire enterprise will be organized according to the demands of this undertaking. . . . In order to attain these results it is absolutely necessary that we be capable of satisfying the entire demands of the Russian kerosine market. And for this it is necessary to construct such a caliber of sea and river faring vessels for the transport of liquid fuel, that it be capable of carrying the entire stock of kerosine needed by the domestic market." ANSSSR, pp. 48-49, (TsGIAL f. 1458, op. 1, d. 1725, pp. 1-9).
40. ANSSSR, p. 753.
41. Fursenko, p. 22; Marvin, p. 283.
42. SSMD (The Saltykov-Shchedrin Manuscript Division of the St. Petersburg Public Library) f. 369, ed. khr. 261, pp. 1-5.
43. A. Argutinskii-Dolgorukov, *Istoriia sooruzheniia i ekspluatsii Zakavkazskoi zheleznoi dorogi za dvadtsat' piat let eia sushchestvovania* [History of the Construction and Operation of the Transcaucasian Railroad in the Twenty-Five Years of its Existence], (Tiflis, 1896), pp. 52-53.
44. RGIA, f. 37, op. 31, d. 483, p. 1.
45. Ibid, at p. 15.
46. Ibid.
47. RGIA, f. 40, op. 1, d. 49, p. 128.
48. Nardova, pp. 82-85.
49. RGIA, f. 40, op. 1, d. 49, pp. 128-129.
50. RGIA, f. 37, op. 31, d. 439, p. 2.
51. ANSSSR, pp. 37-39.
52. NAII Azerbaidzhana (Archives of the Institute of History of the Azerbaijan Academy of Sciences), Inv # 2267, vol. 3, p. 63.
53. ANSSSR, pp. 121-122 (TsGIAL f. 20, op. 5, d. 582, pp. 3-4).
54. Ibid., pp. 146-147.
55. RGIA, f. 560, op. 26, d. 205, pp. 5-6.
56. NAII Azerbaidzhana, Inv # 2267, vol. 3, pp. 81-83.
57. Marat. Dzh. Ibragimov, *Neftianaia promyshlennost' Azerbaidzhana v period imperializma* [The Oil Industry of Azerbaijan in the Period of Imperialism], (Baku: Elm, 1984), pp. 39-43.
58. NAII Azerbaidzhana, Inv # 2267, vol. 3, pp. 171-175.
59. RGIA, f. 1263, op. 2, d. 5766, pp. 339-340. According to the governor of Baku, the workers' demands were usual requests such as increase in pay, shorter working days, better living quarters, and clean drinking water; however, it was evident that these requests were not inspired as much by the workers' condition, but by pressure from revolutionary circles.
60. ANAzSSR, *Rabochee dvizhenie v Baku v gody pervoi Russkoi revoliutsii: dokumenty i materialy* [The Workers' Movement in Baku in the Years of the First Russian Revolution: Documents and Materials], (Baku: Akademii Nauk Az SSR, 1956), pp. 34-36.

61. ANAzSSR, p. 75.
62. Ibid., pp. 77-78.
63. Senator Kuzminski, *Vsepoddanneishii otchet o proizvedennoi v 1905 gody po Vysochaishemu poveleniiu Senatorom Kuzminskim revizii goroda Baku i Bakinskoi gubernii*, [Senator Kuzminski's report on his official appointment to investigate the city and province of Baku in 1905], (St. Peterburg, 1905), pp. 1-9.
64. RGIA, f. 1276, op. 2, d. 149, p. 82.
65. RGIA, f. 560, op. 26, d. 31, p. 162. According to Luigi Villari, an Italian historian of Russia wrote: "I visited the premises of several oil companies at Bibi Eybat soon after the fires, and the spectacle presented simply defies description. . . . Out of the 200 derricks of Bibi Eybat, 118 had been destroyed, and the majority of the other buildings were heaps of black ruins. The whole area was covered with debris and wreckage, thick iron bars snapped asunder like sticks, or twisted by the fire . . . broken machinery, blackened beams, fragments of cogged wheels, pistons, burst boilers, miles of steel wire ropes. . . . Everywhere streams of thick-oozing naphtha flowed down channels or formed slimy pools of dull greenish liquid; the whole atmosphere was charged with the smell of oil." Luigi Villari, *Fire and Sword in the Caucasus* (London: Fisher Unwin, 1906), pp. 205-206.
66. RGIA, f. 40, op. 1, d. 59, p. 102.
67. Ibid, pp. 102-103. Kokovtsev's assertion is supported by the letter from the Baku oil producers to the viceroy, in which they describe the abysmal state of the working force four months after the February events: "The exodus of responsible employees, the mass flight of panicked workers, has made the continuation of industrial activity entirely impossible." TsGIA Azerbaidzhana, f. 484, op. 1, d. 41, p. 1.
68. Kuzminski, p. 11. Villari, too, observed that both the Armenians and the Muslims agreed that "the authorities promoted, or at least encouraged, the feud on their old principle of *divide et impera* [divide and rule]." Villari, p. 169.
69. Ibid.
70. John P. McKay, "Entrepreneurship and the Emergence of the Russian Petroleum Industry, 1813-1883," in *Research in Economic History* 8 (1982): 88. McKay concludes that this hypothesis would form the basis for a "more systematic investigation."

CHAPTER FIVE

Oil and Gas: Fuel for Caspian's Economic Development

Jean-François Seznec

Introduction

This chapter argues that the oil and gas reserves of the Caspian region as being developed presently will not make a significant contribution to the welfare of the Caucasian and Central Asian states. The cost of extracting and transporting oil to the open seas is prohibitive. Countries like Kazakhstan and Azerbaijan will end up with only a fraction of the oil price obtained in European and United States markets. The Central Asian countries will get less than the $20.29 per barrel (bl) that the producers belonging to the Organization of Petroleum Exporting Countries (OPEC) received on the average in 1996. As contrasted with the production and transportation costs obtained for Persian Gulf oil, the huge costs of transport, foreign capital, and field development would cut into the revenue of Central Asian producers. For example, in a recent time-frame when the OPEC basket was at $20.29/bl, Kazakhstan would have netted only about $7.67/bl. Similarly, in 1997, prices per barrel again started to decline, falling $4.00/bl in the first quarter; at world prices of $16.00/bl, the Central Asian countries would see their net revenues decline to less than $3.39/bl.

This chapter argues that the former Soviet republics bordering the Caspian should move away from the policy of extracting as much oil as possible, or negotiating for as many pipeline deals as possible. Instead. the Central Asian countries should develop their oil/gas reserves with an eye to fuel energy-intensive industries such as petrochemicals, steel, and aluminum. For example, Kazakhstan, with one of the largest deposits of bauxite, would be better off in the long run to use its oil and gas to generate the electricity needed to produce aluminum locally.

This change in emphasis would allow a greater amount of money to be invested internally. Shorter pipelines will save a great amount of capital. Using the energy locally will also create jobs for more people than mere oil extraction, and will add substantial value to local production. Aluminum and many petrochemicals are easier to export than oil or gas. True, they will need to be transported to their markets; however they could use all manners of alternative/existing routes through roads and rail links to Russia, and the Near and the Far East. Such exports are much less dependent on the availability of one or two very expensive pipelines and the goodwill of the states in which they run.

Region's Oil and Gas Profile

Because it holds large oil and gas reserves, the Caspian region, inclusive of Kyrgyzstan, Tajikistan, and Uzbekistan, has been described as the next Persian Gulf. Not quite, at least, not yet. The estimating of the region's oil and gas reserves is not based on exact science. Ranges on estimated reserves vary between 29,300 million barrels (mbls)[1] to 200,000 mbls.[2] While these reserves are substantial, they are still much lower than those of the Persian Gulf. Proven reserves in the Persian Gulf stand at about 645,000 mbls.[3] Similarly, the Caspian region's production in 1995 averaged 815,000 bls per day (bls/d), with more than one-half coming from Kazakhstan alone; production in the Persian Gulf in December 1996 averaged approximately 18,180 mbls/d.[4]

The region is also rich in non-associated natural gas. Total reserves are estimated by the U.S. Department of Energy's Energy Information Agency (EIA) at about 282 trillion cubic feet (tcf). At 1996 production levels, this would correspond to 87 years of production. For comparison, Iran has 735 tcf of gas reserves.

The figures above are not meant to diminish the importance of the oil and gas fields of the Caspian Sea region. Reserves of 29 billion barrels are important by any standard. However, they do point out that the Persian Gulf has substantially larger reserves and vastly lower production costs. The Caspian Sea basin will not replace the Persian Gulf, largely because the latter has two natural advantages over the former. First, the marginal cost of pumping crude in the Persian Gulf is the lowest on Earth. Marginal costs go from $0.5 per barrel on shore in Saudi Arabia to about $1.25 per barrel offshore in Abu Dhabi.[5] This compares to the cost of approximately $8 to $10 per barrel in the former Soviet republics, including the ones on the Caspian Sea, and of up to $12.00/bl in the North Sea.[6]

Second, the Persian Gulf countries can ship their oil at very little cost. The pipelines needed to pump crude from the fields to the sea need only be very short. Therefore, the cost of land transport is quite minimal, and the political leverage of the neighboring country is shortened/lessened. Saudi Arabia, the UAE, Kuwait, Oman, and Qatar all have their major exporting harbors within a few miles of their oil fields. Saudi Arabia did build a pipeline across its territory to the Red Sea. However, the Transarabian Pipeline was built to allow the Kingdom to sell at an advantageous price on the Red Sea and to protect itself from potential strategic problems in the Strait of Hormuz.

The energy resources of the former Soviet republics bordering the Caspian Sea and the Central Asian countries may be summarized as follows:[7]

AZERBAIJAN: Oil reserves: 11,000 mbls; oil production: 176,000 bls/d; years of oil production at present rates: 171. Gas reserves: 10,978 billion cubic feet (bcf); gas production: 229.80 bcf/y; years of gas production at present rates: 48. Coal reserves: not available; coal production: not available.

KAZAKHSTAN: Oil reserves: 16,000 mbls; oil production: 473,000 bls/d; years of oil production at present rates: 93. Gas reserveO)Os: 82,955 bcf; gas production: 159.90 bcf/y; years of gas production at present rates: 519. Coal reserves: not available; coal production: 119 million tons per year (mt/y).

KYRGYZESTAN: Oil reserves: 3,00 mbls; oil production: 2,000 bls/d; years of oil production at present rates: 411. Gas reserves: 529 bcf; gas production: 0.00 bcf/y; years of gas production at present rates: not applicable. Coal reserves: not available; coal production: 3 mt/y.

TURKMENISTAN: Oil reserves: 1,400 mbls; oil production: 73,000 bls/d; years of oil production at present rates: 53. Gas reserves: 97,957 bcf; gas production: 1,259.15 bcf/y; years of gas production at present rates: 78. Coal reserves: 800 mt; coal production: not available.

TAJIKISTAN: Oil reserves: 300 mbls; oil production: 2,000 bls/d; years of oil production at present rates: 411. Gas reserves: 988 bcf; gas production: 0.00 bcf/y; years of gas production at present rates: not applicable. Coal reserves: not available; coal production: not available.

UZBEKISTAN: Oil reserves: 300 mbls; oil production: 89,000 bls/d; years of oil production at present rates: 9. Gas reserves: 87,967 bcf; gas production: 1,588.50 bcf/y; years of gas production at present rates: 55. Coal reserves: 4,560 mt; coal production: 6 mt/y.

In Azerbaijan, the oil fields are located both onshore and offshore in the Caspian Sea. The onshore fields were discovered, in the early part of the twentieth century and were the first major oil resource of Russia. The offshore fields are being developed in joint venture with Chevron of the United States, Total of France, Statoil of Norway, and a number of other Western companies.

Azeri reserves are the second largest in the region after Kazakhstan. The EIA estimates them (both on- and offshore) to be at 1 billion barrels. This estimate may be conservative and seems to be on the low end of the various numbers mentioned and may eventually be revised upward substantially. It should be noted that proven reserves tend to increase as price of crude petroleum increases. Azeri production presently (176,000 bls/d) is low; Bahrain, for example, the smallest of the Persian Gulf producers, pumps about 140,000 bls/d from both its off- and onshore fields.

Azeri exports follow the Russian route. The only pipeline leaves Baku for Russia, passing through Chechnya. Thus, Azeri oil cannot be sold to the major oil companies and traders for hard currency without the approval of Russia and the goodwill of the Chechen. Most Azeri joint ventures with foreign oil companies, whether for prospecting, extraction, or pipeline, are likely to be concluded with substantial involvement of Russian oil companies.

In Kazakhstan the oil fields are located mainly in the northeast corner of the Caspian Sea area, the Tengiz field, and in the central part of the country around Kumkol. Daily production is at about 480,000 bls, approximately equivalent to Qatar's production, and has been in decline by 5 percent per year since 1990, hitting bottom in 1995 before rebounding as the result of new joint ventures with Chevron, Total, and others. The general decline in Kazakh production has been due to the same reason as the decline of oil production in all the former Soviet republics: shortage of equipment, lack of spare parts, poor technological and industrial base of the oil and gas sector, scarcity of funds, and low investment levels.[8]

Kazakh exports are effected through three small pipelines, all routed through the Russian Federation. Kazakhstan is limited by the ability and willingness of Russia to take its oil. In 1995, Kazakhstan was allocated a total export quota of 6.0 mt (120,000 bls/d). The existing pipelines and users could not take more. Presently, the Russian link limits the ability of Kazakhstan to export and develop joint ventures with Western companies.

Turkmenistan, Uzbekistan, and Kyrgyzstan share a major oil field, the Feragana Basin, estimated to have recoverable reserves of about 4 billion barrels.[9] Turkmenistan is a major gas producer and has substantial reserves. Its gas production is principally exported to Russia and Kazakhstan by pipeline. With 97 tcf, Turkmenistan has the fourth-largest gas reserves in the world, after Russia, Iran, and Qatar. It is now producing 3.5 bcf per day. Turkmenistan is not moving as rapidly as Azerbaijan and Kazakhstan in developing joint ventures with foreign companies; it lacks a consistent legal system and that makes potential investors nervous.[10] Turkmenistan is also saddled with receivables from its main clients, Kazakhstan and Russia, who only pay intermittently. These payment delays hobble the Turkmen government and economy.

Finances of Production and Export

The present emphasis on production and export of crude from Azerbaijan and Kazakhstan would be worthwhile if it could provide funds for the long-term development of the exporters. Oil and gas reserves are non-renewable. States must be satisfied that they get long-term benefits for permanent losses in reserves. Based on the agreements that Kazakhstan has signed with Chevron and Russian companies for the development of the Tengiz field,[11] it is possible to approximate the type of income that can be expected from the present projects in the Caspian Sea region.[12]

Because of geography, the competitiveness of the fields in all the countries of the Caspian Sea region, particularly the countries farther east of Kazakhstan and Turkmenistan, is limited. Exporting crude implies building long pipelines through terrain that is both physically and politically very difficult. Although the differentials in price of crude vary according to the supply and demand of refineries in each of the major areas of the world and other factors, by and large, crude that are shipped by pipeline then loaded onto ships for long distances sell at a great discount. For the purposes of this discussion, the price of the Caspian Sea oil at $12.00/bl up to $24/bl assumed as realistic.

The cost of transport from the Tengiz field to the Mediterranean is one of the main factors in evaluating the Central Asian projects. Transporting crude oil from the Caspian Sea is a complicated affair. Oil has to be piped through a number of different political entities, all with their own agenda. The politics and geo-politics of access and pipelines has been discussed elsewhere in this volume. Here, without prejudice to the pros and cons of the various projects, it is assumed that the Tengiz oil will be exported through Russia to Novorossiysk, from Novorossiysk to Turkey by sea, and through Turkey by pipeline to the Mediterranean.

There are three major costs associated with using pipelines—capital costs, transit fees, operating costs. At an average of $1 million per mile, the total capital cost of a modern pipeline from the Tengiz field to Novorossiysk and across Turkey would be about $1.5 billion in less demanding terrain. A premium of $500 million may be added to this for the more difficult conditions, and for the marine terminals needed to manage exports.[13] The transit "fee" includes payments to the countries that allow the pipeline through their territory, the rental fee for the terminals at Novorossiysk, and fees for the general pumping and maintenance of the pipeline. Operating costs include the cost of the actual day-to-day running, repairing, and maintaining of the pipeline, pumping the oil, and the like. These costs are estimated at between $11.5 and $13.5 per ton between Kazakhstan and Novorossiysk and at $1.00 to $3.00 per ton from Novorossiysk to the Mediterranean, corresponding to between $1.67 and $2.20/bl.[14]

Developing any oil field to increase production takes large amounts of capital. In the absence of detailed data on the costs of oil field development in the former Soviet republics, one may use the cost of developing oil resources in the Persian Gulf, selecting, for example, the highest in the cost range provided by the EIA (January 1996) at $4,866 to increase production by 1bl/d, to which one may add 30 percent in additional costs to account for the extra costs due to the distance to major harbors and the difficulty of operating in the area. New technological development in oil exploration, especially 3-D seismic research and horizontal drilling, have allowed substantial costs savings to field developments, rendering, for example, the Tengiz field more attractive to oil companies.[15]

The near-term cost of production in Russia is estimated at between $8 and $10/bl, which includes exploration, appraisal, and development outlays and also a 10 to 15 percent real rate of return, but excludes taxes and operating costs.[16] As applied to a $9/bl at a base price of $21/bl, these costs of capital, depreciation, and royalties would amount to $1,651 million per year.

In the case of the development of the Tengiz field, Chevron will have a 45 percent equity ownership, Kazakhstan 25 percent, Mobil 25 percent, and LUKoil of Russia 5 percent. Chevron and LUKoil are partners in the Caspian Pipeline Consortium (CPC) which will be a 1,500-km pipeline from Tengiz to Novorossiysk.[17] It is expected that Chevron will either find or provide the capital necessary to develop both the field and the pipeline. The total amount of capital to be expended by the year 2000 for the development of the fields and the pipeline could total $5.6 billion. Whether this capital is raised on the world financial markets or directly provided by Chevron, it will have a cost. Indeed, Chevron itself will borrow the money to pass it on to the two consortia. One can expect that the cost of capital will be not less than 8 percent per annum for repayment over 15 years.

The cost of running the day-to-day production and maintenance in the oil fields of the Persian Gulf region is estimated by the EIA (January 1996) at 5 percent of the original investment plus $0.25 to $1.00/bl. In the case of the Caspian region, it is realistic to expect the cost of production to amount to 5 percent of the original investments of $3.1 billion for a production of 500,000 bls/d corresponding to $0.86/bl, plus the average of the spread per barrel suggested by the EIA, or $0.63/bl. Therefore, the total cost would be about $1.49/bl.

Some of the largest elements in the costs of oil to the Central Asian republics are the royalties and management fees that the joint venture will have to pay to the foreign oil companies for developing the fields. The fee structure of the agreements has not be released for public information. Therefore, one can only estimate what range will be demanded by the oil companies for providing their technology, know how, and capital. To estimate these amounts, one must assume that the oil companies, like any other firm, will want a certain rate of return before tax on their use of capital.

If the foreign oil companies were to bear the risk and opportunity costs of financing the development of fields and pipelines, they would only invest if they could match their minimum required return on the total amounts they expect to spend for a certain expected risk level. One may assume, realistically, that the amount of capital needed, about $5 billion by the year 2000, will carry an interest rate of 8 percent and that the foreign oil companies will charge the said amount to the joint venture, therefore requiring a return that includes such interest. It would be normal practice for a corporation to require a rate of return on investment of about 30 percent per year. On a total capital of $6,083/bl of production provided by the foreign oil companies, this would correspond to a minimum required annual return of $1,285/bl, or $5.00 bl/d.

In the Tengiz field, Chevron will own 45 percent equity in the joint venture. In case oil prices climbed to $21.00/bl, Chevron would obtain about $2,677/bl/y corresponding to $7.33/bl as its share of the total net consortium income; Chevron's minimum expected return therefore would be met easily.

Cash Return to Producer Country

At the price of $21.00/bl, gross income from oil would be around $3,825 million per year on 500,000 bls/d production rate. The total of the costs discussed above indicate that the cash outflow for this size production would be about $2,380 million per year, leaving to the Caspian countries $1,453 million, or on the basis of a 10-year production average of $8.39/bls, about 40 percent of ultimate sale price. It should be noted, however, that a $21.00lb price assumed here is very high.[18] At prices below $13/bl, the Caspian countries would actually lose cash on their production. Their income would remain at 30 percent of world prices until oil reaches $18/b.

Naturally, it can be argued that chances of oil prices falling below $13/bl are very slim and that any amount of money made above and beyond break even is cash in the state's coffer that would not otherwise be

Estimate of State Oil Income 500,000 bl/d Production

At OPEC basket price ($/bl)	14	17	18	20	21	22	23
Income to State ($/bl)	1.39	4.39	5.39	10.39	11.39	12.39	13.39
total production ($million)	254	801	984	1,896	2,079	2,261	2,444

there. This argument has value. Indeed, with income of $1 billion and more, a state can invest in education and industrial programs that will benefit the country, create jobs and infrastructure. However, the old argument that "a barrel of oil once exported is gone forever" still applies. If production in Kazakhstan were to increase by 1 mbl/d to 1.5 mbls/d, the actual number of years of oil production would decrease from 93 to 29. Proven oil reserves in new fields tend to increase as prospecting improves, but the decline in the number of years of production is irreversible. It forces governments to diversify very rapidly and to somewhat unnaturally embark on unviable industrialization programs.

The concept of oil money being used to develop other types of economic ventures and industries is often unrealistic. The extraction of natural resources is indeed a mixed blessing.[19] By bringing substantial amounts of foreign exchange into the country, oil allows the local currency to remain relatively high compared to the currencies of its neighbors and others. An overvalued currency allows the local population to buy imported goods rather than manufacture locally and this, in turn, hinders efforts to develop a local productive industry. This problem, sometimes called the "Dutch disease," has been seen in the Persian Gulf region, where cheap foreign labor was imported in lieu of using the more expensive local labor. The existing factories and service providers are now addicted to this cheap labor and find it very difficult to switch to local sources, paradoxically creating high unemployment rates in otherwise very wealthy states.

The large inflows of cash from oil sales are more often than not used for nonproductive investments. Military forces, luxurious palaces, unneeded roads and a bloated bureaucracy seem to eat up the funds as they become available. The Caspian countries that are suffering from economic recessions are strapped for cash. The amount of cash obtainable from oil return may be much lower than generated by the oil pumped in the Persian Gulf. But, cash is cash, and its lure may prove irresistible. If world prices were $18.00/bl, the Caspian states would make only $984 million instead of the $3.2 billion that a Persian Gulf state would make, but payments of close to $1 billion are better than what the states obtained before.

One could question the ability of states in general to handle a rent-type of yearly income of $1 billion or more. Too many countries have ill-used their windfall from oil. Nigeria has spent many times what it receives in oil revenues. It has borrowed beyond its ability to repay without much benefit to its citizens. Venezuela is undergoing a very tough economic restructuring program that is not even sure to succeed in putting the country back on the right track. Very wealthy states tend to spend lavishly on military programs. For example, Saudi Arabia spends some 35 percent of its income on the military and, yet, the kingdom is all the more dependent on foreign powers for its defense.

Development Alternatives

In general, there are three economic development alternatives to straight-out production and export of crude oil and natural gas. These are export substitution, export-led growth, and energy intensive industrialization. All three cases do demand some level of oil and gas production; however, they do not demand export.

Export substitution was tried in Saudi Arabia in the early 1970s. The state established the Saudi Industrial Development Bank to provide zero-interest loans to industry. The state also provided companies with very cheap prepared industrial land at one Saudi rial per square meter per year and free access to utilities and subsidized rates. Industrialization increased drastically, with over 2,000 factories built between 1975 and 1985.

Generally, however, export substitution based on investment from oil extraction income creates more problems than it solves. The Saudi factories produced all manner of goods, but always found it difficult to compete with imports from other countries, unless supported by tariffs that increase prices for the local consumer and create tension with trading partners. Many of these plants survived only because of large subsidies and importation of cheap skilled labor. One could only question the wisdom of building Mercedes trucks in the Kingdom of Saudi Arabia using mainly imported Turkish labor, or of assembling air conditioners using Asian labor. The social cost of having up to 4 million foreign workers in the Kingdom is huge. The subsidies cost the state a great deal of money. The local population gets used to non-factory jobs, which are worked by imported labor.

The "success" of the so-called Asian tigers—that is, Singapore, Hong Kong, Taiwan, South Korea, and Thailand—have made export-led growth one of the favorite models of development among political economists. However, the tigers started their growth using low pay, highly skilled labor, and had a geographic advantage of being able to receive raw materials easily, and export finished goods rapidly. They also greatly benefited from the United States wishing to see strong economies surround communism's evil empire, even if it were to cost the United States the loss or erosion of a part of its own industrial base. In contrast, the former Soviet republics in the Caspian region have none of the characteristics of the tigers; they have relatively smaller populations, uncompetitive wage scales, and poor skill levels. Furthermore, they are landlocked. The costs of importing raw materials, transformation/production, and export to the nearest harbors are prohibitive. Only the type of product that can be made from materials imported by air and products exported by air would be competitive. These products, mainly in the domain of electronics, pharmaceuticals, software engineering, and the like, are rare and by and large require highly skilled labor and engineers.

Some aspects of the two aforementioned economic development models could work in landlocked energy-rich countries by capitalizing on a particular country's natural advantages. For example, there are products that can be made primarily from local raw materials, by low-skilled workers, for some domestic consumption, but mostly for export. Most Persian Gulf countries are trying to use their easy access to plentiful energy to develop along these lines. One example of this hybrid approach is found in Saudi Arabia's experience with the Saudi Arabian Basic Industries Corporation (SABIC), a holding company that controls eleven large petrochemical companies, two steel companies, and one very large fertilizer company. In 1996, SABIC produced over 23 million tons of chemicals, plastics, fertilizers, and other products, posting a net profit of $1.18 billion for the year.[20] Its success is due in great part to: (1) some $20 billion in capital investments: (2) access to locally produced raw materials, mainly oil and natural gas at one of the world's lowest production costs: (3) a modern, efficient, integrated, streamlined industrial/ production infrastructure: (4) skilled engineers: and (5) proximity to the fastest growing markets for petrochemicals, South East Asia and China.

There are also more modest undertakings, which may find a greater relevance to economic development of the former Soviet republics in the Caspian region. In Saudi Arabia, for example, both local clay and limestone have been combined with cheap fuel to create active brick and cement industries. The other countries in the Persian Gulf have developed large cement plants as well. Bahrain, Dubai, and Qatar have built large aluminum smelters by applying electricity produced from their own cheap natural gas to alumina imported from Australia. Developing industries based on energy would create added value to local raw material, whether oil or gas. It would create local good-paying jobs. It would limit the dependence of the local states on their neighbors, namely, Russia, Iran, and Turkey.

Using their large reserves of natural gas, Kazakhstan and Turkmenistan would have a natural advantage in producing methanol, ammonia, urea, ethylene, and ethylene derivatives. These chemicals could be exported by truck or rail to China, which presently has the biggest demand for them,[21] as well as to their other usual trade partners. Kazakhstan could also use some of its energy to develop further its aluminum industry. Kazakhstan, with some of the largest bauxite deposits in the world, could supply China, Russia, India, Iran, and others with competitive aluminum and derivatives of aluminum.

Azerbaijan has less natural gas but could also develop a petrochemical industry based on oil, using refined products, such as naphtha, as feed-stock for ethylene and numerous other chemicals. Local production of chemicals, aluminum, steel, and the like would result in diversification, political independence, profits, and jobs. However, the costs are high. In order to develop any kind of petrochemical or other energy-intensive industry, there have to be some basic prerequisites. First, the oil and/or gas fields need to be developed. As was seen above, this can cost $0.63 per new barrel per day. However, the ultimate returns from petrochemicals are much higher than in just selling crude, which would allow the local oil/gas companies to invest less in production and limit extraction.

Second, as is the case in the Persian Gulf, large investments are required for developing petrochemical plants. A sizable petrochemical industry may require many billions of dollars. A precise price tag is hard to ascertain, but some estimates may be given: The BinLaden/SABIC joint venture to produce annually 330,000 tons of ammonia and 570,000 tons of urea in Saudi Arabia is expected to cost $480 million. At present prices, this production would provide yearly sales of about $190 million, using about 81 million cubic feet per day of natural gas, which would correspond to 18 percent of Kazakhstan's natural gas production and 2.3 percent of Turkmenistan's production. If sold at world market prices, the same gas production would fetch about $58 million, one-third of the petrochemical value.

The new extension of the Al-Sharq facility in Saudi Arabia to produce an additional 450,000 tons of ethylene glycol to be made in the same plant into 300,000 tons of low-density polyethylene for export will cost about $1.2 billion. Sales of this product would bring in about $300 million per year at today's prices. Such a plant itself would require a source ethylene cracked from liquid petroleum gas, naphtha, or natural gas. The products that would be made in similar plants in the Caspian region would not require expensive and politically sensitive pipelines. They could be transported by rail or truck to the ultimate users. The transport infrastructure would have to be improved over time, but imports could be started immediately, without additional investments upfront. Crude oil and gas pipelines with all their political ramifications would not be needed, thereby saving billions of dollars in capital expenditures and daily transit fees on oil throughput. In addition, the new plant and production would create extensive collateral industrial activities in maintenance of equipment, transportation, and downstream plants such as aluminum foil plants, specialized plastics plants, and others.

The computation of the total cost-benefit figures for switching the investment emphasis to developing a substantial energy-based industry is complicated. However, simple examples suggest that the heavy capital costs would be vastly overshadowed by the long-term and even the short-term benefits. For example, a plant similar to the aforementioned SABIC ethylene glycol/LDPE plant, requiring a $1.2 billion investment against sales of $300 million, would need an ethylene source that would be produced in a plant of not less than $600 million. Thus, to get sales of $300 million per year, the basic investment would have to be about $1.8 billion. Using industry estimates, going from oil to naphtha to ethylene to ethylene glycol, the production of 450,000 tons of ethylene glycol would require about 56,380 bls/d of oil.[22] Some 43 percent of this oil could be refined into naphtha for the ethylene production and the balance sold as other oil byproducts. The naphtha portion of such oil production, if sold at $18/bl, would represent about $159 million per year, about one-half of what would be returned on petrochemicals.

A similar computation can be made with respect to returns from an ammonia plant. Ammonia's main raw material is natural gas. A 500,000-ton/y ammonia plant requires 45 mcf/d of gas.[23] Assuming the price of ammonia at $227/ton (as of March 1996), the cost of gas for the production of m/t/y of ammonia would equal $32 million,[24] against sales of $113.5 million. In other words, the same amount of gas, if used as feed stock for locally produced ammonia, would return close to four times its market value as gas.

Natural gas can also be used to produce methanol, propylene, butadiene, and others. The aforementioned ammonia production is used to make urea and other fertilizers. Petrochemicals based on byproducts from refined crude oil include ethylene and all its byproducts (including PVC and ethylene glycol), benzene (itself used in producing styrene and polystyrene, phenol, and synthetic fibers), toluene (the base for TDI and urethane rigid and flexible foams), and numerous other high-value products.

Comparative Cost Analysis

Whether oil is exported or used as feedstock for a petrochemical industry, it still needs to be pumped out of the ground, for which reserves have to be developed. Therefore, the argument goes, there would be little savings in establishing a large-scale petrochemical or energy-intensive industry. Both efforts would require the involvement of foreign partners at huge costs. It should be said, however, that the amount of oil and gas required to achieve the present expected cash flow generated from 500,000 bls/d of oil exports from the Tengiz field or others would be between one-half to one-sixth if most of the production were used for creating petrochemicals and other energy-based industries. Substantially lower production would require lower capital expenses. Instead of spending $3.2 billion producing 500,000 bls/d, $500 million could be spent to build or upgrade a refinery to produce more naphtha as base for ethylene production. An additional $1 billion could be spent to build an ethylene cracker, which would end up producing exportable ethylene glycol or LDPE or similar products.

If new pipelines for oil or gas exports are no longer needed to be built, then other types of transportation infrastructure will be needed to export ethylene byproducts, ammonia, urea, and the like. Further, if other energy-intensive industries such as aluminum or direct reduction steel are promoted, new electricity generators will be needed, and facilities would have to be built to transport bauxite or iron ore to the source of energy. The pipeline from the Tengiz field to the Black Sea and through Turkey to the Mediterranean will cost $2 billion. This pipeline will render Kazakhstan even more dependent on Russia. Only by building pipelines through other countries, such as Turkey and Iran, can Kazakhstan reduce its dependency on Russia. This would imply many more billions of dollars. Spending the same amount of money on improving the railroad and road ties to its neighbors would allow Kazakhstan and the other countries of the region to export much more easily through many routes, especially to China.

Financing

Unless the Caspian countries can find large sums of cash or borrow heavily in the European financial markets, the financing of these plants and refineries will have to come from joint venture partners in those industries. One of the advantages of going downstream from pure oil production is the ability to negotiate with many more companies. Most of the large oil companies have petrochemical divisions, but there are also a good number of large chemical companies with the know how, savvy, and financial means to develop sizable plants anywhere in the world. Naturally, these firms will charge for sharing their know how, finding the capital and building the plants, and, just like the oil companies require royalties, they will require either a share of production or special sale prices. On the other hand, these companies will teach the local petrochemical engineers and managers how to market the products worldwide, especially in Asia.

Other Benefits

New developments in energy-intensive activities will benefit a much larger part of the population. The making of petrochemicals and other energy-based products will create a demand for engineers and highly paid skilled labor. In the first stage, many of these skills may have to be imported. However, if the Saudi example holds in the Caspian region, within 20 years many such skills will be developed and supplied locally.

Further, there will be a great demand for collateral services—transportation, maintenance of machinery, construction, and others—which will be provided locally, all creating substantial employment at all skill levels.

Turkmenistan has greatly suffered from its inability to export its gas to anyone other than the former Soviet republics. The state is under great financial stress because neither Kazakhstan nor Russia are paying for their purchase of gas from Turkmenistan. Turkmenistan is totally dependent on the goodwill of Kazakhstan and Russia. Turkmenistan is married to its pipelines. In order to export and be paid in a timely manner, it must either build more pipelines through third countries like Iran, or through the Caspian Sea and Azerbaijan to Turkey. The pipeline solution is expensive and politically difficult. It would make much more sense for Turkmenistan to emphasize developing its petrochemical infrastructure and export its semi-finished and finished products in multiple directions and/or swap them with its neighbors.

Conclusion

Oil resources in Central Asia are not a blessing. The race for exporting crude oil is making the countries very dependent on foreign capital and on foreign oil companies. The Central Asian countries are as dependent on Russia as they were before the breakup of the Soviet Union. Foreign companies are interested in maximizing their revenues in the shortest time possible and to minimize their risks. Geography also renders the Central Asian republics very susceptible to pressures from their neighbors in whose territory pipelines, pumps, and loading terminals are/will be located.

Prospecting, drilling, pumping, and exporting crude oil is extremely expensive in the Caspian region. To maintain a production that is competitive with the price of oil in the Persian Gulf, prices at the wellhead will have to be discounted heavily by the costs of capital and royalties. In the final analysis, the Central Asian and Caspian countries will only receive money if world prices stay above $13/bl and if the countries receive one-third of the final price paid by the users.

The mere export of crude taps a non-renewable resource. Foreign oil companies will always find oil somewhere in the world. However, once a barrel of oil has been exported from Kazakhstan it decreases the oil stock of the country forever. Thus, it is vital for the states of Central Asia and the Caspian region to maximize their return on each barrel of oil and cubic foot of gas. Mere oil production produces few jobs, little money, and it is a disincentive to the development of other industries because it keeps the foreign exchange parity too high.

The Central Asian and Caspian countries need to consider the alternatives to oil production and export. Their natural advantage on the world market being their access to plentiful energy, they must emphasize investment in energy-intensive industries. Granted, petrochemicals, aluminum, direct reduction steel, and others do demand large investments, but the return on that investment to the countries would be often two to four times more than revenues from the mere export of crude oil and gas. At the present, Central Asian and Caspian countries need capital from abroad, but they can also be selective from among many sources of capital other than the oil companies. This increase in the source of capital may give the capital-importing countries a better leverage in their negotiations first among the sources themselves and also with respect to the oil companies. However, in all cases, energy-intensive industries create a much larger multiplier effect for the developing economy than a straight-out "produce and export" program.

Notes

1. This figure is based on the information contained in U.S. Department of Energy, Energy Information Agency, *Country Studies* on Azerbaijan, Kazakhstan, Kyrgyzstan, Turkmenistan, Uzbekistan, and Tajikistan (Washington, D.C. 1997) (<www.doe.eia.doc.gov/emeu/cabs>). Hereinafter referred to as EIA.
2. This figure is obtained from the internet site identified as <mfs.gov.tr:80/Grupf/caspian1>.
3. *International Encyclopedia of Petroleum* (Tulsa, OK: PennWell Books, 1995).
4. *Middle East Economic Survey* (Cyprus), vol. 40:4 (1/2797). Hereinafter referred to as MEES.
5. See EIA, *Oil Production Expansion Costs for the Persian Gulf* (Washington, D.C.: U.S. Department of Energy, January 1996).
6. See Thomas R. Stauffer, *Indicators of Crude Oil Production Costs: The Gulf vs Non-OPEC Sources* (Boulder, CO: International Research Center for Energy and Economic Development), Occasional Papers, no. #19 (1993).
7. These figures are obtained from EIA (1997). The figures for gas are converted from EIA's cubic meters into cubic feet using an equivalent of 35.3 cubic feet per cubic meter.
8. OPEC *Bulletin* (Vienna), September 1995, p. 10.
9. See EIA, *Oil and Gas Resources of the Feragana Basin* (Washington, D.C.: U.S. Department of Energy, December 1994).
10. EIA (www.eia.doe.gov/emeu/cabs/turkmen.html) (1997).
11. Some of the terms of these agreements were disclosed to MEES and are reported in MEES, 40:4 (1/27/97).
12. The estimates set forth in this discussion are based on the model developed by the author and presented at Caspian Associates' First International Conference on "Caspian Oil, Gas & Pipelines," held at the Mariott Marquis in New York City on May 29-30, 1997. The model—not an econometric one—assumes a joint venture with a major western oil company to develop an existing field and produce 500,000 bls/d by the year 2000. Production will increase by 50,000 bls/d in 2001 and 2003;

increasing by a further 100,000 bls/d every year from the year 2004 to 2009 and a further 50,000 bls/d in the year 2010, reaching a total production of 1,250,000 bls/d. The price of crude on the world market is assumed to be the average OPEC basket price ($20.29/bl in 1996), and that it would apply to Central Asian production only after the oil reaches an open sea for shipment. The model assumes a discount or premium to equalize the cost to the main users, which includes the cost of the pipeline and the cost of shipping based on what is called "Worldscale." The shipping discount is assumed to be the same as the OPEC basket, that is, once crude has been delivered to the Mediterranean by way of Novorossiysk on the Black Sea and either the Bosphorus or by pipeline to Ceyhan, the differential in price to the average of all the crude from both the Persian Gulf, Nigeria, Algeria, and Venezuela will be similar. The model assumes other discounts such as API degrees (the higher the degree the lighter, and more expensive, the oil), the sweetness (level of sulfur), and the like.

13. According to Petroleum Finance Company, *Pipeline Politics (II)* (Washington, D.C., April 1994), pp. 19-20, a 42-inch, 300,000 bls/d pipeline from Tengiz to Novorossiysk would cost $1.5 billion. The figure includes port facilities and pumping stations. The cost of facilities needed to cross Turkey is estimated to add another $573 million for 500,000 bls/d.

14. Anne Bingham, "Costing Kazak Oil Exports: An Economic and Political Analysis of Transporting Kazak Oil to World Markets," unpublished student paper, Columbia University, New York (December 1996) (citing an oil economist at the World Bank).

15. See, Bingham, ibid. But see J. E. Hartshshorn, *Oil Trade: Politics and Prospects* (Cambridge: Cambridge University Press, 1993), p.50 (the Tengiz field, even though very large, was not economically viable).

16. Stauffer, pp. 8 and 15.

17. MEES, vol. 40:4 (1/27/97).

18. The OPEC basket price in October 1995 was $15.85/bl; it had been as low as $12.87/bl in December 1993. OPEC *Bulletin* (3/95) and (2/97).

19. *The Economist* (London), December 23, 1995, p.87.

20. MEES, Vol. 40:11 (3/17/97).

21. See SRI International (Menlo Park, California), "Polypropylene in the Early 21st Century" (1995) and "Ethylene, The Next Decade" (1995), papers presented to the Schroder Wertheim Chemical Industry Conference, New York, December 10-11, 1996.

22. These ratios are based on SRI International, *PEP Yearbook 1995* (Menlo Park, CA, 1995) and SRI International *Chemical Economics Handbook* (Menlo Park, CA, 1995) and were rounded. Crude will yield 43 percent naphtha. Naphtha will yield 48 percent of ethylene, ethylene 89 percent of ethylene oxide, ethylene oxide 84 percent of ethylene glycol.

23. MEES, vol. 40:11 (3/17/97).

24. 45mm cf/d x 365 days x $1.96/mcf.

CHAPTER SIX

The Caspian's Environmental Woes

Siamak Namazi

Introduction

Following the dissolution of the Soviet Union, the Caspian Sea is increasingly the subject of political and academic attention. The sudden increase in the number of littoral states from two to five, the discovery of vast oil and gas reserves, and contention over the legal status of the world's largest lake have all drawn much expert attention. Yet, as the littoral states seek to gain a greater share of the Caspian's hydrocarbon reserves, and as multi-nationals rush to the area in search of business opportunities, the environmental problems are often neglected. Meanwhile, the nearby Aral Sea stands as a stark reminder of the consequences of such inattention.

Environmental pressure on the Caspian is intensifying: The rise of the sea's level has resulted in coastal inundation; pollution is on the rise; and over-exploitation of the marine resources, especially the sturgeon and seal, is widespread. Yet, most academic work is drawn to the legal and political aspects of contemporary issues encompassing this body of water. There is a distinct lack of attention toward understanding the Caspian's environmental challenges.

The purpose of this chapter[1] is to bring together the scarce literature on the Caspian environment in an attempt to paint an overall picture of the situation and to expose the challenges that must be overcome. In that sense, this is much more a descriptive, rather than prescriptive, study.

Less apparent than the environmental problems of the Caspian Sea is the solution to these problems. The Caspian remains devoid of a defined legal and institutional structure capable of identifying the responsibility of each littoral state toward the lake's environment. Nonetheless, the impact of these challenges to the Caspian Sea does not allow the time it takes the diplomatic process to resolve the legal issues. No doubt, finding a way to prevail over environmental problems amid the geo-political upheaval surrounding this lake is an extremely difficult task. The only unmistakable guide to facing the impacts of the Caspian's environmental problems is that while improvement on the national level is definitely needed, regional cooperation is crucial.

This chapter will describe the Caspian Sea's environment in terms of geographical data, look at its legal regime, and give a picture of its biological and hydrocarbon assets. This is followed by an examination of the pressures that human economic activity and resource exploitation place on the region's environment, followed by a review of recent attempts at environmental salvation.

Geographical Data

The Caspian Sea, the largest inland body of water in the world, is situated on the far end of southeastern Europe, bordering Asia. It covers 393,000 square kilometers, with a water volume of 78,000 cubic kilometers, which accounts for some 40 to 44 percent of the total lacustrine waters of the world. The Caspian occupies a deep continental depression within the largest catchment basin in Europe. The coastlines of this lake are shared by Iran, Azerbaijan, Russia, Kazakhstan, and Turkmenistan.[2]

On the basis of physio-geographic characteristics, the Caspian Sea is divided into three areas: Northern, Middle, and Southern Caspian. The border between the Northern and Middle sections runs over the edge of the Northern Caspian shelf, the Mangyshlak threshold, though the conventional border is taken as the line connecting Chechen Island with Cape Tiub-Kargan. The Aspheron threshold constitutes the border between the Middle and Southern sections, while conventionally the line connecting Zhilloi Island and Cape Kuuli delineates the two. Water volumes in the three regions vary to a great degree. The Northern area, with some 28 percent of the total area and average depth close to 62 meters, accounts for only 1 percent of the total water volume. The Middle area covers 36 percent of the Caspian area and accounts for 33 percent of the volume, with an average depth of 176 meters. The Southern area, with an average depth of 330 meters, also covers 36 percent of the sea and accounts for 66 percent of the total volume.[3]

Salinity ranges from 0.3 percent near the mouth of the Volga to over 300 percent in the Kara-Boghaz-Gol Bay, though in most parts salinity lies between 12 and 14 percent, compared to approximately 35 percent in the world's oceans.[4] Kara-Boghaz-Gol, located on the shores of Turkmenistan, is the largest gulf in the Caspian Sea. It plays a major role in the water's equilibrium, given that it is a source for the extraction of many salts formed by the evaporation of the water. Nearly 140 large and small rivers flow into the Caspian, including the Volga, Ural, Terek, Sulak, and Kura. The major inflowing system is by far the Volga, which provides annually close to 88 percent of the lake's total water input from rivers.[5] The river network is distributed very unevenly within the catchment; all the main rivers flow into Northern Caspian.

The Caspian is not connected to any ocean; it is artificially connected to the Black Sea via the Volga and Don Rivers, by the Volga-Don Canal. It has also been joined to the Sea of Azov, which is essentially a branch of the Black Sea. For years, Iran has wanted to connect the Caspian Sea with the Persian Gulf by means of a giant navigation channel which would stretch over 1,500 kilometers. Currently this plan is being studied at the Ministry of Construction Jihad, which claims that its design is complete. Nonetheless, this project seems far from being ready for initiation.

Legal Regime

Prior to the breakup of the Soviet Union, the management of the Caspian was regulated primarily by the 1921 treaty between Iran and the Russian Socialist Federal Soviet Republic, and the 1935 and 1940 treaties between Iran and the Soviet Union. After the dissolution of the Soviet empire, considerable disagreement over the classification of the Caspian Sea has surfaced, affecting each riparian state's share of resources.

Broadly speaking, Azerbaijan and Kazakhstan are leading the effort to classify the Caspian as a "sea," which by the United Nations Convention on the Law of the Sea would allow each state independent exploitation of resources within defined nautical limits of sovereignty. Iran, Russia, and Turkmenistan, however, are in favor of keeping the status quo and argue that the Caspian is indeed a closed body of water, or a lake, therefore subject to joint use, including joint development of its hydrocarbon resources.

The current contention over the legal regime of the Caspian has resulted in many problems affecting its environmental condition. There is a pressing need for a consensus between the five riparian states to regulate their current activities. Navigation, environmental management, conservation of fish stock, and the position of permanent installations and pipelines all require regional cooperation and a mutually respected regime.

Biodiversity

As a result of 5 to 7 million years of isolation from the oceans of the world, the Caspian has turned into a safe haven for a number of very ancient indigenous species, including a unique flora. Some 75 percent of the species inhabiting the Caspian are endemic to the Caspian; 6 percent are from the Mediterranean; 3 percent from the Arctic; and the balance are freshwater species. It remains the only body of water with large quantities of the sturgeon. Some six species of that fish in the Caspian account for 85 percent of total reserves and commercial fishing equal to some 90 percent of the world's total catch. Other important fish include the volba, zander, carp, and kilka. Estimates of total annual fishing range between 500 and 1,000 tons. Some 575 types of marine plants are found in the Caspian and nearly 2,000 tons wash up annually along the shorelines. The southern coast of the sea includes valuable wetlands located on the migration path of millions of birds. The Caspian is known to be the breeding ground for ducks, moorhens, cormorants, bitters, and herons, among others. There is only one marine mammal in the lake, namely, the Caspian seal *(phoca caspica)*, with an estimated population of 400,000.

Hydrocarbon and Other Mineral Resources

The Caspian basin is thought to have the third-largest world oil reserves, behind the Persian Gulf and Siberia. The marine extraction of oil was introduced for the first time in history in the Caspian Sea. According to a 1994 estimate, there are some 150 million barrels of proven oil reserves along with 15 trillion cubic meters of natural gas in the Caspian Sea. If these figures are accurate, the Caspian holds nearly 16 percent of the world's proven oil reserves and 53 percent of the proven natural gas reserves. According to the same source, it is estimated that the Caspian's marine fields presently yield close to 9 million tons of oil and 14 billion cubic meters of natural gas, with much greater yields anticipated in the future.[7] Moreover, there is a unique possibility for the development of chemical industries, including the world's largest sulfate deposit in Kara-Boghaz-Gol Bay.[8]

Within the existing legal frameworks, the hydrocarbon reserves falling in each of the former Soviet republics bordering the Caspian are as follows: Azerbaijan with 4 to 5 billion cubic meters (bcm) of oil and 0.6 trillion cubic meters (tcm) of gas: the Russian Federation with 2.3 bcm of oil (including condensate) and 1.3 tcm of natural gas; Kazakhstan with 3 bcm of oil and 1.7 tcm of recoverable natural gas; and Turkmenistan with 0.5 bcm of oil and 0.2 tcm of natural gas.[9] At the moment, Azerbaijan is the only country extracting significant amounts of oil from the Caspian Sea, largely from two fields lying on the Aspheron coast. In 1995, the total oil extraction in Azerbaijan amounted to nearly 9.2 million tons and 6.6 million cubic meters of natural gas. It is estimated that since operations began, some 430 million tons of oil and 300 billion cubic feet of gas have been extracted by that country.[10]

Tourism and Marine Transportation

Historically, the Caspian has been vital in meeting the food and economic requirements of the millions on its shores. Besides its valuable marine resources, the sea acts also as a magnet for tourism; in 1991, the Caspian attracted some 190 million visitors.[11] Moreover, its marine transport facilities are of paramount importance. The Caspian serves as the nexus between its riparian states and ports on the Black Sea, the Baltic, and the northern basins. Marine transport between the littoral states is currently considered the most economical medium. In fact, the Caspian route can decrease the transport time between Iran and Europe by up to ten days. Presently, the ships that operate in this waterway have a carrying capacity up to 5,000 tons. They provide the only means for transporting heavy machinery and equipment that are too large or heavy for the imperfect roads and rail systems of the littoral states.[12]

Pressures on the Environment

The challenges to the Caspian environment are largely rooted in three distinct sources: flooding, pollution from onshore and offshore sources, and the over-exploitation of resources.

Despite the obvious importance of the resource-rich Caspian Sea to all five riparian countries, the state of its environment is far from desirable. It is hard to establish the overall environmental status of this lake, given that much of the available data is largely site- or country-specific. The Soviet system was notorious for encouraging the under-reporting of problems, a legacy that appears to have devolved intact to the successor republics. Among this legacy is the lack of communication between various government institutions and the absence of classification of findings, all of which make it extremely difficult to gather reliable information. Again, the need for a regional body whose responsibility would include the amalgamation of information is greatly felt.

The data collected in 1991 reveals that the Caspian Sea received effluent consisting of more than 3,000 tons of oil products, 28,000 tons of sulfites, 315,000 tons of chlorides, and 25,000 tons of phenols. Furthermore, some 200,000 tons of tar entered the Caspian annually. Concentrations of oil products and phenols in the Northern Caspian and along its western coast were reported to be 4 to 6 times the maximum admissible values. Along the coast of Azerbaijan these figures reached 10 to 16 times the maximum. The three out of the six rivers considered as major spawning grounds for the sturgeon run in Daghestan, a republic within the Russian Federation. The potpourri of pollutants found in these rivers in 1991 was alarming. The quantities of heavy metals, pesticides, phenol, arsenic, boron, selenium, among others, found in them exceeded the maximum permissible values for fisheries by 60 to 100 times and 2 to 10 times the maximum for sanitary-hygienic parameters. In fact, some studies indicate that nearly all 140 rivers that flow into the Caspian carry a myriad of pollutants including sewerage and agricultural and industrial wastewater.[13]

It has been reported that the data generated by the United Nations Industrial Development Organization indicates that along parts of the Azerbaijan coastline, pollution levels exceed international standards by a multiple of 16. Along the Volga in Russia this figure is 4 times that set by international standards; in Kazakhstan twice the set level; and in Iran and Turkmenistan close to the maximum allowable.[14]

It is crucial to realize and appreciate that given the gyration of the Caspian's currents, environmentally unsound activities in one riparian state is quick to reach the others. This and the fact that the sturgeon and other Caspian species are highly migratory—that is, they feed off the coast of one riparian state, spawn in the river of another, and are caught elsewhere—render futile any unilateral effort by a riparian state at environmental protection and resource conservation. Meanwhile, the population of the sturgeon in the Northern Caspian alone has diminished by nearly 7 million and the population of the endemic seals is rapidly diminishing.[15]

Flooding

In the mid-1970s, the Caspian Sea reached its lowest levels in recent times. As the sea withdrew, human activity expanded into the newly exposed lands. Fearing the example of the dying Aral Sea, the Soviet Union responded with engineering solutions aimed at bringing water to the Caspian from other parts of the country. However, unexpectedly, in the late 1970s, the lake's water level began to rise and has continued to do so at an alarming rate of about six inches per year.[16]

The fluctuation in the Caspian's water level is not a recent phenomenon. During the past 2000 years, variations of up to eight meters have been reported. From 1930 to 1977, the major concern was the drop in volume of that lake; this decline corresponded to the period in which the Soviet Union, especially Russia, undertook mega development projects in agriculture and industry, which included the building of artificial lakes and large dams. After 1977, as the sea continued to expand, the drastic rise in the Caspian's water levels became the source of increasing concern. Today the littoral countries are trying to cope with the adverse consequences of the flooding that has followed.[17]

Scientists confess that the source of fluctuation is not fully known. Generally three hypotheses have been advanced to explain the phenomenon: probabilistic (or cyclical), tectonic, and climatic. The majority of experts currently support the last mentioned theory. A number of other phenomena are thought to contribute to the fluctuation. These include the reduction of water evaporation surface as a result of oil pollution, diversion of the routes of rivers to and away from the Caspian Sea, and the artificial creation of links between the Black Sea, Aral, and the Caspian.[18]

The increase in water levels began manifesting its devastating effects in 1986, a year that was also marked by torrential rains and high winds, causing financial loss, destruction of agricultural land, disruption of the haul-out sites for seal and sturgeon hatcheries, and loss of human life. Some 10,000 houses were destroyed in Iran alone. Water, electricity, and fisheries installations and port facilities were damaged. The Iranian government estimated that the losses due to flooding amounted to some $2 billion. There has been further damage in terms of lost revenue on the part of beach businesses and recreational facilities, among others.[19]

In Kazakhstan, approximately 20,000 square kilometers of land was inundated between 1978 and 1995. It is estimated that over 300,000 people incurred some sort of damage as a result of this flooding. Nearly 357,000 hectares of agricultural land were reportedly lost, and 200,000 livestock died due to lack of forage. The European Bank for Reconstruction and Development (EBRD) has pledged some $54 million (42.5 Ecu) in loans to rescue Kazakhstan's only major port, Aktau, from flooding.[20]

The Russian share of the Caspian coast spreads across the territories of Astrakhan, and the republics of Daghestan and Kalmukia. In Astrakhan, 10 percent of all agricultural land were lost in 1978-1995 as a result of the rising water, which inundated 400,000 hectares of valuable coastal land. Half of one meter of water now covers the once inland Martyshi oil fields. In Daghestan, 40 percent of production facilities are at least partially flooded in four urban areas and some 150,000 hectares of land were submerged in the sea. The Republic of Kalmukia lost some of its most valuable forage land when the Caspian flooded approximately 70,000 hectares of its shoreline.[21]

In Turkmenistan, the coastal recreation area of Kheles has been inundated and the Cheleken peninsula has turned into an island. The port city of Turkmenbashi, where the Krasnovodsk Oil Refinery is located, is now threatened by the rising water, and some settlements experienced flooding. The EBRD has plans to extend credit to Turkmenistan to redevelop this port. Meanwhile, in Azerbaijan, rising water cut off the Sara peninsula from the mainland; the Kazylagachskii Reserve was completely inundated and water overwhelmed the Sumagait and Kurilskaya bays.[22]

The rising of water causes further concern because the land in the immediate proximity to the sea is situated among some of the most polluted areas of the former Soviet Union. Examples of high risk areas include 20 of the 32 oilfields in Kazakhstan's Atyrau region, including Tengiz and Korolevskoye.[23] Interestingly, inhabitants of the Caspian, aware of the water fluctuations, refrained from building large facilities close to the shore, but, of course, modern centralized planning changed all that, as recent planning oversights resulted in the destruction of the sand dunes that once insulated the prudent inhabitants from the flooding.

Offshore Pollution: Hydrocarbon Resources

Experience has shown that oil revenues are never used toward improving the marine environment that generates them. Rather, exploitation of hydrocarbon resources is largely responsible for ecological degradation and even environmental disaster. In fact, transportation of oil by tanker fleets, oil exploration, extraction, and refining are the main causes of offshore pollution in the Caspian Sea. Accidents and the natural seepage of oil constitute an additional potential hazard associated with hydrocarbon reserves.

Annually, some 200,000 to 6 million tons of oil are estimated to seep naturally into the sea; this is caused in part by the high rates of vast volcanic activity in the sea and, especially, in Azerbaijan. The offshore loading of oil introduces three additional adverse environmental impacts: (1) spillage; (2) tanker movements near oil platforms; and (3) contamination through discharge of ballast water. Furthermore, initial steps in oil exploration involve the use of heavy charges that disrupt the delicate marine ecosystem and pollute the water. The drilling operation introduces another set of environmental concerns; the rig itself is a foreign element disturbing the marine environment and there is always the possibility of a blow-out during drilling operations.[24]

In the future, the increased financial importance of the Caspian region, as well as better infrastructure capabilities resulting from increased earnings from oil and gas resources, will play a big role in bringing greater industrialization and urbanization to the coastline, causing yet greater pollution and straining even further the Caspian's fragile and unique ecosystem.

Land-Based Pollution

The intensive industrial and agricultural development of the Caspian area already poses a severe threat to the sea. Over 200 large enterprises and 100 large urban areas are situated in the Caspian Basin. Together, these areas are responsible for disposing more than 8 cubic kilometers of waste into the sea, containing toxic, chemical, and organic substances.[25] The run-off water from the surrounding agricultural land drains into the sea; it includes high doses of nitrogen and phosphorus. Moreover, agricultural pesticides enter the Caspian directly by air or mixed with ground and surface water. Fertilizers and pesticides are known to be used in excessive and increasing amounts in the surrounding agricultural activities. Meanwhile, some studies have indicated that a trace of 5 parts per million of DDT in the eggs of freshwater sturgeon can result in 100 percent fatal-ity;[26] data on the effect of DDT on the Caspian sturgeon apparently is yet to be compiled and made available.

The direct dumping of household waste into the Caspian and its adjoining rivers is commonplace. For example, four urban areas in Iran alone are responsible for the dumping of 400 million tons of waste into the Caspian annually.[27] Nearly 6 million Iranians inhabit some 120 urban and rural areas along the Caspian coast; their sewerage is dumped into the sea without any treatment whatsoever.[28] There are practically no treatment facilities for the industrial waste and sewerage generated elsewhere around the Caspian either. In Azerbaijan, about 10,000 cubic meters of Baku's wastewater are discharged directly into the Caspian.

The Caspian region supports many industrial facilities. These include tanneries, textile plants, food processing centers, slaughterhouses, metal plating, and pulp and paper plants, among others. The waste from these industries is dumped frequently into the Caspian. In Iran alone, these industries are known to account for a biochemical oxygen demand load equal to 104,000 tons.[29] More disconcerting are the Russian nuclear facilities along the banks of the Volga and the chemical plants at Sumqaieet.

Over-Exploitation

Russia and Iran annually release some 63 and 15 million young sturgeon, respectively, and yet the population of the caviar-producing fish is in constant decline. In the period between 1990 and 1995, the population of sturgeon in the Caspian declined from some 200 million to 50 to 60 million. Today, the caviar produced in the Republic of Azerbaijan is only 5 percent of the production of a decade ago.[30] Besides over-fishing, a number of other factors are responsible for this decline. Improper netting, for example, is responsible for the death of nearly one million young sturgeon annually. Nonetheless, over-fishing is the major threat to the population of the Caspian sturgeon. The Caspian seal is also in severe danger. Some 30 to 40 thousand seals are hunted annually, which makes the Caspian the world's second biggest hunting ground for these mammals.[31]

Agreements between the Soviet Union and Iran had helped perpetuate the sturgeon population at sustainable levels. In the post-Soviet era, restrictive quotas are either lacking or are not being enforced; 90 percent of the sturgeon that are now caught are caught without allowance for adequate spawning time. Pressed by economic hardship, the newly independent republics bordering the Caspian hope to get a larger share of the profitable caviar trade; they are the major source of over-fishing. While the development of productive industries and infrastructure requires time and capital, the valuable sturgeon has proved to be a ready and highly valued source of quick cash. The caviar-producing fish has become tantamount to "hard currency with fins."

The rampant nature of illicit exploitation of the Caspian's living resources may cast doubt on the efficacy of a regulatory solution to the problem. The authorities in the riparian states are aware of the causes and practice of illicit exploitation. By one account, observed at what was called a research trip by the Caspian Fisheries Research Institute of Russia, official bodies are not only aware of such activities, but they themselves are involved. In this particular case, the Institute collected 40 percent of the profits from the slaughter of seals. The hunters engaged in illicit exploitation were said to believe that reduction in their catch would only benefit those living across the Caspian in the other littoral states.[32] Strictly enforced catch and hunting limits remain the only hope, and their success will depend on regional cooperation.

Far from being adequate, the recent efforts toward cooperative action and the show of concern over environmental issues by the governments of littoral states have been at least encouraging. Such efforts include the placing of environmental conditions on oil exploration activities, calls for regional cooperation, and soliciting the help of the United Nations.

Regulating Oil Exploration

A number of steps can be taken in order to reduce the impact of pollution coming from drilling operations and to minimize the risk of harm to the environment. These include the study of previously successful models, strict implementation of safety procedures, and the use of appropriate engineering such as accurate approximation of the drilling rig to the sea lanes.[33] According to one industry observer, a zero pollution level in the Caspian may be obtained by using appropriate technology, such as is applied in the Zuidwal field in the Netherlands.[34] Although this is most probably an overstatement in order to win business in the region, it goes to show that many hazards may be avoided if proper steps are taken before initiating operations.

Environmental conditions have been placed already upon oil companies interested in the exploitation of the Caspian's hydrocarbon reserves. In 1994, the State Oil Company of Azerbaijan Republic (SOCAR) signed an agreement with 11 oil companies representing a total of 6 nations for the development of the Azeri, Chiraq, and Gunashli oilfields. This agreement created a body known as the Azerbaijan International Operating Company (AIOC), which is essentially a partnership of the foreign companies and SOCAR.

Under the AIOC agreement, the AIOC is charged with conducting an environmental baseline study (EBS). This task was subcontracted to an international environmental consulting group called Woodward-Clyde International (WCI). WCI set out to determine the environmental con-

ditions of the contract area in a four-step approach that included:
(1) review of existing literature and available data; (2) review of relevant
international environmental standards for similar bodies of water; (3)
audit of existing operations and practices; and (4) collection of environmental data. The data collected is archived supposedly for general future
use in a computerized database, though it remains unclear how to gain
access.[35] While the EBS exercise was an encouraging development, many
factors, however, discredit much of WCI's methodology and the study's
findings. For example, samples were taken from very selected locations
of the contract site. Thus, even if there were no measuring or processing faults, the results cannot be used toward the assessment of the situation in the Caspian as a whole, or even the entire contract area for that
matter.

Regional Cooperation

In 1992, during the Caspian states' conference, held in Tehran, Iran called
for the establishment of the Organization for Regional Cooperation of the
Caspian States (ORCCS). The draft by-laws of the organization made provisions for the formation of a number of committees on areas of cooperation in the Caspian Sea, including the Committee on Coordination and
Supervision of Utilization of Resources of the Sea Bed. The idea behind
the organization was received enthusiastically by all littoral states, as the
organization represented a desire essentially to separate the seemingly
intractable legal issues from those that require urgent and necessary
regional cooperation.

In 1995, Turkmenistan praised the ORCCS plan and commented that
the organization could be a moral guiding force for the countries as well
as to carry out technical and research projects in such areas as ecology, utilization of natural resources, and navigation.[36] However, as late as November 1996, during a meeting of foreign ministers of the Caspian states in
Ashkabad, all five littoral states were still paying lip-service to the importance of the quick signing of the ORCCS treaty,[37] without actually doing
so. It appears that the Caspian riparian states were all hoping for international funding for this project, which so far has not been realized.

A number of other attempts at bilateral and multilateral cooperation
have been made in recent years. In September 1995, Iran and Russia set
up in Moscow the Center for Caspian Research Studies, expecting other
littoral states to join. This body is to carry out research on topics such as
managing the water resources of the Caspian Basin, balancing the level of
the Caspian's water level, and identifying the high-risk points in the
coastal areas of the lake.[38]

Role of the United Nations

In 1993, Iran and Russia requested assistance from the United Nations (UN) with respect to the environmental woes of the Caspian region. In that year, a resolution relating to the Caspian was passed by the Governing Council of the United Nations Development Program (UNDP), which resulted in that agency's efforts to bring the various United Nations bodies, particularly the three organizational members of the Global Environment Facility (GEF)—the UNDP, the World Bank, and the United Nations Environmental Program (UNEP)—to forge a joint response to the environmental challenges of the Caspian Sea. Realizing that the UN knew little about the world's largest lake, in early 1995, ten GEF officials set out on a three-week fact-finding mission. The mission concluded that the Caspian is indeed a "unique environment and of a unique global environmental concern under the Biodiversity Convention."[39] It was further declared that this lake deserves the attention and funds of the UN community, in particular the GEF.[40] With that, the UNDP, UNEP, and the World Bank created the Environmental Caspian Sea Initiative (ECSI).

The ECSI drafted a document calling for addressing the consequences of the Caspian's water fluctuations, planning for the long-term protection of its biodiversity, creating an information system for the management of natural resources, and fortifying the institutional, legal, and regulatory frameworks in the riparian countries.[41] All five Caspian states quickly ratified the document. But, the GEF was quick to learn firsthand of the difficulty in dealing with an international lake whose legal regime remained in dispute. When UNDP referred to the Caspian as the largest saltwater lake in the world, Iran was quick to object in a harsh cable, claiming that the Caspian is in fact a freshwater lake, given the very low salinity levels.[42] The UNDP relented and identified the Caspian as "the largest land-locked body of water in the world," avoiding altogether reference to its water as saline or sweet.[43] By labeling it a "land-locked body of water," the UN essentially has confirmed that the Caspian is a lake.

The GEF had hoped to raise some $10 million to fund the activities prescribed by the ECSI. However, it has encountered a myriad of daunting issues, including the fact that not all of the Caspian littoral states are members of the GEF, nor have they all ratified the international Biodiversity Convention. Under the circumstances, it is argued that it is impossible to apply for GEF funding. These difficulties were not addressed and the initiative lay dormant.

Conclusion

The Caspian Sea is a unique ecosystem. Flooding, pollution, and over-exploitation of natural resources are the three major factors contributing to the Caspian's environmental degradation. While the fluctuations in the water levels are a historically documented phenomenon, their cause is not yet fully understood by scientists. The rising water is leading to economic loss, and even greater environmental consequence as the highly polluted adjoining lands are inundated. The pollution levels in the coastal waters off the littoral states, while varying by multiples, are all higher than they should be.

Hazards associated with hydrocarbon resources have been the main cause of offshore pollution. The dumping of industrial waste and untreated sewerage have further contributed to the degradation of the Caspian's delicate ecosystem. Meanwhile, economic hardship and lack of regulations and enforcement have contributed to unsustainable levels of exploitation of the Caspian's indigenous sturgeon and seal populations.

All governmental, international, and independent assessments of the Caspian Sea's environment acknowledge the need for the development of an accurate and comprehensive information system and objective statistics as a first step to addressing environmental issues. This and the inability of any one country to alleviate the Caspian's environmental woes make it obvious that regional cooperation is all the more imperative.

The need for a combined effort is well appreciated by the littoral states. However, contention over the legal regime and geo-political considerations has led to an atmosphere of distrust. While the Caspian countries all may seem to be genuinely interested in cooperating on environmental issues, without the mediation of a funded and mutually trusted international body, such cooperation remains illusive.

Notes

1. The author wishes to thank all those who helped in finding resources for this chapter, with special thanks going to Dr. Michael Glantz of the National Center for Atmospheric Research and Dr. Liz Rogers of the Azerbaijan International Operating Company.
2. There are some 65 international lakes in the world, including Lake Alberta, Lake Chad, Lake Constance, Lake Geneva, the Great Lakes, and Lake Victoria.
3. A. N. Kosarev and E. A. Yablonskaya, *The Caspian Sea* (The Hague: SPB Academic Publishing, 1994), p. 1.
4. See J. Shaygan and A. Badakhshan, "Iranian Research on the Caspian Sea Level Rise," a paper presented at the Workshop on Scientific, Environ-

mental and Political Issues of the Caspian Circum Region, held in
Moscow, May 13-16, 1996. Hereinafter referred to as "Caspian Workshop
(Moscow, 1996)."

5. Kosarev & Yablonskaya estimate that in the period 1900-29, the Volga
contributed an annual surface water input of some 250.6 billion cubic
meters, from a total input of 335.7 bcm (74.6 percent); for the period
1942-69, the annual input for Volga was at 241.2 bcm of a total of 285.4
bcm (84.5 percent); in the period 1978-93, the annual input for Volga
was 274.3 bcm of a total of 310.9 (88.2 percent).

6. The data on Caspian's biodiversity is culled and synthesized from the following sources: B. Kristofferson and L. Rogers, "Operating in the Caspian
Environment," a paper presented at the International Conference on
Health, Safety & Environment, organized by the Society of Petroleum
Engineers and held in New Orleans, Louisiana, June 9-12, 1996, p. 2
(hereinafter referred to as "Health & Safety Conference [New Orleans,
1996]"); Hadi Manafi, "Challenges and Prospects," a paper presented to
the International Conference on Oil and Gas Prospects in the Caspian
Region, sponsored by the Institute for International Energy Studies and
the Institute for Political and International Studies, Tehran, December
10-11, 1995 (hereinafter referred to as "Caspian Conference [Tehran,
1995]"); Parviz Kardavani, *Ecosystemhay-e Abiy-e Iran* (Iran's Marine
Ecosystems) (Tehran: Ghoms Press, 1996), p. 178; M. Ghaheri and
J. Shaygan, "Iranian Share in Polluting Caspian Sea," Caspian Workshop
(Moscow, 1996); B. Amirahmadian, "The Geography of the Caspian Sea,"
in *Central Asia and the Caucasus Review*, vol.5, no. 14 (Summer 1996), p. 24.

7. Mohsen Hashemian, written remarks at the Caspian Conference (Tehran,
1995). Hashemian, President of the Institute for International Energy
Studies in Iran, estimated that by the year 2000 production could reach
12 million barrels per day.

8. Kosarev and Yablonskaya, p. vii.

9. "Caspian Club of Five, To Be or Not to Be," in *Caspian Sea Bulletin for
Decision-Makers*, no. 1 (1995) (hereinafter referred to as "CSB"). Accord-
ing to one source, Turkmenistan may have much greater hydrocarbon
reserves: 3 billion tons of oil and 4.8 trillion cubic meters of natural gas.
See Hekim U. Eshanov, witten remarks at the Caspian Conference,
(Tehran, 1995).

10. Khoshbakht Yousefzadeh, written remarks at the Caspian Conference
(Teheran, 1995).

11. Kosarev and Yablonskaya, p. vii.

12. Kardavani, p. 178.

13. For details, see Kardavani, pp. 199-225. See also "Ecology of the
Caspian", in *CSB*, no. 1 (1996), but it does not identify the standard used
for establishing the "admissible" threshold.

14. A. Maleki, written remarks at the Caspian Conference (Tehran, 1995).
This source does not cite the source for the UNIDO data.

15. For details, see Shaygan and Badakhshan; *Key Findings of the NATO
Advanced Research Workshop*, held in Moscow, May 13-16, 1996,
www.dir.ucar.edu/esig/new_nato.htm).

16. M. Glantz, "In Central Asia: A Sea Dies, A Sea Also Rises," in *Fragile-cologies* (1995), www.sni.net/~mglantz/oct_95.html.
17. For details, see I. Zonn, "The Caspian: 18 Years of Continuous Water Rise," in *CSB*, no. 2 (December 1, 1996).
18. Manafi, ibid., posits that the rise in water levels as well as the linking of the Caspian to other bodies of water are responsible for the presence of non-indigenous species now found in the Caspian.
19. See generally Shaygan and Badakhshan. Given the instability of the Iranian currency, it is hard to convert the value given in Iranian rial to United States dollars. Currently there are two official rates for the rial: 1,780 rials to the dollar for internal transactions and 3,000 rials to the dollar for exports. Black market rates are higher.
20. See generally, Zonn, p. 2; EBRD, "Longterm EBRD Financing to Rescue Kazak Caspian Port" at <www.ebrd.com/opera/pressure1/pr1996/27april96.htm> (April 27, 1996); "Caspian's Drama," in *CSB*, no.1 (1996).
21. "Caspian's Drama," p.5.
22. Ibid.
23. Zonn, p. 2.
24. Shaygan and Badakhshan.
25. Zonn, p. 2. See also *Focus Central Asia*, no. 17 (1995)
26. Kardavani, p. 208.
27. Ibid.
28. Ghaheri and Shaygan.
29. Ibid.
30. See generally Amirahmadian, p. 23; R. Bundy, "Legal Aspects of Protecting the Environment of the Caspian," in *Central Asia and the Caucasus Review*, vol. 5, no. 14 (Summer 1996), p. 127.
31. Kosarev and Yablonskaya.
32. L. Yampolski, "The Caspian Seal and the Legal Status of the Caspian Sea," in *Ecostan News*, vol. 4, no. 4 (April 1, 1996).
33. Shaygan and Badakhshan.
34. Theo Happel, written remarks at the Caspian Conference (Tehran, 1995). Happel represented the Elf oil company.
35. Kristofferson and Rogers, p. 4.
36. Eshanov, p. 15.
37. Zonn, p. 2.
38. Islamic Republic News Agency (IRNA), "Iran and Russia Set Up Joint Center for Caspian Sea Research Studies," via World Wide Web (October 5, 1995) (quoting the Director of Research Studies of the Caspian Sea). At this writing, it could not be confirmed whether this center has started its activity, though it appears that it lacks adequate funding.
39. Michael Schulenburg, written statement at Caspian Conference (Tehran, 1995), p.86.
40. The World Bank, *Mainstreaming the Environment* (Washington, D.C.: The World Bank, 1995), p. 60.
41. "U.N. to Protect the Caspian Sea Environment," in *Iran News* via World Wide Web (April 23, 1995).

42. The Caspian is a far cry from being a "fresh water" lake. The Iranian government is trying obviously to distance the Caspian from factors that help identify it as a "sea."
43. Schulenburg, p. 86; The World Bank, *Mainstreaming the Environment*, ibid., p. 60.

CHAPTER SEVEN

Organizational Response to Caspian's Environmental Needs

Hormoz Goodarzy

Introduction

In recent years, the environmental issues facing the Caspian Sea region have been the subject of increasing international and regional attention. So far, they have also remained largely unaffected because international organizations are interested mainly in issues of global relevance, instead of working on regional or local problems. That task has befallen largely on the regional actors, who may not agree on the nature of an environmental crisis or solutions thereto. This chapter examines the scope and content of international and regional institutional responses to the environmental challenges of the Caspian region.

Background

In the last quarter of a century, there have been great strides in unifying or coordinating the roles of the many United Nations agencies with respect to the environment. The first step in that direction was taken with the 1972 UN Conference on the Human Environment in Stockholm. The conference laid the foundation for the United Nations Environmental Program (UNEP). But, the UNEP had no executive powers and had to depend on other national and international organizations to implement its programs. In 1989, the UN General Assembly established the United Nations Conference on the Environment and Development (UNCED) and charged it with the task of "strengthening national and international efforts in promotion of sustainable and environmentally sound development."[1]

The United Nations Conference on Environment and Development ("Rio Earth Summit"), which took place in Rio de Janeiro in 1992, was in many ways a "Stockholm II," because like its predecessor it dealt with the larger global issues. It produced four conventions on climate change, biological diversity, biotechnology, and forests. The recent fifth year reunion of the Rio summit, which took place in Kyoto in 1997, once again dealt with the "larger picture," this time focusing on the abatement of air pollution and the equitable distribution of pollution quotas among countries.[2]

The Rio summit led to Agenda 21, a plan designed to redefine the roles and responsibilities of the various UN agencies with respect to environmental issues, to provide the financial mechanism necessary to sustain UN environmental programs, and to pave the way for transfer of environment-friendly and pollution-abating technologies from the developed countries to developing countries. The plan meant to create a consensus among the UN member countries to balance environmental protection and the requirements of economic development activity, with economic, social, and environmental development forming a triad.

Under Agenda 21, environmental initiatives would come from the United Nations Development Program (UNDP) and its affiliates with input from the countries involved. In addition, it would emphasize the achieving of integration, participation, and sharing of information through strengthening existing capacities and investing in local resources and knowledge. The financing of programs under Agenda 21 would come principally from a trust fund founded by the partner countries, and from the financial resources of the Global Environment Facility (GEF), a joint World Bank and UN facility. The financing would be made available for programs through UNDP, other specialized agencies, and non-governmental organizations (NGOs).

Over the years, the UNDP has placed a great deal of emphasis on "capacity building" as a part of Agenda 21 initiatives. As part of Capacity 21, countries are encouraged to enhance their abilities to "manage" development. The UNDP stresses what it calls the "cross-sectoral" approach, a method by which countries can view environmental management as an integrated all-sectors process. Designed to give as many actors as much ownership in what eventually emerges as a policy, the program uses the basic ingredients of participation, information sharing, monitoring, and reporting in formulating conservation strategies and designing national plans.

The bringing of actors from various institutions together is hoped to lead to formulation of policy by consensus which, in turn, is hoped to translate to legislative and enforcement action. The underlying logic behind this approach is that in making everyone a user and provider of information, it would be possible to chart the course of development more effectively. UNDP's Sustainable Development Networking Program serves as the vehicle for dissemination of information and improvement of communication by establishing networks of information on the one hand, and linking them to electronic resources such as the internet on the other.

Regional Disconcert

Regional efforts at addressing the Caspian's environmental problems have not fared any better than the ECSI. In 1992, the Caspian Sea Conference, held in Teheran, led to the establishment of the Organization for Regional Cooperation of the Caspian States (ORCCS). Ostensibly, the organization was to have a number of committees, including one on the coordination and supervision of utilization of resources of the seabed, even though there was no consensus among the littoral states as to the legal regime governing the exploitation of the sea's natural resources. Iran and Russia believe in the joint exploitation of the sea's resources, while Azerbaijan, Kazakhstan, and, to a lesser extent, Turkmenistan, favor dividing up the sea into exclusive national sectors. It came as no surprise, therefore, when in 1995 the vice president of Turkmenistan could speak of the potential contributions by the ORCCS in the conditional tense. A similar aspirational tune serenaded the ORCCS at the November 1996 meeting of the Caspian foreign ministers gathered in Ashkhabad. Severe financial constraints and a clear lack of political will have scuttled any sustained regional concert on the Caspian.

The Environmental Caspian Sea Initiative

While there have been large-scale efforts by UNDP and its affiliates under Agenda 21 and Capacity 21 programs, the local environmental issues have yet to become subject to substantial planning and financing. The focus on the larger global environmental problems is preventing the more immediate and pressing local and regional concerns from being redressed. The responsibility for these is being shifted increasingly to local actors. One case in point is the Environmental Caspian Sea Initiative (ECSI).

In 1993, Iran and Russia requested assistance from the UN with respect to the deteriorating state of the Caspian environment. Chief among their concerns were the rising water levels, flooding and run-off from polluted land areas, pollution, and diminishing fish stocks. The UNDP's governing council adopted a resolution whereby various UN programs combined their efforts to meet the sea's environmental challenges. A visit by ten GEF officials on a fact-finding mission to the region culminated in the statement that the Caspian was a unique global environmental concern under the Biodiversity Convention and that "the lake" deserved the attention of the UN in general, and the GEF in particular.

This led to the creation of the ECSI by the UNDP, UNEP, and the World Bank.

The ECSI seemed like a major step forward; it set into motion an organizational response to an array of the Caspian region's environmental issues, ranging from the geological categorization of the sea to the increasingly devastating effects of its recent fluctuations. The littoral countries—Azerbaijan, Iran, Kazakhstan, Russia, and Turkmenistan—signed onto the ensuing agreement calling for coping with the consequences of water fluctuations, protection of the sea's biodiversity, creating an information system for the management of natural resources, and strengthening of the national institutional, legal, and regulatory frameworks within "Capacity 21."

With an initial funding of $500,000, supported by a Japanese grant administered through the World Bank, the program's ambitious objective was to focus first on building institutional capacities at national and regional levels.[3] On its long agenda figured programs to reduce the overall population burden on the Caspian basin, monitoring biodiversity, managing the coastal zones and wetland areas, and developing self-financing mechanisms. Yet, the agenda did not reflect the local sense of priorities. For example, in 1995, the World Bank's report on the Caspian's environmental challenges made no mention of the critical issue of sea-level rise.[4]

From the start, the ECSI ran into trouble. First, the question of the legal status of the sea as a "lake" or "sea," regardless of the geography, pitted Iran against the UNDP's characterization of the sea. Second, as it turned out, not all the littoral states were members of the GEF, nor all had become parties to the Biodiversity Convention. This latter issue made the funding of the ECSI a legal impossibility.

In a case such as the Caspian Sea region, the situation is further complicated by the fact that a common resource is being simultaneously shared by parties with mutual and diverse interests and needs. It is also more and more evident now that demands set forth by these organizations in providing solutions to the complex issues of the Caspian Sea littoral states, in particular, require a sophisticated level of planning, if only for the effective use of local technical expertise, that is often in short supply and at a high premium when imported.

In terms of priorities, allocation of financial resources, promotion of technology transfer, and capacity building are the basic building blocks of the international organizations in addressing what are predominantly global environmental concerns. The World Bank's current long-term strategy for protecting and managing the Caspian environment is based on the Regional Sea and River Basin Model. The global financial mechanism of GEF, and the Multilateral Fund for the Montreal Protocol (MFMP) facilitate transfer of these resources to governments. Yet, in nominative terms, even the World Bank's assistance for global environmental conservation is small relative to the Bank's other activities, with a cumulative portfolio value of less than one billion dollars, a figure comprised of client governments and donor contributions put together.[5]

Regional Constraints

Reducing global environmental solutions to suit regional needs is often a challenging and risky venture. Regional negotiations among large countries, especially those with different ecological, economic, and cultural circumstances, provide important clues as to how one might handle global environmental treaty negotiations more effectively.[6] The opposite case, namely applying general methods or global models to specific situations, involves challenges and risks that may outweigh the benefits. The geographical boundary of the Caspian poses the additional challenge of a region for which the developments of the past few years have created exceptional circumstances, ranging from creation of new states to realignment of economic, political, and strategic interests.

The change in the historical identity of the Caspian littoral from a two-state, Iran–Soviet Union entity to a five-nation one, has elevated the geo-political equation to a much more complex matrix. The developments have carved a new map of alliances. Prior to the break-up of the Soviet Union, the Caspian Sea had been the exclusive preserve of the Soviet and the Iranian interests, regulated by treaties and protocols between the two countries. The present situation requires a far greater degree of sophistication in negotiation skills.

The environmental issues facing the Caspian are articulated by the web of governmental leaders, unofficial or nongovernmental interest groups, including business associations, scientific associations, and multi-lateral agencies. The role of international organizations has indeed been critical in bringing to light the issues of the Caspian Sea. Still, the region and its unique environment demand a level of engagement beyond the immediate measures for a greater coordination among the UN and its sister institutions, and for improvements in handling resource management questions exclusively.[7]

The unique geography of the Caspian region as a critical variable cannot be emphasized enough. In a recent cross-country model of economic growth for some 25 transition economies, a component such as physical geography has climbed to the very top of the list of determining factors for economic performance. In the case of transition economies that are landlocked, such as Azerbaijan, Kazakhstan, and Turkmenistan, it is concluded that apart from the traditional variables such as economic policies, political and economic institutions, and demography, economic policies that neglect their physical geography would misread the opportunities for economic growth and would feel the negative impact on their trade liberalization plans.[8]

Economic and Technological Currents

The Caspian Sea and its immediate surroundings are rich in marine life as well as subsoil mineral resources. Apart from the sturgeon, it is blessed also by an abundance of other species of fauna, important to the human and animal life alike. Besides fisheries, the Caspian's seabed contains hydrocarbon resources. As a geological entity, the Caspian has been going through what is perhaps the most dramatic natural development of its recent history. The current sea level is 2 meters higher than a decade ago, which has resulted in flooding and the destruction of coastal structures. Responses to the environmental and economic disaster caused by the flooding have been on a short-term, country-to-country basis.

The development of the Caspian's oil and gas reserves is another source of anxiety for the environmental future of the sea. Fueled by financial incentives, policymakers in different countries are acting as brokers in making multi-national contracts in absence of clear legal standards aimed at protecting the environment. Where there may be environmental protection clauses in contracts or clear and enforceable legislation against polluters, the prospect of multi-nationals wiggling out from under their obligation, by favor or corruption, is very real.

Azerbaijan's position of dividing the whole sea into national sectors has been perceived as one device to minimize liability for the international oil and gas polluters. This country's Soviet legacy of rampant and negligent oil extraction practices have not gone unnoticed. With onshore and offshore wells exceeding ten thousand in number, contamination of the sea, groundwater resources, and marsh habitats have had tremendous negative impact on the marine life.

The desperate situation faced by the Caspian region, where opportunity costs of engaging in developmental schemes are being debated from conflicting perspectives, could find the needed attention in encouraging the rapid emerging of technological innovations and free and liberal exchange of information.

At the Rio Earth Summit, the need to improve environmental information for decision making had been explicitly articulated in chapter 40 of Agenda 21. The UNEP and the UN Environment Assessment Program (EAP) stated the need to provide the world community with improved access to meaningful environmental data and information, and to help increase the capacity of governments to use environmental information for decision making and action planning for sustainable human development.

There are various international and national organizations that are now involved in the establishment of environmental information networks worldwide. These networks are expected to improve assessment of the positive and negative effects of development activities nationally and internationally. The term "environmental information" is used to form all sorts of information products, such as databases in the areas of geographical information, technical guidelines and laws, news services, films, and the like.

Conclusion

The Caspian's environmental problems will get worse if a serious effort is not made to bring the global initiatives to the regional level and to force the regional actors to cooperate for the good of the region as a whole. What may help in that regard is the gradual increase in local awareness to address the ecological issues, prompting the provincial and national governments to propose concrete actions with a dual purpose—that is, to alleviate the immediate problems at a national level, much like self-help, and to tailor a viable regional solution for the long term.

In many countries, including in the Caspian region, the disparity in acquired technology vis-à-vis utilized technology still remains the stumbling block in taking full advantage of what information technology might offer today in the service of environmental protection. Where the information technology has been present, it has remained in the hands of the few, thereby making it more of a novelty than a practical tool.

A general search of the internet for topics related to the Caspian Sea reveals a handful of sites with topics ranging from geo-political to historical, and scientific/technical research. What seems to be lacking is a coordinated forum for accumulation of a "shared" resource on regional information, in the form of databases, and preferably, interactive internet resources, to cover critical environmental/economic concerns of the region, including the monitoring of the sea-level fluctuations. The establishment of a joint Iran-Russian center for Caspian studies in Moscow, in 1992, was hoped to fill the void in the area of environmental data collection and dissemination; it seems that the venture has died for want of resources.

The key to raising the public's awareness and consciousness about environmental problems would be best served if the environmental data collected on the region is shared openly by all of those wishing to access such data. However, multi-nationals whose projects in the Caspian region would normally generate useful environmental data are not willing to share such data, largely because the data is deemed proprietary and confidential, business secrets, or outright scandalous and embarrassing. To the extent that these operators share the collected data with government agencies, the governments themselves sit on the data without any obligation to share the same with the public.

Notes

1. Lawrence E. Susskind, *Environmental Diplomacy: Negotiating More Effective Global Agreements* (Oxford: Oxford University Press, 1994), p. 37.
2. The World Bank, *Mainstreaming the Environment: The World Bank Group and the Environment Since the Rio Earth Summit* (Washington, D.C.: The World Bank, 1995), p. 60.
3. Ibid., p. 60.
4. Ibid.
5. Ibid., p. 63.
6. Susskind, p. 8.
7. Ibid.
8. Jeffrey D. Sachs, "Geography and Economic Transition," a paper at the Central Asian Forum, held at Harvard University, Cambridge, Massachusetts, December 9, 1997, reprinted in the Harvard Institute for International Development, *Emerging Asia* (ADB, 1997).

Part III

Pipelines and Outlets

CHAPTER EIGHT

By Way of Iran: Caspian's Oil and Gas Outlet

Narsi Ghorban

Introduction

In the aftermath of the Soviet Union, the former Soviet republics in Caucasia, Central Asia, and the Caspian have realized that their newly found political independence and their economic prosperity depend on regional economic and political cooperation. One of the main areas for such cooperation is petroleum resource development, which, in view of the interest shown by international oil companies, promises to inject new blood into the battered economies of these countries. However, these countries do not enjoy free access to the open seas. Therefore, logically, the exit routes for Kazakh and Turkmen oil and gas would have to pass north through Russia or go south through Iran; Azerbaijan has the additional option of exporting its crude oil west through Georgia and Turkey. This chapter will discuss the proposed pipeline schemes and will assess the role of Iran in the export of Caspian oil and gas in the twenty-first century.

Export Routes

While reported estimates vary, the extent of Caspian's proven recoverable oil reserves are estimated generally at 30 billion barrels. The oil export from this region by the year 2010 is estimated to be nearly 2 million barrels per day (mbls/d). Naturally, one of the factors that would determine the ultimate level of oil production in this region will be whether sufficient export pipelines are in place in order to carry the oil out.

The Azerbaijan International Operating Company (AIOC) has decided to export its early oil from Azerbaijan (around 80,000 bls/d) to the Black Sea by way of either one or both of the routes for which the infrastructure is already in place. The northern route runs to Russia using the existing pipelines from Baku through Grozny, in Chechnya, leading to Novorossiysk, a Russian port on the Black Sea. The AIOC is to spend over $50 million to repair the section of the pipeline in Azerbaijan alone. Russia's Transneft will pay for the repairs needed on the Russian side. This route is expected to deliver 4 million metric tons (mmt) of oil (80,000 bls/d) in 1997 and 5 mmt (100,000 bls/d) in 1998. Reportedly, a transit fee of $2.20 per barrel would apply to the trans-shipment.

The second exit route for Azerbaijan oil would use the existing pipeline going westward through Georgia to Supsa on the Black Sea. Crude oil will move from Sangachal to Kazi Magorned in Azerbaijan, where it would enter the existing pipeline at Kazakh, near the Georgian border. The AIOC plans to lay a 117-kilometer long, 40-inch gauge pipeline between Kazakh and Tbilisi. An existing oil pipeline would carry the crude for another 352 kilometers to Supsa. The overall length of this pipeline is nearly 950 kilometers and the cost, excluding a new terminal facility that must be built at Supsa, is estimated at $300 million. The oil exports through Georgia are scheduled to commence by the end of 1998.

Altogether, nearly 210,000 bls/d of oil will be transported through the two Azerbaijani routes. However, around 500,000 bls/d of oil are expected to be produced by the year 2000 and 1 mbls/d in the first decade of the twenty-first century. Additional export routes may be necessary. One proposed route is to pass through Georgia to Turkey's port of Ceyhan on the Mediterranean Sea. Another proposed plan calls for connecting the Azerbaijan line to the Caspian Pipeline Consortium (CPC) pipeline, which is expected to take Kazakhstan's oil to Novorossiysk.

Kazakhstan has great potential for exporting oil in the coming years. Production from the Tengiz field by a joint venture between Kazakhstan and the international oil companies is expected to rise from 65,000 bls/d presently to nearly 700,000/d by the year 2010. The amount of oil for export from Kazakhstan in that year could exceed one million barrels per day. The only crude oil pipeline out of this country is presently through Russia; the movement of large quantities of oil through this is constrained by the capacity of the pipeline network, as well as Russian export quotas. The new CPC pipeline route through Russia will increase Kazakhstan's dependence on Russia. Furthermore, the CPC would be piping large quantities of oil for transportation through the Bosphorus, which Turkey is unlikely to appreciate because of the effect on safety, environment, and volume of tanker traffic in the already burdened and congested straits.

There are three export pipelines/routes being proposed to carry out Kazakhstan's oil. First, there is the CPC pipeline through Russia to the Black Sea. Presently, its shareholders are Russia (24 percent), Kazakhstan (19 percent) and Oman (7 percent). The other 50 percent of shares is divided among Chevron, LUKoil, Mobil, British Gas, Agip, Rosneft, Oryx, and Munaigaz. Second, there is the proposed line under the Caspian Sea to Azerbaijan, extending through Turkey to the Mediterranean Sea. Third, there is the proposed line from Kazakhstan south through Turkmenistan to the Gulf of Oman via Afghanistan and Pakistan.

The oil exports from Turkmenistan by the year 2010 are estimated at over 200,000 bls/d. This does not justify the construction of a separate pipeline to the Black Sea, Mediterranean, or the Gulf of Oman. The export of oil would be through hook-ups to the lines emanating from Kazakhstan, for which purpose Turkmenistan would have only to construct lines linking its fields to the Kazakh lines. Another export route for Turkmenistan could include a line under the Caspian Sea to Azerbaijan, connecting with the aforementioned AIOC pipeline

Oil Exports via Iran

The experts and international oil companies operating in the Caspian region recognize the advantage of oil swaps with Iran as a method to export Caspian's oil production. Equally significant is the recognition that Iran offers a desirable land route for exporting Caspian oil. But due to political pressure by the United States, the use of this route has been undermined considerably.

The factors responsible for making Iran a central actor in Caspian oil exports may be summarized as follows. First, there is Iran's unique geographical position between Caucasia, the Caspian Sea, Central Asia, the Persian Gulf, Gulf of Oman, Turkey, Iraq, Afghanistan, and Pakistan. Second, the northern part of Iran has a total oil refining capacity of some 650,000 bls/d, which could be adapted with relatively low cost for oil swap arrangements with Azerbaijan, Kazakhstan, and Turkmenistan. Third, Iran has a number of crude and product pipelines within 50 to 150 kilometers of its ports on the Caspian Sea, with a combined capacity of one million barrels per day; that capacity could be used for transportation of oil to its refineries. Fourth, Iran has extensive export facilities in the Persian Gulf, capable of exporting over 2.5 million barrels per day above its present export levels.

Iran's crude oil presently is pumped from the oilfields in the south of the country and piped to the refineries in Isfahan via a 32-inch pipeline and Tehran via a 24-inch pipeline. From Tehran, two parallel 26-inch pipelines carry crude to the refinery in Arak, while a 1-inch and a 16-inch pipe carry crude from Tehran to Tabriz.

Trans-shipment of oil through Iran would entail the following three phases. In the first phase, oil from the Caspian will be used in the existing refineries at Tabriz and Tehran, with 100,000 bls/d and 200,000 bls/d capacity, respectively, replacing the current flow of crude from the oilfields in the south of Iran. This stage can be implemented in less than two years and can absorb easily all the early oil produced in the Caspian region (300,000 bls/d) in exchange for Iranian oil delivered in the Persian Gulf.

In the second phase, pipelines between Tehran and Isfahan (450 kilometers) and between Tehran and Arak (340 kilometers) would be reversed so as to bring oil from the Caspian basin to the refineries at Isfahan and Arak, each with a capacity to handle 220,000 bls/d and 150,000 bls/d, respectively. This will involve the construction of some new pipelines as well as increasing the capacity of the existing lines by looping. Substantial investment will be required, but it would be considerably less than the amount needed to build an alternative pipeline in any direction passing through several countries. The project could be implemented in less than three years and could accommodate a significant amount of the oil exported from the region by the year 2005.

The third phase would entail the construction of new pipelines in order to carry additional quantities of oil from Kazakhstan, Turkmenistan, and the Caspian in general to Iran's export terminals in the Persian Gulf. This project could be completed in less than two years, provided projections would argue in favor of its creation. Finally, a pipeline may be built to carry oil from Central Asia to the Gulf of Oman by way of Afghanistan and Pakistan.

Details of the Swap

The swap arrangement for the export of crude oil from the Caspian basin entails the exporting of crude produced in the Caspian region to the nearby Iranian refineries in exchange for Iranian crude of equivalent value delivered at Iran's export terminals in the Persian Gulf. The Caspian oil from Azerbaijan, Kazakhstan, and Turkmenistan would be shipped/piped to points/refineries in Iran and/or piped into Iran's crude/product pipeline networks.

Presently, the two Iranian ports of Anzali and Neka are best situated to receive swapped oil by sea. Most of the crude oil to be produced in the Caspian region is either offshore or very near to the Caspian coast. Shipments to Iranian ports can be made in 3,000 to 5,000-ton tankers. This would require 3 to 4 shipments per day for the estimated 80,000 bls/d of early oil produced in Azerbaijan by 1998.

Anzali is less than 150 kilometers from the two crude oil pipelines that deliver over 100,000 bls/d of oil to the Tabriz refinery. The construction of a pipeline capable of carrying up to 250,000 bls/d of oil from Anzali to these two pipelines will result in delivery of some 100,000 bls/d of crude to the Tabriz refinery, with some going to the Tehran refinery in a reversed flow. The cost of the Anzali pipeline and the required infrastructure would amount to about $80 million, of which sum more than

60 percent will be paid in Iranian rials. Such a project could be financed easily and implemented by the Iranian government and/or private sector in about a year's time.

Neka is situated at the southeastern part of the Caspian Sea. The facilities at Neka were being upgraded to handle up to 50,000 bls/d in 1997 and nearly up to 100,000 bls/d by the year 2000. Presently, the oil swap agreement between Iran and Kazakhstan moves through Neka. By the end of 1997, the Neka-Tehran pipeline would have handled about 40,000 bls/d of Kazakh crude. The oil reaches the Tehran refinery via a pipeline that had been constructed previously to move petroleum products from Tehran to the cities on the Caspian coast. This pipeline has been modified to be able to move crude oil in the opposite direction. In order to increase the volume of the oil swap to 100,400 bls/d, the capacity of the pipeline has to be expanded considerably. The cost of the upgrade and additional port facilities would be about $40 million, most of it in Iranian rials.

With the necessary improvements to and expansion of pipelines, port facilities, and other infrastructure at Anzali and Neka, the refineries at Tabriz and Tehran could run for the most part on crude from the Caspian basin, resulting in a total volume of swapped oil in excess of 300,000 bls/d by the end of 1998. This would be the most logical and economic approach for the export of early oil produced from the Caspian basin.

The export of crude oil from Iran on behalf of Azerbaijan, Kazakhstan, and Turkmenistan would take place at no extra cost to the exporters at Iran's terminal on Khark Island, in the Persian Gulf. Iran would not have to produce extra oil under this plan, as the oil supplied by Iran is equivalent to the amount received from the Caspian basin. All proposed alternative routes for transporting some 210,000 bls/d of Kazakh, Turkmen, and Azeri oil would require long lead-in times, would be less reliable, and would be considerably costlier than swap.

The volume of the swapped oil exported in the manner described above could be doubled easily at a reasonable cost. With an infrastructure capable of handling 650,000 bls/d of oil, the refineries at Arak and Isfahan also can receive Caspian oil by way of Tehran. This will mean that the port facilities at Anzali and Neka have to be expanded considerably in order to handle 650,000 bls/d of Caspian crude. Further, a second pipeline parallel to the existing Neka-Tehran pipeline would be required. On the Anzali side, the pipeline from Anzali joining the Tehran-Tabriz pipeline must be designed so that it may handle the additional crude. This would require an expansion in the pipeline between the junction of the Anzali pipeline and Tehran itself, ideally, to handle 350,000 barrels per day.

The costs associated with the expansion of the lines from Anzali, Neka, and Tehran could be reduced if plans for these lines are incorporated in the construction plans for the earlier lines. The cost of an additional pipeline from Neka to Tehran, running some 300 kilometers, is estimated at less than $60 million. The cost of an additional line from the junction of the Anzali pipeline with the Tehran-Tabriz line first to Zanjan and then to Tehran would be another $60 million. The cost of expanding and/or improving port and infrastructure relating to the line from Anzali to the Tehran-Tabriz line would be about $70 million. The total cost of moving 650,000 bls/d of crude oil from the Caspian basin to the open seas through Iran will be around $340 million, which is far less than the projected cost of moving just 210,000 bls/d of Azerbaijan's oil from Baku to the Black Sea by way of Russia and Georgia.

There are other routes proposed for swap purposes. One is the possibility of using the IGAT I gas pipeline from Baku to Ardebil in order to bring large quantities of oil to Iran. From Ardebil, a pipeline would carry the oil to the refinery at Tabriz, whence it may flow to Tehran along the Tehran-Tabriz line. Another route would have a pipeline carry Turkmen oil to Neka and further to Tehran. This pipeline would link up eventually with Kazakhstan.

Gas Export Pipelines

Most of the gas export pipelines in and around the Caspian basin are proposed with Kazakhstan and Turkmenistan in mind. Azerbaijan does not have much potential as a gas exporter. While it has the largest gas reserves after Russia, Iran consumes its gas production internally, including using it for reinjection into oil wells. Turkmenistan, on the other hand, exported nearly 72 billion cubic meters (bcm) of gas in 1990. Presently, it exports around 23 bcm/year, having therefore a large underutilized export potential. Its limited gas exports are caused by Russia cutting back on Turkmen access to Russian gas networks and by consumers not willing to pay for their gas imports from Turkmenistan. Therefore, finding reliable export routes and access to paying customers is an overriding objective of Turkmenistan's gas sector. Iran, by necessity, will play a central role in Turkmenistan's gas exports.

Among the proposed pipelines from Turkmenistan there is a 1,120-kilometer line to Pakistan via Afghanistan; this is promoted by Bridas, an Argentinean company, and its rival, a consortium consisting of Unocal, Delta, and Gazprom. This pipeline, which is expected to carry 20 bcm/y of gas and would cost nearly $3 billion to construct, cannot be built unless peace and stability return to Afghanistan. Furthermore, the mountainous terrain of Afghanistan creates hazards and risks for the construction process itself. Another proposed pipeline consists of a 6,700-kilometer line from Turkmenistan to China. The leading proponents of the Turkmenistan-China pipeline are Exxon, Mitsubishi, and China National Petroleum Company. This pipeline would pass through Uzbekistan, Kazakhstan, and across China to the markets in the Far East. The cost of the project is estimated at over $12 billion.

Kazakhstan is the other producer of gas. In 1996, it produced nearly 5 bcm; British Gas/Agip operation is responsible for more than one-half of this production, mostly obtained from the Karachaganak field. The annual domestic demand for gas is around 9.5 bcm/y, however; the balance is imported from Turkmenistan. A 4,000 to 5,000-kilometer pipeline from Kazakhstan to western China would provide an outlet for Kazakh and, eventually, Turkmen gas to the Far East. At present levels, Kazakhstan's gas reserves and production by themselves do not justify the construction of a Kazakhstan-China line.

The only logical and economic route for transporting Turkmen gas is by way of Iran, bypassing altogether the Russian network. The present Iranian gas network is within 100 to 200 kilometers of Turkmenistan, Armenia, Nakhchevan, and Turkey. The network can be linked to these places in a relatively short time and without foreign investment. Presently, there are two pipeline projects underway between Iran and Turkmenistan. The first is a 200-kilometer line connecting the gas pipeline networks of the two countries. This pipeline is under construction by the National Iranian Oil Company and is near completion. Turkmenistan is bearing its part of the cost by paying Iran in gas, which would be used in Iran's northeast provinces. Another pipeline, approved by Turkmenistan, Iran, and Turkey, is designed to get the Turkmen gas to Europe. In Iran, the line will extend to Gorgan and Semnan, passing south of Tehran and heading north toward the Turkish border. Work on several sections of this route is underway; it will be a part of the delivery system of Iranian gas to Turkey under a separate Iran-Turkey gas deal.

Iran's own gas export projects are taking shape as well. On one front, the existing Iranian gas network will be connected to the lines in neighboring Turkmenistan, Armenia, and Nakhchevan. On another front, the gas exports to Turkey and Pakistan will commence in all likelihood by the turn of the twenty-first century and the year 2002, respectively, which would require an additional development of gas reserves and 1,600 kilometers of pipeline at a cost of over $3 billion. On yet a third front, Iran plans to sell gas to India and Europe, which would involve huge investments in development of gas resources and construction of new pipelines.

Conclusion

The objective of the oil and gas producers of the Caspian region is to export their hydrocarbon resources. To that end, quite a number of oil and gas pipeline schemes are either proposed, approved, or under consideration. Obviously, not all these pipelines emanating in the landlocked countries of Caucasia, the Caspian, and Central Asia are economical or practical. While the multi-national oil and construction companies lobby the region's governments for the sake of their own schemes and profits, the governments in the region should choose the options that speak directly and logically to the interests of their respective nations in the long term. Most of the oil pipelines presently under construction from the region will end up on the Black Sea and/or on the Mediterranean. There are four basic problems associated with this choice. First, they are passing through areas that have been and will remain politically unstable in the foreseeable future. Second, the outlet to the Black Sea will be at the mercy of Turkey because of environmental and traffic concerns through the Bosphorus and Dardanelles. Third, the market served by these lines will be saturated in the future. Fourth, the high cost of these pipelines will reduce considerably the income of the oil-exporting countries.

The only logical approach to the export of Caspian oil and gas is by way of swap arrangements with Iran. The required pipeline network is either in place or may be put in place in a relatively short time at reasonable and very competitive cost. Not only may the Caspian oil and gas be exported cheaper in terms of transportation cost, but it can also can reach both Europe and the huge Asian market. Swap arrangements would also create interdependency among the counties in the region, providing valuable synergies for the long-term development of Caspian's oil and gas resources.

The choice for the landlocked countries of Azerbaijan, Kazakhstan, and Turkmenistan are obvious; their oil and gas exports either move north through Russian territory, or south through Iran, or perhaps they move in a diversified fashion north and south. To export exclusively through either country gives the transit country leverage; the decision therefore rests on which route costs less, is the fastest, and which is equally dependent on the exporting country. The swap arrangement with Iran works to the maximum mutual advantage of the exporting countries and Iran.

Instead of pursuing individual pipeline projects, the Caspian countries must cooperate in the creation of the most economically efficient oil and gas outlets to as broad a market arena as possible. A comprehensive multi-lateral study followed by the drawing up of a regional energy master plan free from obscene self-interest and political acrimony is needed.

CHAPTER NINE

Transportation of Caspian Oil Through Russia

Felix N. Kovalev

Introduction

The countries of Armenia, Azerbaijan, Georgia, Kazakhstan, and Turkmenistan are counting on the development of the Caspian Sea's energy resources to buoy their economies and improve the living conditions of their people. To that end, a considerable amount of attention has been devoted to the role that Russia plays and can play in transporting the Caspian's oil and gas exports to the outside world. While it shares in the hopes and aspirations of its neighbors, the level of Russia's cooperation would depend on its neighbors taking into account Russia's interest as well. This chapter examines briefly Russia's interests and concerns with regard to development and exportation of Caspian oil and gas.

Legal Status of the Caspian Sea

Closer cooperation among the Caspian countries is being impeded by the lack of noticeable progress in the drafting of a mutually acceptable convention on the legal status of the Caspian Sea. This unsettled state of affairs translates into risk for the huge investments that have been earmarked for oil and gas development in the Caspian. For example, the total of Azerbaijan's first three international contracts for the development of Caspian oil called for $14 billion in investments; just between $800 to $900 million of the amount called for has been committed, however. The monies actually invested in Azerbaijan and Kazakhstan in 1996 amounted to 20 percent of the funds promised. The tensions between Azerbaijan and Turkmenistan over offshore concession areas makes the legal environment for investment in the Caspian more risky.

The Russian view regarding the legal status of the Caspian Sea is grounded in the Soviet-Iranian agreements. According to these agreements, the Caspian is regarded as a joint Soviet and Iranian sea. Since there are no longer two but five states bordering the Caspian, the sea has become a joint Azeri-Kazakh-Russian-Turkmen-Iranian sea. Accordingly, the Caspian and its resources are open to exploitation by all Caspian states. Unless some agreement to the contrary is reached to deal with the matter, exploitation may take place by the Caspian states either in combination or individually anywhere in the sea, resulting often in overlapping activities with respect to specific deposits, which could lead to greater dangers of confrontation. This in turn will have a negative impact on the level or commitment of investment and technology.

Possibly better than any other state in the region, Russia understands the dangers and disadvantages of a unilateral or maverick approach to exploiting Caspian's resources. Russia, therefore, remains the most active among the Caspian states in developing a regional consensus to a new legal regime for the Caspian, one based on mutual recognition among the five littoral states of one another's interests as the first step to arriving at an overall agreed-upon compromise. This stance, which is significantly more accommodating than Russia's earlier proposal, was put forth by the Russian foreign minister, E. M. Primakov, at the meeting of Caspian foreign ministers in Ashkabad in November 1996. However, it was dismissed flatly by the Azerbaijan representatives.

Russia's effort to devise a compromise solution to the legal status of the Caspian Sea dates back to 1993. At every turn, however, this effort has been thwarted by the suggestion that the sea be divided into national sectors. The voice of the regional actors advocating the division of the sea has been joined by that of representatives from influential quarters outside the region. The interference by the extra-regional actors is likely to undermine the harmony that would be otherwise taking shape. Publicly, of course, these outsiders pay lip service to the notion that the Caspian's new legal regime be founded on mutual agreement among the littoral states and without outside interference.

Russia does not support the idea of dividing the Caspian, particularly its superjacent waters, into national sectors. In a Caspian divided into national sectors, the environment would become inevitably a victim of mindless exploitation of its resources, resulting in decimation of the valuable sturgeon stock and pollution from oil development and production activities. While each state would develop its own environmental protection and abatement procedures, the requisite basis for joint action and responsibility would be much harder to come by. The Caspian Sea is a landlocked body of water. It is accepted universally by world renowned scientists and experts and, what is more, by influential politicians, that it does not take much for an environmental mishap in the Caspian to turn into a first-class ecological disaster, affecting all the littoral states, some more than others, perhaps. Another casualty of the division of the sea into national sectors will be the freedom of navigation as presently enjoyed by the littoral states over the whole sea. Freedom of fishing, too, will become confined to national sectors.

While it has not sought to have a veto over the development of Caspian's resources by the other littoral states, Russia would be justified in insisting that, insofar as practicable, resource development agreements be drawn up in a manner so as to reduce the threat of ecological disasters and preserve the unique biological resources of the Caspian. Such a procedure is adopted throughout the world, even in those marine areas where the threat to the ecology is not as great as in an enclosed basin such as the Caspian.

The Road to Novorossiysk

From its own economic perspective, Russia is interested in transportation of Caspian oil through Russia to Novorossiysk and from there by way of the Black Sea to Europe. It would be pointless to deny this. Some investment would be required to construct some parts and modernize the Novorossiysk pipeline in general. However, economic analysis of the Novorossiysk route confirms the viability of this export outlet, resulting in the cheapest and most convenient export route for Caspian oil. The route is endorsed also by Caspian oil operators and exporters for political considerations, chief among them being the relative safety and security of operating pipelines on Russian territory.

Russian preferences aside, the Novorossiysk route is favored also by some of the Caspian exporters themselves for their own reasons. The Azerbaijan International Operating Company (AIOC) is on the record for favoring the Novorossiysk pipeline because of its economic advantage for Azerbaijan. The diameter of the Novorossiysk pipeline is 700 millimeters and has an annual capacity of 17 million metric tons. This greatly exceeds the capacity required by the exports of Azerbaijan's "early" oil and, in considerable measure, it covers also the capacity needed to handle Azerbaijan's "late" oil from the Caspian. With agreements between the Chechen authorities and Russia in place as of June 1997, the previously perceived risk of transit through Grozny is no longer an impediment to exports. The restoration of the Chechen section of the pipeline would cost only $1.2 million and take between 20 to 30 days to complete. In September 1997, Russia and the Chechen authorities signed an agreement launching the repair process, in anticipation of Transneft and AIOC scheduled exports of crude from Baku in October 1997.

The Caspian Pipeline Consortium (CPC), whose aim is to supply the world markets with oil exported from the Tengiz field in Kazakhstan, is expending great efforts to modernize the pipeline from Kazakhstan to

Novorossiysk. Chevron, who recently acquired a major shareholding in CPC, has stated for the record that it sees no other economically advantageous route than the route over Russian territory to the port of Novorossiysk. The Kazakhstan state oil company is on the record as stating that the Russian route is and will be a major export route for Kazakh oil regardless of any problem that may arise between the two countries.

The annual capacity of the CPC pipeline will be 67 million metric tons, large enough to meet the needs of Kazakhstan's oil exports. It will be 1,500 kilometers in length, costing $1.9 billion to construct. In December 1996, Kazakhstan and the international consortium operating in its territory received 49 percent of the shares in the CPC project; Russia and its companies, LUKoil and Rosneft, received 44 percent, and Oman received 7 percent of the shares. The deal, in effect, resulted in majority foreign ownership of a project that will be operating on Russian territory. This demonstrated Russia's flexibility in dealing with issues of regional concern in a mutually advantageous manner. The forming of entities such as the CPC project with equity participation by public oil and gas enterprises of the former Soviet Union may become in the future a model for solving problems of oil and gas transportation from countries bordering Russia.

Oil producers and exporters, including the western companies active in the Caspian region, are influenced foremost by financial and economic factors and these argue in favor of the Novorossiysk route. By favoring this route, exporters favor also Russia's economic interest, but no one could accuse the major Western companies operating in Azerbaijan and Kazakhstan of picking the Novorossiysk route in order to help Russia maintain its influence in the Caspian region.

The Potholes in the Road to Ceyhan

Among the alternatives to the Russian route is the pipeline that is to carry Azerbaijan's late oil from Baku to the Turkish port of Ceyhan on the Mediterranean. The perils of having this line traverse numerous mountainous points aside, the cost of its construction is prohibitive. In October 1996, the AIOC estimated that the construction of such a line would cost at least $3.5 billion; to pump one metric ton of oil through it would cost no less than $6, whereas the cost of sending the oil through to the Georgian port of Supsa would cost $1 to $2 per metric ton and only slightly more when pumped through to Novorossiysk. Nevertheless, the decision to build this pipeline is not Russia's to make; the government of Azerbaijan, the Azerbaijan State Oil Company, and the AIOC would have to make that decision for themselves. One cannot help but be concerned by the security risks associated with a Baku-Ceyhan pipeline. The line will pass through areas either under Kurdish control, and therefore Kurdish leverage, or through areas where Turkish control may not be firm due to local political conditions.

Turkey, for its part, endorses the Baku-Ceyhan line. To boost its prospects, Turkey undertook in 1994 to restrict tanker traffic from the Black Sea through the Bosphorus in order to dampen the enthusiasm for the rival Novorossiysk route. The tanker restriction is being justified by Turkey on ecological grounds. While tanker traffic and navigation can stand improvement in the Bosphorus and Dardanelles, ecology per se is a false issue. In the opinion of oil-producing companies, including the AIOC, the Bosphorus has been purely a political problem; there are no real insurmountable technical or ecological problems associated with increased tanker traffic. Turkey simply wishes to force the oil community to choose the Baku-Ceyhan alternative for Azerbaijan's late oil exports.

Russia does not oppose the role of Turkey as a conduit for Caspian oil and gas. In the course of a visit in January 1997 by the deputy head of the government of the Russian Federation to Tbilisi, Russia and Georgia signed an agreement that provides for the construction of a major gas pipeline from Russia to Turkey through Georgia. Earlier, in July 1995, Russia supported the development of two pipelines for carrying Azerbaijan's early oil from Baku; one leading to Novorossiysk, but the other one extending from Baku to the port of Supsa on Georgia's coast on the Black Sea. The latter line is incorporated in the Georgian government's plans to revitalize the country's economy. While Russia fully understands the contribution of this pipeline to Georgian economy, it expects Georgia to recognize in return and take into consideration Russia's interests in the Transcaucasus.

The Dire Straits

Undoubtedly, the navigation system in the straits of Bosphorus and Dardanelles needs to be modernized. According to the studies by the authoritative Lloyds Register, the problem in the straits does not stem from the narrowness of the straits or extremely congested tanker traffic. The problem stems from the lack of an effective traffic control system and the inadequacy of coastal navigation equipment. In view of these findings, Russia supports the recommendations of the relevant bodies of the International Maritime Organization to ensure free and unimpeded passage through the straits in strict accordance with the Montreux Convention. To that end, Russia supports a joint investigation in the straits in order to ascertain what in fact needs to be done to improve the safety of navigation and what expense it would require. Neither Russia nor another interested state, nor, for that matter, the oil-producing companies, will refuse to cooperate in implementation of the measures that are found to be necessary.

According to estimates, by 2010 the annual flow of oil to the Mediterranean will increase by roughly 29 million metric tons to some 70 million metric tons. This translates into an increase of at the most one tanker of 100,000 to 175,000 metric tons dead weight passing per day in one direction. This estimate is substantially lower than the one calculated by Lloyds Register in assessing the throughput capacity of the Bosphorus. One can only hope that the Turkish government will meet its obligation to freedom of navigation by implementing the recommendations of international organizations in this regard.

Improvement of navigation through the straits will be in the interest of Turkey itself. It is well known that other routes are being developed for the export of oil from the Caspian region to world markets. On the occasion of the ministerial meeting of Black Sea Economic Cooperation in Moscow, in October 1996, discussions were held among the Russian, Greek, and Bulgarian ministers regarding an "orthodox" project to transport oil from the Bulgarian port of Burgas to the Greek port of Alexandrupolis. Among those prepared to participate in this project are the European Bank for Reconstruction and Development and international oil companies such as Chevron. Naturally, the double loading and unloading of oil from tanker to tanker that would be involved under this project would lead to an increase in the cost of transportation. However, it is still not known whether such an increase would be greater than the increase in cost of pumping oil over mountain routes from Baku to Ceyhan.

There have been discussions also with respect to the use of sea-river tankers in order to transport Caspian oil from the Romanian port of Constanca along the Danube to the very heart of Western Europe. Interest has been shown by oil companies in the existing systems of oil pipelines all over Russia and the other members of the Commonwealth of Independent States. The western companies involved in developing the oil resources of the Caspian region are beginning to show increasing interest also in the possibilities of supplying oil to the Gulf of Finland on the Baltic Sea. As for Kazakhstan and Turkmenistan, pipelines have been proposed or planned extending eastward and southward for carrying oil and gas to China and Pakistan, respectively.

Unwarranted Russophobia

Not only in Turkey but elsewhere some still worry about transportation of oil through Russian territory. This was evident in the views of some participants at a number of conferences on Transcaucasus, the Caspian, and Central Asia, held in Washington, D.C., variously in October and November 1996. Much debate was devoted to the development of alternative East-West transport corridors bypassing Russian territory. Reminiscent of the worst periods of the cold war, the proponents of alternatives

operated on the proposition that everything that is economically beneficial to Russia is harmful to the West.

Expressive of the deep Russophobia gripping some opinionmakers is a
certain article by S. F. Starr published in the spring 1997 issue of *The
National Interest.* There, the author accused the new, democratic Russia of
pursuing an "aggressive and threatening strategy" with respect to Azerbaijan and Central Asian countries. Unfairly and without recourse to any
evidence supported by facts, the article concluded that, by exerting its
leverage on them, Russia was jeopardizing the ability of its southern
neighbors to develop as sovereign entities.

Fortunately, the mainstream political and business opinion in the
United States does not share in the foregoing example of Russophobia. In
regard to the Novorossiysk pipeline, for example, the cold economic and
business factors compel the conclusion that the route is advantageous
both for the West and Russia. Honest and conscientious business cooperation will in the end prevail over the sentiments of some who are still har-
boring Cold War attitudes.

Conclusion

To conclude, one would reiterate that at the present time the most expedient route out for Caspian oil is the Novorossiysk terminal through
Russian territory. Bearing in mind the vast reserves of oil and gas in the
Caspian region, the need may arise in the future to construct other routes.
Russia and Russian companies will be ready to participate jointly. Even
at this stage, it is important to create favorable conditions for business
cooperation, taking into account parties' mutual interests.

CHAPTER TEN

Pipeline Politics in the Caspian Region

Hooshang Amirahmadi

Geo-politics of pipelines from Central Asia and the Caucasus to markets in Europe, Asia, and elsewhere has become a major foreign-policy issue for the United States in the last few years. Countries like Iran, Turkey, and Russia are competing to gain a piece of the great pie. However, Washington favors Turkey for political reasons and against the will of its business community. At stake for Iran is strategic, not just economic, gains or losses. No wonder that the United States is not letting the business executives and the states in the Caspian region play the pipeline game among themselves. The winners of the pipeline game will reap strategic benefits while losers will become marginalized for some time to come. It is in this context that I assess the political risks of various pipeline routes and suggest an alternative.

To assess the political risks associated with various pipeline routes from Central Asia and the Caucasus to markets in Europe, Asia, and other world Regions, one must account for a multiplicity of often paradoxical factors at national, regional, and global levels. Yet the current positions held by the major players involved often ignore this complexity in favor of narrowly defined strategic and economic interests largely informed by shortsighted political animosity, rivalry, or alliances. To advocate for particular routes on the basis of a policy that excludes some players and includes others in the so-called great game that has ensued in the wake of the oil and gas rush in the Caspian basin is haphazard at best. The current approach is equally dangerous, for it remains oblivious to internal political and economic developments of the countries involved.

In what follows, I shall first provide a description of various routes and their advocates and then give an outline of the major risk factors involved, including the extent that they are ignored or accounted for in the positions held by the regional players. I shall conclude by proposing that decisions concerning oil and gas pipelines should recognize the need for multiple routes as dictated by political, economic, technical, and strategic realities, and that a grand cooperative and win-win strategy is preferred over the current alliance-making and win-lose games. Yet, the most important preconditions for a sustainable transport of Caspian energy are national political and economic developments. The proposed framework is based on the assumption that the long-term prospect for every player is much richer than what it can achieve by maximizing its short-term gains.

Pipeline Routes and their Advocates

Currently, five pipeline routes are available, proposed, or contemplated. They include Northern routes, Southern routes, Western routes, Eastern routes, and Southeastern routes. As we shall shortly see, some are extensions of existing pipelines while others are altogether new and have to pass through untested and contested geographies. Rough terrain, ethnic violence, bureaucratic infighting, and individual ambitions need to be accounted for along the way. They are also distinguished in terms of their strategic significance, economic feasibility, and technical complexity. More importantly, these routes involve uneven political and environmental risks, as explained in the subsequent sections of this chapter, and are viewed within a framework of win-lose and alliance-making strategy.

Yet the real difficulty with pipeline politics is that it must find a solution to often opposing business and strategic interests. For example, while Iran provides the most economical routes, the United States opposes the alternative in an attempt to curb Iran's future regional influence. The fact is pipelines offer more than economic benefits and trade possibilities; they form strategic cores of power along which communications, transportation, and other infrastructure corridors develop. The nation or alliance that controls such corridors would supposedly hold sway over the region. The U.S. policy aims for an East-West axis, the so-called New Silk Road, which excludes Iran and Russia. As this U.S. policy goes against the economic logic of the companies involved, it has become a stumbling block to pipeline construction; meanwhile other countries have used the U.S.-Iran tension to push for less than optimal routes through their territories with a view to gain business and strategic advantages.

Northern Routes

Advocated by the Russians, both Kazakhstan and Azerbaijan could join existing Russian pipelines by building extension or new pipelines that would take their oil to Novorossiysk on the Black Sea. The Caspian Pipeline Consortium (CPC) is already busy developing the line. For the

Kazakh oil, the pipeline will be built as it encounters no rival or opposition at present. In 1977, CPC signed a $2 billion contact for the construction of the pipeline; another $2 billion is needed for the Azeri part. For the Azeri oil, however, the routes will have to pass through the insecure Chechnya territory or near it, a rather unpleasant possibility for prospective investors. Chechnya has a ruined economy with no real prospect for future growth and is seeking political independence from Russia. These conditions, combined with elite rivalry and a growing drug trade, call for continued political violence there. Besides, both Azeris and Kazakhs remain concerned about Russia's continued dominance of their political life; for the Azeris, the concern is elevated to fear by their Turkish, Israeli, and American allies. The real problem is this: Russia as the holder of the world's largest reserves of natural gas can hardly be excluded from the growing world gas market, thus making the Northern route a real option.

Western Routes

Preferred by the United States, Turkey, Azerbaijan, and Georgia, these routes are intended to bypass the Russian territories and Iran. The less expensive alternative ($1.5 billion) is to build an upgraded pipeline to the Georgian port of Supsa on the Black Sea; from there oil will have to be taken by tankers through the Bosporus to Europe. One immediate problem is the current political instability in Georgia: the Abkhazia separatists would have to be suppressed or co-opted first. Even then there is the problem with the rebellious South Ossetia. According to one report, the people living in the vicinity of the pipeline going to Supsa made some 800 holes in the line, forcing the Azerbaijan International Operating Company (AIOC) to build a whole new line for its early oil. The other problem is environmental. Turkey claims, rightly, that the Bosporus is already too congested and that further tanker traffic will endanger Istanbul's safety. Despite these problems, this route seems to be on schedule for construction given a lack of better or more politically acceptable alternatives. The Bosporus problem could be addressed in a number of ways including a $1 billion Bosporus bypass from Bulgaria to Greece or a pipeline from Supsa to Ceyhan.

Turkey instead has pushed, with American and Israeli support, for another pipeline from Baku to its Mediterranean port of Ceyhan, a direct one. A Trans-Caspian pipeline will then feed Kazakh oil and Turkmen gas to these routes. The United States is currently lobbying the Kazakh and Turkmen governments to support the pipeline. The United States' attraction to the Ceyhan route and the Trans-Caspian line emanates from its desire to build an East-West axis of influence and commerce in the Eurasia region. But this alternative is too expensive and passes through the Kurdish-dominated territories. The proposed Trans-Caspian pipeline poses additional environmental hazards to the sea. No wonder that despite serious U.S. effort, the Ceyhan alternative is resisted by the companies. Even AIOC has argued against it. The companies had asked for American subsidies at a substantive level. The U.S. originally resisted, but after it became clear that the line may not be built by the companies, the United States offered to assist with $823,000 toward a projected cost of $3-$4 billion (*New York Times,* October 22, 1998). Turkey, which originally did not come up with the promised incentive package, was also forced to bow, but its offer remains as yet vague and unattractive. Financial difficulty aside, companies also complain that "the instability of Turkey's government has made it a difficult partner"(*New York Times,* October 11, 1998).

Southern Routes

Favored by Iran and oil companies, the Southern routes make economic and commercial sense. They are cheaper to build (under $1 billion), pass relatively safer territories, and pose no serious environmental hazard. Significant pipeline and port infrastructure also exists. A gas pipeline extends from Turkmenistan to Iran, which they hope to extend it Turkey via a new pipeline to be constructed by Shell. Extensive oil pipelines to south of Iran also exist, as do port facilities in the Persian Gulf, from where both European and big Asian markets could be efficiently served. Most notably, the Southern routes also offer the swap option, something no other routes have offered as yet. Oil companies and governments worry that the Southern option increases the world's reliance on the Strait of Hormuz, a concern that can be addressed by linking the pipelines, from Central Asia, to the port of Jask on the Oman Sea.

Certain geologists have also argued against the line because of possible seismic problems in Iran. Yet, in the last several decades earthquakes have not posed problems for the pipelines in Iran. The United States is opposed to the Southern routes for obvious political reasons and has made it a policy to prevent its realization. Opposed to the routes is also Azerbaijan, who remains wary of Iran's intention, a fear largely instigated by its allies, the United States, Israel, and Turkey. Yet, the United States and others may find it hard to advocate exclusion of Iran, which holds the world's second-largest gas reserves and is its fourth-largest oil producer.

Eastern Routes

China is increasingly energy-hungry and needs to seek new markets. The Kazakh option is attractive because it is, comparatively speaking, the most accessible. Thus, China signed a contract with Kazakhstan in September of 1977 to build a 2,000-mile long and extremely expensive pipeline ($3.5-5.0 billion) from two fields in Kazakhstan that China has proposed to purchase, to its western territories. The deal, commercially unattractive, can only go if China is to continue viewing Kazakh's option as a new strategic necessity. All indications to date point to China's commitment to the proposed pipeline. However, financing the project can prove much harder than the Chinese had originally anticipated. This is the only route that seems to have no rival or enemy despite the fact that it can cause China's influence to rapidly grow in the Caspian region.

Southeastern Route

Favored by Pakistan and Afghanistan, UNOCAL, an American oil company, with Saudi Arabia Delta Oil, has been promoting a pipeline to transport oil and gas from Turkmenistan and possibly Kazakhstan, through Afghanistan, to Pakistan and eventually India. The line also had the tacit approval of the U.S. government until last summer when, following the bombing of American embassies in Africa, Washington had to bomb Osama bin Laden's terrorist camp inside Afghanistan. The Taliban's identification with bin Laden has also forced UNOCAL to withdraw its proposal for the time being. The Afghan geography presents some difficulty, and cost ($1.9 billion) could make fund-raising a bit too difficult. Yet, the real obstacle is a political one. As long as the Afghan civil war is not fully ended and the Iran-Afghanistan crisis continues, any attempt to build the line will prove futile.

Major Political Risk Factors

Alternative pipeline routes are exposed and vulnerable to uneven political risks and involve risk factors of national, regional, and global origins and significance. It must be noted that these risk factors are interdependent, given that national and regional borders are increasingly at the mercy of global forces. The risk factors are also dynamic due to the fact that the region as a whole is in a state of transition to a new political-economic future. Clearly, the transient character of the region makes any short-term strategic alliances unstable as political changes will make loyalties shift and national interests change. For example, Russia, whom Americans thought would become a strategic partner of the United States in the wake of the Soviet collapse, has already become a strategic rival to it. Another case in point is Afghanistan. Iran thought encouraging Islamic movements there would help its cause, but the Taliban victory can jeopardize Iran's national interest. Indeed, the Taliban will not serve even the strategic interest of Pakistan and Saudi Arabia, the group's main political patron and cash register, respectively.

National-Level Risks

As *The Economist* (February 7, 1998, p. 6) noted: "Oil companies take a more relaxed attitude to political risk than many other firms. They are used to dealing with violent or unstable countries. Because oil is simply pumped out of the ground and can be speedily exported, they can tolerate economic mismanagement, civil disobedience and even isolated violence in the host country more easily than other industries." Nevertheless, political independence, stability, and certainty concern them to a great extent. Pipeline security will particularly depend on the political stability in countries of origin and transverse. These include Azerbaijan, Turkmenistan, and Kazakhstan as countries of origin, and Russia, Iran, Afghanistan, Georgia, Armenia, and Turkey as countries of possible transverse.

The fact that proposed pipelines would have to pass through two or more countries makes the situation even more complicated. Additional political sensitivity arises due to the multi-ethnic and socially polarized character of the countries involved. Another risk-contributing factor is the undemocratic nature of the ruling elite in most these states, in the countries of origin in particular. Political independence is not fully assured given the harsh political-economic conditions in the countries of origin and due to the growing Russian interventionism in its near abroad. The explosive population growth is generating increasing and urgent needs for jobs and economic growth. Meanwhile, the new generation is demanding more and has higher expectations than its predecessor generation. They will hardly accept the continuation of current political repression and backwardness for long. They will demand democracy and development.

Equally unacceptable will be the current social inequalities in income and wealth distribution and in geographic distribution of national expenditures, leading to extreme territorial imbalance. Such inequalities generate abject poverty and lead to environmental degradation and unsustainable growth. Just in the case of Persian Gulf states, oil economies in the Caspian region tend to overestimate the role that oil revenue can play in sustainable growth of their economies. Any undue over-reliance will result in unstable economic policies and unwise spending. Finally, ethnic disparity and the unresolved nationalities' question are additional sources of national cleavage in the Caspian states. The desire of the ethnic elite to gain economically and politically from pipelines adds to the volatility of the political situation.

Regional-Level Risks

The Caspian pipelines face a series of risks that originate from regional and inter-state conflicts. Ethnic movements are quite prevalent in the region. The Kurds in Turkey dominate the eastern mountains of the

country and are a major source of worry for Ankara, which wishes to promote the Ceyhan route. The current crisis between Turkey and Syria is an
indication of Ankara's deep concern with its Kurdish question. The Lez-
gins are struggling for national unity and independence. At present, they
are divided into two parts in the Russian Federation and Azerbaijan. The
Abkhazia separatist movement has destabilized the government in Georgia and continues to remain a source of national cleavage in that country.
Chechnya and Dagestan continue to remain potential violent spots in
Russia's Caspian frontier. The Chechens are particularly adamant about
their independence and possible gains from future pipelines through their
territories. Conflict among more then a dozen other ethnic groups in the
Caucasus can flare up if their political and economic demands are not met.

The conflict between Armenia and Azerbaijan over Nagorno-Karabakh
is on hold and could lead to renewed fighting in the future. Presently, 25
percent of Azeri's territories are occupied by the Armenians and this
makes pipelines from Baku less than safe. In Afghanistan, while the Taliban seems to have secured its position as the dominant force, the
endgame remains less than certain there given the crisis with Iran. Other
regional or international crises can develop given the alliance-making
policy that some states follow. Meanwhile two other important sources of
tension remain: dispute over the legal regime of the sea and a possible
environmental conflict that can follow from the unwise utilization of
resources in the Caspian by the oil-producing states, Azerbaijan in particular. Iran and Russia have little oil and gas resources around the sea;
instead, they depend on its clean environment for fishery and caviar. In
addition, agriculture, forestry, and tourism are vital for the people on the
Iran side of the sea. Any degree of pollution can harm Iran's interests and
become a cause for political conflict.

Global-Level Risks

Another set of political risks that face a rational decision regarding
pipeline routes emanates from global games for strategic gains. The East-
West axis strategy followed by the United States proposes to exclude Russia and Iran while including Turkey along with states in the Caucasus and
Central Asia. The rivalry this strategy generates can prove unproductive
for the independent and democratic development of the very states the
United States wishes to promote. Both Iran and Russia are central to the
coherence and well-being of the larger region. Their cooperation can help,
while their antagonism is sure to hurt. To view the Caspian solely as a
linchpin of an American global game for strategic gains is unwarranted.
The current power game and alliance-making policy goes against the
need for cooperation in building, utilizing, and safeguarding the
pipelines.

As part of its global and regional games, the United States has tried to cripple Iran economically and isolate it politically. The economic impact of the containment policy has been simply devastating for the Iranian people, who continue to suffer from declining income and employment opportunities. Politically, too, the U.S. policy has hurt Iran by making it enemies to its otherwise natural allies. At present, three sets of regional alliances are organized around and against Iran: the Turkey-Azerbaijan-Israeli alliance in the northwestern and western borders, the Iraq-Mojahedin-UAE alliance in the western and southwestern borders, and the Pakistan-Taliban-Saudi Arabia alliance in the southern and southeastern borders. These destructive attempts notwithstanding, the United States has not been able and will not be able to bring Iran to its knees as the country benefits from a rich history of national dignity and regional role.

Aside from destructive rivalry between the states, the present policy discourse in the Caspian region will increase the tendency toward anti-externalism, reviving the largely outmoded anti-imperialist political culture so embedded in the minds of the Caspian people. If current outside intervention leads to failure of the states to develop their respective societies, alternative social systems and ideologies will come to challenge the current drive for establishing liberal political and economic systems. Extremism and national fascism are candidates for such an eventuality. From this perspective, pipeline routes need to be decided with a view to regional development, not as tools to serve strategic interests of particular states or groups of states. The key to altering the present discourse is a change in U.S. policy toward Tehran, and a good starting point will be for the United States to release Iranian assets in a symbolic gesture and to drop its opposition to pipelines going through Iran.

Toward a Cooperative and Win-Win Strategy

The above analysis indicates, among other considerations, that the best solution to the current stalemate may be found within a cooperative framework that emphasizes four principles: a win-win strategy, a multiple-pipelines approach, political reform, and sustainable economic development. A win-win strategy will de-politicize decisions with regard to pipeline routes, seeks a balance of national interests, and includes existing and prospective investors in determining the optimal routes. Meanwhile, the states with high stakes in the region, namely the United States, Russia, Iran, and Turkey, must realize that their current rivalry or hostility serves no one in the longer term and that a more amicable solution to outstanding claims will serve all parties involved. In this regard, a solution to the current standoff between the U.S. and Iran is the most critical. Here, as in the case of other rivalries, visionary leadership is called for in which sensationalism is subordinated to rationalism and long-term gains.

A multiple-pipelines approach is another most optimal and logical solution to the current stalemate. First, a multiple-pipelines approach will reduce dependency on a few countries and avoids concentration of world energy in a few hubs such as Baku, the Strait of Hormuz in the Persian Gulf, and Novorossiysk or Ceyhan and Supsa on the Black Sea. Second, the multiple-pipelines approach has economic logic. For example, while Iran offers a cheaper alternative than Russia or Turkey for pipelines from both Central Asia and the Caucasus, its comparative advantage lies in providing swap arrangements and easier access to fields in Turkmenistan and Kazakhstan. Turkey and Russia could have offered attractive routes for pipelines originating from points in Azerbaijan but, cost aside, feasibility is seriously constrained by political and environmental difficulties. Third, the political logic of multiple-pipelines is equally attractive as it reduces sensitivity to political instability in a given country or countries along a given route. The current largely political logic applied to pipeline routes not only goes against the economic logic but is counterproductive because of its superficial treatment of regional politics. For example, while bilateral political and strategic alliances are emphasized, internal political development is ignored.

Fourth, the technical feasibility of various routes also increases with a multiple-pipelines approach. This is due to the varying geographic and environmental difficulties that certain proposed routes pose. Within a cooperative framework, pipeline routes can be modified to avoid country-fixed solutions and possible inter-state environmental conflicts. Finally, the multiple-pipelines approach lessens possible strategic losses while increasing potential strategic gains. Pipelines are long-term commitments to a nation, and countries that lose will do so strategically. While there is no guarantee that the gaining people will remember the favor, the losing people will surely develop a structural animosity toward countries it considers the culprits. Given the deep-rooted anti-imperialist political culture or nationalistic tendencies in most of these nations, such a structural anti-externalism can lead to social upheavals. Needless to say that in that case, the domestic and international players in the pipeline and energy games of the Caspian region will encounter a lose-lose situation.

Political reform is key to political stability in the Caspian region and as such is the most important precondition for the safe operations of the pipelines in the long run. The reform must be genuine and lead to political participation, elite circulation, and the role of law. It must also end corruption, create discipline within the state, and increase its accountability. Current personality-based and exclusionary politics will not last for long. The new generation expects freedom, participation, and development and in that regard it looks to the developed world as a model to emulate. The future leaders of the Caspian states will have difficulty justifying under-development and dictatorship in an increasingly democratizing and developing world. As a basic requirement of political reform, the states must prepare the ground for expansion of civil society institutions and alternative discourses in all spheres of national life.

Political development is a precondition for sustainable economic development. Yet, the states in the region cannot afford to postpone the latter until the former has been achieved. Here lies the challenge facing the emerging economies in the Caspian region: They must develop as they democratize. The one and only possible alternative in this direction is a balanced development strategy: one that allows for economic growth and diversification, provision of basic needs, expansion of civil institutions, and circulation of the elite. As states promote political liberalism and free-market principles, they must also devise visions for re-inventing the government, expanding social networks, promoting political competition, and reducing reliance on oil income by encouraging export-oriented industrialization. In short, political and economic pluralism along with a complementary social policy is the key to state-building and national development in the region. Unless and until these conditions are met, pipelines will continue to remain vulnerable to domestic violence and inter-state conflicts.

In closing, I wish to invite attention to the consequences of the Iraqi invasion of Kuwait for the pipelines that used to take millions of barrels of Iraqi oil to international markets. Those who promoted the pipelines on the basis of economic, technical, or strategic criteria never thought that Saddam Hussein's dictatorship and undemocratic approach to the states in the region would become a cause for their closure. Had the Iraqi people and the states involved insisted on a balanced political-economic development of the country, Iraq would not have lost billions in oil revenue investment, and the pipelines would have been carrying more oil to the world energy markets for the benefit of all involved (except for the Caspian states!). As it is rightly said, democracies do not fight, internally or externally; but dictatorships do. I am only hoping that this and similar other examples will make current decision makers think deeply about the multiple conditions that need to be satisfied for a sustainable Caspian energy transport.

Part IV

Security and Geo-Politics

CHAPTER ELEVEN

The Geo-Politics of the Caspian Region

Pirouz Mojtahed-Zadeh

Introduction

This chapter provides a general perspective on the geo-political factors that have been influencing international relations in and around the Caspian littoral, including Caucasia and Central Asia. The emergence of Caspi-centric forms of "regionalism" centered around the Caspian Sea, or Central Asian "bloc politics" in this area is owed as much to the requirements of geo-economics as to the new political geography of the region. The chapter will discuss the central role played by Iran in any Caspi-centric and/or Central Asian regional alignment as a matter of geo-economic and strategic necessity.[1]

At the time of the dissolution of the federated Union of Soviet Socialist Republics, the Caspian was the exclusive concern of the USSR, including its constituent republics and territories, and Iran. Since the de-federation of the Soviet Union, the Caspian's littoral has come to be shared also by the independent states of Azerbaijan, Kazakhstan, and Turkmenistan. The importation of foreign capital and technology, aimed primarily at exploration and exploitation of Caspian's oil and gas resources, has infused into this basin extra-regional economic and political influences from every direction.

Geography and Culture

The Caspian region may be seen as and including the vast expanse stretching from the Black Sea in the west to Central Asia in the east, at the same time divided and connected by the Caspian Sea. Caucasia, including parts situated inside the Russian Federation, consists of a mix of Muslim and Christian nations with limited economic resources; chief among them are Georgia, Armenia, and Azerbaijan, of which the last-named

176 • Pirouz Mojtahed-Zadeh

borders the Caspian Sea. In contrast to Caucasia in general, the Caspian Sea littoral, on the other hand, is resource rich. Its gas reserves boast an estimated 57.1 trillion cubic meters, and its estimated 59.2 billion barrels of proven oil reserves ranks third in the world.

Central Asia includes the countries of Turkmenistan, Kazakhstan, Uzbekistan, Tajikistan, and Kyrgyzstan, with the first two countries also belonging to the Caspian littoral. Only Iran and Russia abut all three of the geopolitical regions, with Iran alone bordering the strategic Persian Gulf and Gulf of Oman, making it a veritable land-bridge connecting Caucasia and Central Asia to one another and the two regions to the Indian Ocean.

The peoples of Iran, Afghanistan, Tajikistan, Uzbekistan, Turkmenistan, and parts of Kazakhstan together with most peoples of Caucasia have shared a common historical experience as part of the various Iran-based political structures in successive periods from the Achaemenid to Qajar empires. The pivotal role that Islam plays in these states provides the cultural glue for the most part of the region. Within the context of Islam, there are other areas of homogeneity: Iran and Azerbaijan are the two countries with majority Shi'ite populations; Iran shares its Persian language with Tajikistan, most of Afghanistan, and parts of Uzbekistan. Central Asian cities like Bokhara, Balkh, Merv, Samarkand, Khiveh, and Kharazm figure prominently in the region's literature as traditional centers of Persian arts and sciences. The Persian new year Nowrouz is celebrated in Central Asia and parts of Caucasia. Meanwhile millions of ethnic Azeri and Turkmen live in various parts of Iran, including in the Iranian territorial provinces of East and West Azerbaijan and in Turkmensahra.

In view of the region's political geography and potential resources, the region's politics has been dominated by two major issues: (1) maritime and land access to international markets, and, to a lesser extent, (2) delimitation of offshore areas of national jurisdiction in the Caspian Sea. The dictates of ideology on the part of some players, and of sheer economic self-interest on the part of others, dominate the substance of regional political alignments and economic groupings. Nowhere is this better illustrated than in the case of the debate over the development of a regional infrastructure, including pipelines, for transportation of oil and gas from the Caspian and Central Asian producers to the outside world.

Geo-Politics of Access, Routes, and Pipelines

The Moslem republics of the former Soviet Union are landlocked countries. Kazakhstan, Turkmenistan, and Azerbaijan, along with Iran and Russia, border the Caspian Sea. This particular geography is conducive to the development of maritime trade among the littoral states. However, any cooperation and trade in the Caspian will not compensate for the lack of maritime access to the outside world. A practical solution to this problem can be sought in the idea of linking all these republics and Afghanistan to the Persian Gulf and Gulf of Oman by road, railway, and pipeline networks. With its two thousand miles of coastline on the Persian Gulf and Gulf of Oman, Iran offers these landlocked countries efficient and secure access to the global markets.

In December 1991, Kazakhstan and Iran signed an agreement providing for the Central Asian republics to extend their railway networks to the Persian Gulf across Iran.[2] Another agreement signed in the same year between Iran and the waning Soviet Union permitted freedom of travel up to 45 miles on either side of the Iran-Azerbaijan and Iran-Turkmenistan borders by ethnic Azeri and Turkmen, respectively. In June 1995, Iran, Turkmenistan, and Armenia signed a tripartite agreement providing for expansion of overland trade among the three.[3] In March 1995, Iran inaugurated the linking of its railway system from the port of Bandar Abbas on the Persian Gulf to its northeastern terminus in the province of Khorassan. The occasion was witnessed by some 50 heads of state and representatives of Asian countries, at which gathering the Kazakh leader proclaimed Iran the main bridge between Central Asia and the outside world.[4] In March 1996, the Iranian rail network was hooked up to Central Asia.

As a land-bridge connecting the Caucasian, Caspian, and Central Asian regions to the Persian Gulf and the Indian Ocean, Iran may also prove pivotal to India's commercial undertakings. Presently, Indian ships use Georgian and Ukrainian ports, and negotiate hundreds of kilometers of road through Russia in order to reach Central Asia. Reportedly, Iran has suggested that India fund a 700-kilometer railway connecting Bafq to Mashhad,[5] thereby shortening the Central Asia–Persian Gulf connection by several hundred kilometers. This would give India a competitive edge over Pakistan in Central Asia.

From an Iranian perspective, a pipeline connecting the Caspian littoral across Iran to the Persian Gulf or Gulf of Oman is unquestionably the most practical. This connection will offer the nearest route from the Caspian to the oil markets of the Far East, particularly Japan, where demand for energy is projected at an ever-increasing rate for the foreseeable future. In addition, Iran's manpower, skilled in petroleum matters, developed transportation and shipping infrastructure, existing ports and refineries, and existing network of oil and gas pipelines offer considerable technical and logistical advantages to the exporters of Caspian oil and natural gas. Iran's existing gas pipeline network is connected to Azerbaijan and is within a short distance from Turkmenistan. A pipeline connecting Kazakhstan and Turkmenistan to this network will be, at least, four times shorter in length and cheaper than any alternative destined for the Black

Sea and the Mediterranean. The alternative networks bound for the West would have to pass through mountainous terrain, as well as negotiate a political landscape marred by persistent insecurity brought about by armed conflict among multiple ethnic and political factions.

The advantages of an Iran-bound network already has gained converts. In September 1994, Iran and Turkmenistan agreed to the construction of a gas pipeline from Central Asia to Europe across Iran. Iran will cover 50 percent of the cost. Scheduled to be completed in 25 years, in the first phase the line will carry annually 10 to 12 billion cubic meters of gas. In the second phase, the line will be expanded to carry annually 28 to 30 billion cubic meters.[6] Agreed to between the parties in July 1995 is the construction of a second gas project connecting Turkmenistan to existing Iranian gas lines in Iran. Some 60 kilometers of the 200-kilometer-long line will be on Iranian soil.[7]

Pursuant to an Iran-Turkmenistan agreement made on January 30, 1995, Turkmen oil is transported initially to Iran overland and later piped to an Iranian refinery and then exported by way of Iranian ports on the Persian Gulf.[8] The first consignment of 3,200 tons of Turkmen oil was shipped to the northern Iranian port of Bandar Nowshahr for onward transit for export on February 1, 1995.[9] Discussions have been held with respect to the pipeline, which will be capable of delivering daily 120,000 barrels of oil to Iranian refineries.[10] Another pipeline capable of carrying daily 400,000 barrels of Turkmen oil to the Persian Gulf is scheduled for completion in the year 2002.[11]

The Iranian Oil Ministry has also presented the Turkmen government with elaborate technical studies, including route maps, detailing a pipeline system for transportation of Turkmen natural gas to Europe by way of Iran and Turkey. The proposed pipeline is estimated at $3 billion in cost and 1,400 kilometers in length, capable of delivering annually between 15 and 30 billion cubic meters.[12] Similarly, companies operating in Central Asia appear eager to transport their production across Iran to the Persian Gulf.

Iran has called for the construction of a pipeline network that would connect Caspian's gas exporters with the Persian Gulf. As sketched by the Iranian Ministry of Foreign Affairs, the project will consist of a gas loop system within Iran capable of connecting the gas reserves of Russia, Iran, and the Persian Gulf, representing some 70 percent of all the world's gas reserves. This would provide Europe, India, and the Far East with adequate gas supplies needed in the twenty-first century. One half of the pipeline dedicated to the loop is currently in existence and operational.[13] In August 1995, Iran and India agreed to establish a network connecting the Iranian and Central Asian gas fields to India through a 2,000-kilometer pipeline.[14] This network has the potential to play a crucial role in linking the Caspian–Central Asia region to the gas consumers on the Indian subcontinent and in the Far East.

Not surprisingly, Pakistan, too, has been pressing hard for a place in the networks connecting the Iranian and Central Asian gas fields to Southwest Asia. Having secured an understanding with Iran in this regard, Pakistan also reached an agreement with Turkmenistan in October 1995 whereby a gas pipeline would connect the former to Central Asia across Afghanistan.[15] This will prove to be a major link in the emerging multi-national gas pipeline networks originating in Iran and Central Asia. However, concerns over Afghanistan's security situation may dim the prospects for the implementation of a trans-Afghanistan pipeline.

The idea of a trans-Afghanistan pipeline prompted the United States to give active support to the Taliban forces in Afghanistan. In their support for the Taliban, a fanatically Islamic fundamentalist group, the United States sought to help into power in Afghanistan a regime that, due to sectarian differences, would be hostile to Iran, while Pakistan, another Taliban supporter, and UNOCAL, an American oil company, would pipe Turkmenistan's gas away from the Iranian route through Afghanistan instead.

The Turkmenistan-Europe pipeline via Iran is under construction. Some 2,000 miles in length, it will have a daily capacity of 5 million cubic meters.[16] The logical geo-economic advantage of it passing through Iran has been such that in July 1997 the United States found it difficult not to ease off on its prior adamant objection to the pipeline passing through Iran.[17] In the meantime, increased tensions in regions of Afghanistan[18] are making the Iranian option all the more attractive.

Piping oil or gas to the international markets through the territory of a third country has proven costly, involving major political and security risks. The case of Iraq is illustrative of the point at hand. Iraq's major oil pipelines pass through Syria, Lebanon, Turkey, and Saudi Arabia, all of which were closed at one time or another for political and security reasons. Even if the Taliban manage to gain full and complete control of Afghanistan, the proposed Afghanistan-Pakistan pipeline would still run risks twice as great as that of an Iran-bound pipeline passing through the territory of one country. Furthermore, given that the Taliban's politics are directed against Western liberalism and influence, coupled with Afghanistan's historical hostility toward Pakistan, there will be no real and lasting guarantee for a secure and uninterrupted flow of oil and gas through the proposed trans-Afghanistan pipeline. In a statement on this subject, Turkmenistan's foreign minister has pointed out that "due to continuation of war and insecurity in Afghanistan, it will be too risky for Turkmenistan to export gas via the war-torn country."[19]

In recent years, China has shown considerable interest in getting involved in the geo-politics of the Caspian–Central Asia region. It has signed a number of political and economic cooperation agreements with Russia and Iran, a gas pipeline deal with Turkmenistan in 1997, and has offered Iran an oil industry cooperation deal that would expand Iran's role in the oil and gas business of the Caspian–Central Asia region.[20]

Iran's potential contribution to the development and exportation of Azeri oil and gas has been noticed by Azerbaijan as well. In 1994, Azerbaijan allocated to Iran 5 percent of its share in the Azerbaijan International Operating Company (AIOC) as the price for Iran's participation in the exportation of Azerbaijan's Caspian oil via a pipeline through Iran and on to Turkey's Mediterranean coast. A cooperation agreement to that effect was signed between the Azeri oil consortium and the National Iranian Oil Company on February 3, 1995, pursuant to which Azerbaijan formally tendered to Iran the allocated percentage.[21] However, in April 1995, as the result of pressure from the United States, the Azeri government withdrew the offer.

The United States and Iran-Azeri Cooperation

The aforementioned United States–inspired withdrawal of the Azeri offer notwithstanding, the United States, too, has recognized, at least as a practical proposition, Iran's pivotal role in the marketing of the region's oil. The trade sanctions imposed by the United States on Iran specifically exempt the swap of Caspian oil for Iranian oil delivered in the Persian Gulf.[22] Nevertheless, the Unites States position regarding Iran's role in the region is premised on the notion that any idea of an Iranian involvement in the affairs of the Caspian–Central Asian region is to be avoided.[23] Consequently, in September 1995, after having sifted through a dozen or more pipeline prospects for over a year, the AIOC, predominantly owned by American and allied companies, chose the Russian pipeline plan for creating a conduit for Azeri oil exports across Chechnya (Russian) and/or Georgia.[24]

There is a limit to United States' influence in the region. Iran, as an outlet for Caspian oil and gas, looms in Azerbaijan's future plans. In May 1995, Azerbaijan announced that the crude oil produced by AIOC will be exported partly by way of the Persian Gulf in an Iranian-Azeri oil swap,[25] leaving open the possibility of future trans-shipment of Azeri oil to the Iranian coast on the Persian Gulf. A similar arrangement will enable the exportation of the Kazakh production from the Tengiz oilfield.[26] Moreover, in addition to its own exploration activities in the Caspian, in 1995, Iran formed exploration companies with Russia and Azerbaijan to prospect for oil offshore.[27] According to one of these contracts, Iran and Azerbaijan agreed to drill jointly for oil in Iran's offshore area.[28]

The Iran-Azeri cooperative relations in the Caspian are a logical extension of the two countries' neighborly relations, which are owed to a variety of factors, least of which involves decades of joint benefit from the
various hydro projects that straddle the countries' border on the River
Alas. Near the two countries, the Caspian oilfields cut across Iran-Azeri
limits. The desire to prevent either party from directional drilling and the
virtue of unitization on its own technical merit may provide the geo-
economic background to the Iran-Azeri joint exploration/ exploitation
efforts in the Caspian.

Regional Organizations

The foremost Caspi-centric regional organization is the Economic Cooperation Organization (ECO) whose members include the immediate
Caspian littoral states of Iran, Azerbaijan, Russia, Kazakhstan, and Turkmenistan, in addition to the Central Asian countries of Uzbekistan, Tajikistan, and Kyrgyzstan, as well as Iran's neighbors Turkey, Afghanistan,
and Pakistan. Based on the tripartite Iran-Turkey-Pakistan model of the
Regional Cooperation for Development (RCD) from the 1960s, the ECO,
with its expanded membership since February 1992, has held annual
summit meetings. The Iranian government has expressed openly the hope
that ECO become a regional economic grouping able to compete with the
other trading blocs in the twenty-first century.[29]

The development of ECO into a fully functional regional organization
depends on forging workable relations among its members. However,
internal factors and external influences have made this difficult to achieve.
Internally, differences divide the members in outlook and political orientation. The six Caspian–Central Asian republics of this organization are
also members of the Commonwealth of Independent States. Turkey, a
NATO country, has subordinated its relations with and within ECO to its
desire to join up with the European Union, expand military ties with
Israel, and to do, often than not, the United States' bidding in the region.
Similarly, the bulk of Pakistan's geo-political attention is focused on ter-
ritorial/strategic disputes with India and the desire to raise Afghanistan
as an alternative pipeline route to Iran.

In 1992, Iran embarked upon the notion of creating a Council of
Caspian Sea Countries (CCSC) which includes the five countries bordering the sea. The council has been slow to define its specific role in the
region. However, given its Caspi-centric focus and membership, CCSC
may prove more of a workable arrangement than ECO. As yet, CCSC has
not been able to formulate an action plan addressing matters of joint
regional concern such as the legal regime of the Caspian and the ecology.
While matters of security and economic integration, including the emergence of a trading bloc or a common market, must await the forging of

closer political ties among the members of ECO, the more pressing issues of resource exploitation and ecological degradation may argue persuasively for a more robust and immediate framework for Caspi-centric regional cooperation in the framework of such an organization as CCSC.

From the very beginning of the emergence of a Caspi-centric international subsystem, the Western oil concessionaires rushed into the region, prompting concurrent claims by the littoral states to the offshore areas of the Caspian. Consequently some concessions straddle disputed areas.

As early as October 1993, an inter-governmental conference among Russia, Azerbaijan, Kazakhstan, and Turkmenistan convened in the Russian city of Astrakhan. It discussed the Russian desire to establish a common/unified position among the CIS membership for presentment to Iran.[30] Russia favored the 1921 treaty between Iran and the Soviet Union as the foundation of a starting point for the development of a new legal regime in the Caspian.[31] The treaty in question had provided for Iran and the Soviet Union equal rights to freedom of navigation on Caspian waters. It followed from it, Russia argued, that the new Caspian regime provide for its common use by its littoral states on equal basis, sharing equally in the economic benefits derived from it. As considerable economic harm could come to the companies that had already invested in the areas off Azerbaijan, Azerbaijan favored a 10, 20, or 40-mile exclusive economic zone (EEZ) for each littoral state.[32] The United States and other Western interests support the Azeri position, arguing that delimitation of the offshore area is the most logical, equitable, and workable resolution of the conflicting national rights to Caspian's natural resources.[33]

Unhappy about the expansion of oil exploration by Azerbaijan and Western companies, Iran and Turkmenistan are gradually moving toward the Russian position.[34] In 1995, the officials of Russia, Iran, and Turkmenistan warned that exploitation of Caspian's resources should not proceed prior to there being installed a common legal regime for the sea.[35] Russia went insofar as to suggest that such activities be considered illegal.[36] The Iranian government warned that unilateral exploitation of the Caspian Sea could have an adverse impact on regional cooperation in general.[37]

The absence of a clearly defined commonly accepted legal regime serves as a potential cause for serious conflict, as well as making it difficult for the littoral states to proceed with development or exploitation of offshore resources. For example, Azerbaijan has granted oil exploration concessions to various companies whose rights extend as far east as areas considered by Turkmenistan within its maritime limits.

Environmental degradation in the Caspian basin is another regional concern that can best be addressed in a regional context. The pollution caused by oil production activities on the northern (Russian) and eastern (Turkmenistan) shores of the Caspian Sea is alarming. On its western coast in Azerbaijan the level of pollution is even more rampant.[38] Furthermore, the rise in the sea level[39] has caused considerable damage to coastal regions of all five littoral states. In Iran's Mazandaran province alone, 9.9 billion rials ($33 million) have been spent on constructing coastal barriers along ports of Bandar Turkmen, Nowshahr, Ramsar, and Tonekabon, with much more left to do.[40]

Conclusion

Iran has not formulated a clearly defined strategy to capitalize on the advantages that its unique location offers to the world's two most important energy deposit centers—the Persian Gulf and Caspian/Central Asia. Somehow in the cacophony of anti-Iranian rhetoric piped from regional and extra-regional mouthpieces, the sheer geo-economic facts have yet to state the case for Iran's value and role in the development of Caucasia, Caspian, and Central Asian regions.

Russia has great influence over the Moslem republics of the former Soviet Union. Russia's economic support, military presence, and sway with the bureaucracies and political establishments of these republics together serve as a formidable obstacle to other power-seeking influences among these republics. This goes for Iran, and for the United States, who seeks influence in the region through Turkey.

In 1995, Iran turned to a series of geo-economic and strategic engagements as a way to reassert itself in regional affairs; as a result it was able to broker a number of cease-fires between Azerbaijan and Armenia in the Nagorno-Karabakh conflict and among the warring factions in Tajikistan. Most notably in the latter, which took place in July 1995, Iran mediated between the country's pro-Russian government and leaders of that country's Islamic opposition. While the Tajik leaders have called for a defense cooperation pact with Iran,[41] Iran has taken the lead to forge a tripartite agreement with Turkmenistan and Tajikistan for expansion of economic, political, and cultural ties.[42] This may serve as a basic building block for an Iran-Central Asian regional integration and development union.

In 1995, Iran and Russia forged a coordinated political front in their approach to issues facing the Caspian and Central Asian regions.[43] Russia confirmed its continued participation in Iran's nuclear and technological projects and in October 1995 the two countries agreed to set up jointly an oil company. On December 2, 1996, the two agreed to establish a joint drilling company to undertake exploration in the Caspian.[44]

In its diplomatic overtures to the countries in the Caucasia, Caspian, and Central Asia, Iran has been emphasizing economics and downplaying its perceived image as one seeking ideological hegemony. Iran's likelihood of success as an emerging significant regional player and an economic powerhouse will depend in large part on the institution of major structural alterations to its foreign and domestic policies. Its continued role as an honest broker in regional conflicts—as opposed to being perceived as a source of tension—will strengthen its image in the region and therefore blunt the negative publicity directed against it from quarters in bed with Washington.[45]

The recent agreement of cooperation among Iran, Armenia, Georgia, and Greece will provide a badly needed balance to Iran's position in Caucasia, creating a counterweight to the United States' inroads through Turkey, Azerbaijan, and its other financially/technologically dependent countries in the region.[46] A larger strategic agreement among Iran, Russia, and China, yet to materialize, will provide Iran with the necessary strategic wherewithal to contain the United States' push for influence into Central Asia.

In order to enhance its status in the region, Iran could offer to Russia, Armenia, Azerbaijan, Kazakhstan, Turkmenistan, Uzbekistan, Tajikistan, Kyrgyzstan, and Afghanistan the use of one of its many ports in the Persian Gulf region, exclusively for international trade and commerce by those countries. This will bind the Caspian and Central Asian regions with the Persian Gulf and create a viable counterweight to the present United States–led geo-political arrangements in that region.[47]

Notes

1. This chapter is based on Pirouz Mojtahed-Zadeh, *The Changing World Order* (London: Urosevic Foundation, 1992); Pirouz Mojtahed-Zadeh, "The Changing World Order and Iran's Geopolitical Regions," in *The Iranian Journal of International Affairs* (Tehran: IPIS), vol. 5, no. 2 (Summer 1993); Pirouz Mojtahed-Zadeh, "The World of Politics," in *Ettela'at-e Siasi-Egtesadi* ("Political and Economic Ettela'at"), Farsi language newspaper, vol. 9, nos 7-8 (Tehran: March-April 1995); Pirouz Mojtahed-Zadeh, "Geopolitics and Hydropolitics: An Iranian Perspective," an address before the International Water Forum at the United Nations University, Tokyo, March 1995.

2. "New Republics—Problems of Recognition," report in *Echo of Iran* (London), no. 12 (47), December 1991, p. 6.

3. "Iran signs trilateral agreement with Turkmenistan and Armenia," in *Ettela'at-e Bayn el-Mellali* ("International Ettela'at"), London, no. 246, June 6, 1995, p. 10. This publication is referred to hereinafter as "Ettela'at".

4. "Kazakh official: Iran, Kazakhstan's strategic ally," in *Ettela'at*, no. 273, June 20, 1995, p. 10.

5. "Iranian port welcomes Indian traffic to Central Asia," in *Ettela'at*, no. 216, March 28, 1995, p. 8.

6. "Iran, Turkmenistan ink gas trunkline agreement," in *Ettela'at*, no. 329, September 7, 1995, p. 10.

7. "Contract with Turkmenistan will have up to $350 million of annual profit," in *Ettela'at*, no. 286, July 7, 1995, p. 1.

8. Ibid.

9. "Turkmen oil in first transit through Iran," in *Ettela'at*, no. 198, February 23, 1995, p. 8 (quoting Energy Compass of London).

10. "Russian No-Go to Turkmen Gas May Open the Way for Iran," in *NEFTE Compass* (London), vol. 3, no. 28, July 14, 1994, p. 7.

11. "Turkmenistan, New Ukrainian Gas Deal Helps Swell Key Annual Revenue," in *NEFTE Compass*, vol. 3, no. 6, February 10, 1994, p. 7.

12. "Turkmenistan, Iran makes a play about pipeline blueprint," in *NEFTE Compass*, vol. 3, no. 30, July 28, 1994, p. 7.

13. Narsi Ghorban, "Middle East Petroleum and Gas Conference: Bahrain, January 16-18, 1994," in *The Iranian Journal of International Affairs* (Tehran: IPIS), vol. 6, nos. 1-2 (Spring/ Summer, 1994), p. 297.

14. "The project for transfer of Iranian gas to India," in *Ettela'at*, no. 304, August 3, 1995, p. 2.

15. "Pakistan oil Minister in Tehran," in *Ettela'at*, no. 354, October 12, 1995, p. 10.

16. "Transfer of Turkmenistan gas via Iran to begin this winter," in *Ettela'at*, July 29, 1997, p. 10 (quoting IRNA news agency).

17. "America has modified the policy of economic sanction against Iran," in ibid., at p. 2 (quoting Reuters news agency).

18. "Taliban wary of opposition groups closing on Kabul," ibid., p. 2 (quoting Daily Telegraph).

19. "Future of Sino-Iranian oil cooperation bright," *Tehran Times*, Tehran, May 15, 1997, p. 4.

20. "Turkmenistan's outlets to high seas," *Tehran Times*, May 18, 1997, pp. 1 and 14.

21. "Iran agrees to transfer of Caspian oil to world market," *Ettela'at*, no. 185, Monday, February 6, 1995, p. 8.

22. Remarks by Vahe Petrosian of the Middle East Economic Digest at CBI Conference on Iranian Economy, held in London, Tuesday, November 19, 1996.

23. Remarks by Glen Race, U.S. Department of State, at the Seminar on the Caspian Sea, held in London, February 24, 1995 (hereinafter referred to as "Caspian Seminar"); "Iran most logical route to export Caspian oil," *Ettela'at*, no. 200, February 27, 1995.

24. "Azeri President: Jeopardizing Baku-Tehran ties not allowed," in *Ettela'at*, no. 331, September 11, 1995, p. 10.

25. "Iran swaps still on option for first Azeri oil exports, says Aliev," *Ettela'at*, no. 251, March 17, 1995, p. 10.

26. "Chevron seeks swaps deal with Iran for Kazakh oil," in *Ettela'at*, no. 331, September 12, 1995, p. 10.

27. "Joint excavation activities to start soon in Caspian Sea," in *Ettela'at*, no. 344, September 28, 1995, p. 10.

28. Ibid.; "Drilling of Iran's Caspian oil well to start," in *Ettela'at*, no. 306, August 7, 1995, p. 10.

29. "Iran's suggestions to ECO," in *Ettela'at*, no. 212, March 15, 1995, p. 2 (quoting remarks by the Iranian president Hashemi-Rafsanjani at ECO Third Summit meeting, Islamabad, Pakistan, March 1995). Iran continues to stress the truly regional nature of ECO at every occasion, including

a proposal for the creation of an ECO university in order to promote regional integration and organization. Ibid., October 29, 1996. Iran also favors establishment of close economic ties between ECO and the Association of South East Asian Nations (ASEAN).

30. "Caspian Carve-up," in *NEFTE Compass*, vol. 2, no. 43, November 24, 1993, p.9.

31. Ibid.

32. Remarks by Alexander Khodadov, Russian Ministry of Foreign Affairs, at Caspian Seminar, op. cit.

33. Race, Caspian Seminar, op. cit.

34. *Salam*, Tehran, July 3, 1995 (reporting on statement made by the Iranian Minister of Foreign Affairs on a visit to Switzerland).

35. "Turkmen minister underlines formulation of legal regime for the Caspian Sea," *Ettela'at*, no. 310, August 11, 1995, p. 10.

36. "Cooperation in the Caspian Sea," in *Echo of Iran*, London, no. 98, August/September 1995, p. 5 (reporting on statement made by the Russian Deputy Minister for Foreign Affairs on a visit to Almaty, Kazakhstan).

37. "Iran, Kazakhstan call for declaration of Caspian Sea legal system," *Ettela'at*, no. 309, August 10, 1995, p. 10.

38. Soviet Business Intelligence Bureau, *Central Asia: Special Report* (Moscow: East Consult, June 1993), p. 47.

39. According to Hamzeh Behnam of Shavak Consultant Engineers, Rasht, Iran (interview, March 19, 1995), the rise in Caspian's water level is a periodical phenomenon, reported to recur in 30-35 and/or 100-year cycles, following a few consecutive wet winters. Remarkably, in the past 18 years, the water has risen by more than 2 meters, without any significant climate change in the region. In 1995, there was a considerable decrease in the salinity of the water. The source of the rising water is yet to be ascertained with certainty. For details, see Ahmad Barimani, *Darya Khazar* ("The Caspian Sea") (Tehran: Rangheen Publications, 1947), pp. 78-90.

40. "Iran to build barriers to stop encroachment of Caspian Sea," *Ettela'at*, no. 338, September 20, 1995. p. 10 (quoting reports from the Governorate-General of Mazandaran Province).

41. "Defence cooperation discussed between Iran, Tajikistan," in *Ettela'at*, no. 369, November 2, 1995, p. 8.

42. "Iran, Turkmenistan, Tajikistan sign trilateral agreement," in *Ettela'at*, no. 333, September 13, 1995, p. 10.

43. "Iran, Russia emphasis on defining Caspian Sea legal regime," in *Ettela'at*, no. 368, November 1, 1995, p. 8.

44. *Tehran Times*, December 4, 1996 (reporting on Iran-Russia oil exploration agreements).

45. On the failure of Turkey's U.S.-inspired approach to Central Asia, see Petrosian, op. cit.

46. "The Tehran quadripartite agreements," in *Ettela'at*, no. 637, December 6, 1996, pp. 1-2.

47. Pirouz Mojtahed-Zadeh, "Mashhad-Sarakhs-Tajan railway: a window of hope to the future world," in *Ettela'at*, no. 496, May 21, 1996, pp. 1-6.

CHAPTER TWELVE

America's Drive to the Caspian

Bradford R. McGuinn and Mohiaddin Mesbahi

Introduction

This chapter examines the United States' drive for influence in the Caspian region and the role that Turkey and Azerbaijan play in that effort. In May 1997, Azerbaijan and Turkey signed a "strategic cooperation" agreement, which included a commitment by Azerbaijan's president Haidar Aliyev to ship Baku's oil to Turkey's Mediterranean port of Ceyhan.[1] "In my soul," he declared, "I have always been on Turkey's side and, therefore, I will see to it that the oil pipeline will go via Turkish territory."[2] The language was more mystical than might be expected from a former politburo and Soviet Government Security Committee (KGB) member, but its meaning was clear: Turkey's ambition to serve as the principle outlet for Azerbaijan's considerable offshore oil would be realized.

The Ceyhan option represents less a triumph of Azeri-Turkish relations or pan-Turanism than the ascendancy in the region of the United States and the financial and security systems it leads. The creation of a tangible link between Turkey, Azerbaijan, and Georgia, a direct connection between the Caspian and North Atlantic Treaty Organization (NATO), is indicative of a broad shift in the regional balance of power. To say, as the Russian secretary for Commonwealth of Independent States' defense and security issues did, in May 1997, that "we are increasingly losing the Caucasus,"[3] is not the exaggeration it would have been a year earlier. Russia's failure in Chechnya, its weakened grip over the CIS, the Yeltsin government's dependence on the Western economic system, and the intensification of outside involvement in the Caspian oil and gas regime, contributed toward this change. This, combined with America's attempt to isolate and punish Iran, have created an opportunity for

Turkey to project its influence in the Caucasus. The pipeline to Ceyhan would be its most tangible expression.

This chapter argues that the prevailing "Great Game" metaphor is misleading in its attempt to make intelligible the region's "pipeline politics."[4] It is flawed at one level in its image of a strictly bipolar confrontation between the United States and Russia. Actually, this relationship is more complex, as are the roles and ambitions of the various local actors. The "Great Game" theme also implies the existence of a non-ideological power rivalry. Such a view is consistent with the general depiction of the post-cold war period as a "post-ideological" era. But the political, economic, and security aspects of the Caspian are thoroughly invested with ideology, from aggressive state-nationalism to ethnic irredentism, Islamist paradigms, variants of the Soviet legacy, and Western neo-liberalism.

Western neo-liberalism is generally associated with "pragmatism" and actually contains a strong messianic impulse and serves as the normative framework for the expansion of Western power into Russia's southern periphery. It serves also as the ideological basis for a Western-oriented "epistemic community,"[5] the network of professionals and technical specialists, in policymaking areas and the petroleum industry, whose views and behavior influence the dynamics of the emerging Caspian security order.

While neo-liberalism's discontents complicate the West's agenda in the Caspian, the region's allure guarantees that the agenda will be vigorously pursued. Having aligned its interests with those of the United States and the West, Turkey awaits its windfall.

The American Push

The scope of the Caspian's oil and natural gas deposits are unknown. The region's oil reserves are estimated to be near 200 billion barrels,[6] with an overall value of $4 trillion. Worldwide, there are thought to be 1,036.9 billion barrels in proved oil reserves, with 261.3 billion in Saudi Arabia, 96.5 billion in Kuwait, 93 billion in Iran, and 112 billion in Iraq.[7] Therefore, if current estimates are reliable, the Caspian deposits will represent a major oil source,[8] making the region a focus for world markets for decades to come.

The Caspian's oil and gas potential has not gone unnoticed. Since the signing of "the contract of the century," the Azerbaijan International Operating Company (AIOC) and other international concerns have signed a variety of contracts with Azerbaijan, bringing the total investment by international oil companies to $18 billion.[9] There has also been a rapid expansion in the involvement of American oil service companies as well as legal and financial firms.

President Aliyev asserted, in August 1997, that the United States alone had invested $25 billion in the Azerbaijani economy.[10] The nature of this involvement, the close association that exists between the oil industry and U.S. security policy, can be discerned by an inventory of the participants, which include Richard Cheney, Brent Scowcroft, Zbigniew Brzezinski, James Baker, Henry Kissinger, Alexander Haig, Richard Perle, Lloyd Bentsen, John Sununu, Robert Oakley, and Zalmay Khalilzad.[11] The composition of this group, a policy-oriented epistemic community,[12] suggests that the American commitment to this region will be substantial and long-term. Azerbaijan and the Caspian are, as Boris Nemtsov, Russia's former Minister of Energy, observed, now within America's "zone of strategic interests."[13]

The potential geo-political rewards for the United States and its allies are considerable. As James Baker suggested, in his call for closer ties between America and Georgia, the dynamic of the Eurasian landmass has been "transformed . . . from the north-south character of the former Soviet Union into the east-west orientation of the new independent nations."[14] The creation of a "new Silk Road"[15] would split Russia from its possessions and provide the West with a highway from Georgia to China, a thoroughfare that happens to run through the Caspian's oil and gas fields.[16]

Dueling Containments

The Clinton Administration has been alert to the possibilities. According to Deputy Secretary of State Strobe Talbott, it is the policy of the United States to press for the creation of democratic, free-market systems in the area, to promote conflict resolution efforts in Georgia, Azerbaijan, and Tajikistan, and to seek the region's integration into Euro-Atlantic economic, political, and security frameworks.[17] In testimony before the Senate Foreign Relations Committee, Stuart Eizenstat, Under Secretary of State for European Affairs, indicated that the United States would support "robust U.S. commercial participation" in the development of the Caspian's energy sources.[18] He observed further that these frameworks would include NATO security structures.[19] As the State Department asserted in April 1997, America seeks "to tie the region securely to the West."[20] In this way, the ideological mission implied in neo-liberalism is tangibly joined to U.S. commercial interests and military power.

However, linking the Caspian to the West necessitated a structural change in the American approach to the region. The previous indifference[21] shown by the United States toward Russian interventions in its southern periphery[22] has given way to a pronounced concern for the "independence and prosperity" of the former Soviet republics in the Caucasus and Central Asia.[23] Strobe Talbott's earlier deference to Moscow's "Monroe Doctrine" toward its former republics has been modified by the growing Western interest in the Caspian region.[24] Earlier anxieties regarding "post-Soviet instability" were replaced by traditional state-centric concerns regarding the ambitions of regional great powers, notably Russia and Iran. How to counter Moscow and Tehran, in an area where they enjoy obvious geographical and historical advantages, engendered a still inchoate debate concerning the merits of "dual containment," "balance of power," and "constructive engagement" strategies.

With regard to Russia, the Clinton Administration has emphasized "integration" as its central objective,[25] whereby seeking the integration of Russia into Euro-Atlantic structures and promoting Moscow's "pacification" while at the same time easing out the countries of the Caspian from the CIS into the Western system. As the stakes in the Caspian have risen, America's previous concern for Russia's sensibilities has diminished.

The American approach to Iran has, obviously, been more confrontational. "We remain," Talbott asserted, "highly suspicious of Iran's motives."[26] "It is essential," Eizenstat insisted, "that the U.S. have a proactive policy to assure that the Caucasus and Central Asia remain free of Iranian influence."[27] Such a view is consistent with the administration's dual containment policy in the Persian Gulf and it represents one of the organizing themes of Clinton foreign policy,[28] a theme reinforced by domestic political factors. Yet, considerable pressure has developed to alter the U.S. policy toward Iran. Much of it emanated from oil industry and former foreign policy officials who are concerned that a doctrinaire American approach toward Iran could have adverse effects on Western interests in the Caspian.[29] "The U.S.," wrote a former CIA analyst in *The Wall Street Journal*, "must ensure that the Central Asian states and their oil and gas wealth, stay free of Russian power."[30] A rigid dual containment policy toward Russia and Iran would, according to this view, serve only to enhance Moscow's relationship with Tehran.[31]

For its part, the Clinton Administration has signaled a tolerance for this perspective. In March 1997, Secretary of State Madeleine Albright told a European audience that "your critical dialogue hasn't worked. Of course, our critical silence doesn't seem to have accomplished that much either."[32] This was followed by the election of the Khatami government in Iran and, in July, by an announcement that the United States would not oppose the construction of a natural gas pipeline from Turkmenistan to Turkey, which involves a 750-mile segment through Iran.[33] The domestic political support for the dual containment effort against Iran remains formidable,[34] though its intellectual bases are weakening as the Caspian question increasingly informs U.S. security discourse.

What the "dual containment," "balance of power," or "constructive engagement" strategies have in common is a recognition of Turkey's critical role in this geo-political space. "Turkey's increased attention and activism has been," Talbott noted, "a source of solace and support to those who rightly worry about the projection of Iranian influence."[35] Richard Burt, an official in the Reagan and Bush Administrations, provided the emblematic view: Turkey is "the key 'front-line' state in confronting the dominant dangers of the post–cold war period: state-sponsored terrorism, the proliferation of nuclear and conventional weapons and the spread of radical Islamic fundamentalism."[36]

The comments by Talbott and Burt reflect a strong continuity in America's interpretation of Turkey's place in the international system. Turkey's role in the American-dominated world-system has been to serve as a "bulwark" against revisionist tendencies, which have emanated from Moscow or the Middle East, and also as a "pillar" of the U.S.-dominated security order.[37] Such a view represents a point of conceptual convergence between American "cold war" conservatives, neoconservatives, Israel, and Turkish elites. Now, Turkey undertakes another assignment. As the United States and its allies become more deeply involved in the Caspian region, Turkey's function as a safe outlet for the region's oil and a "bridge" for NATO expansion into the Caucasus and Central Asia has become ever more critical. However, all of this could come to nothing should Turkey's domestic balances radically change.

"Into an Endless Darkness"

"The people and the country are relieved," noted the commentator Sedat Ergin in *Hurriyet*.[38] Although the Western media were moved to express concern over the manner in which Necmittin Erbakan's government was overturned, there could be no mistaking the widespread relief upon the Refah Party's fall on June 18, 1997.[39] The reaction of Western governments was similar.

Of course, concern over Refah's progress had served as a point of convergence between Turkey's secular forces and the West. There was agreement that Erbakan's government, left unchecked, could overturn Turkey's Kemalist political culture and fundamentally alter the nation's relationship toward the West. Statements by members of Turkey's secular establishment reflected the anxiety. "The Republic is facing an extremely serious threat," asserted General Fevzi Turkeri, head of Turkey's military intelligence, "political Islam is working closely with Iran and some other Islamic countries to pull Turkey into an endless darkness."[40] Despite the pressures against Turkey from within the American political system, the U.S. government was willing to support the government's actions against internal challenges. On May 14, 1997, Turkey launched a large-scale operation designed to destroy the presence of the Kurdistan Workers Party (PKK) in southeastern Turkey and northern Iraq. The constraints America imposed on Turkey, in the form of delays in military equipment, did not reveal a fundamental divergence between the United States and

Turkey on the Kurdish issue.[41] An American official asserted that the Clinton Administration "understood Turkey's right to self defense."[42] The operation against the PKK concluded on June 27, but concerns regarding the security of the proposed Caspian pipeline had not been allayed.

Of even greater concern regarding Turkey's long-term viability, as the main outlet for Azerbaijan's oil, was the Islamic question. The formation of Erbakan's government in June 1996 caused tangible worries for the military and its secular allies within the Turkish political and financial sectors, and the West. The reaction of the Turkish elite and the Western, neo-liberal, epistemic community was aptly captured by an officer with the Western investment firm, SBC Warburg: The rise of Refah fostered "the worst possible environment" for business activity.[43]

The actual power of the Refah was limited by its involvement in a coalition with Tansu Ciller's secularist True Path Party. However, its success in advocating an "Islamic agenda" in areas such as education, as well as Erbakan's diplomatic activities with Iran, aroused much concern.[44] More broadly, Refah's electoral success and the growing Islamist trend within the Turkish civil society suggest that the country's Kemalist framework had entered into systemic crises of legitimacy and authority.[45] It was the prospect of a long-term erosion of Turkish secularism and Western orientation, as expressed in the celebrated Turkish-Iranian natural gas deal,[46] that prompted interest in a rapid termination of Refah's government.

By early 1997 the Erbakan government faced increasing pressure from the Turkish military and the business sector.[47] The "creeping coup" began with several dramatic and open confrontations between the military-dominated National Security Council and Erbakan, in which the latter was obliged to sign documents reaffirming Turkey's secular traditions and orientation.[48] Pressure continued to build through the spring as members of parliament defected from the REFSOL coalition. As discussion of a putsch became more common, the United States indicated that it wished for Turkey to proceed "in a secular, democratic way,"[49] clearly indicating its preference for a new government.

The end came on June 18.[50] By June 30, Mesut Yilmaz, head of the Motherland Party, succeeded in fashioning a government,[51] declaring that he would "meticulously guard the basic principles of the republic."[52] The ascendancy of Yilmaz enhanced Turkey's ability to position itself within the "dominant order."[53] The nature of the Turkish economic crisis lent itself to the neo-liberal reforms favored by the United States. And, the policies of Yilmaz were generally consistent with this agenda.[54] By July 1997, the World Bank announced that a $1.5 billion loan program would become available to the new government.[55] The Turkish equity markets responded with a sustained rally throughout that summer,[56] though doubts remained concerning the government's willingness to engage in deep structural reforms.[57]

Strategic Denial

The establishment of the Turkish port of Ceyhan as the main outlet for Caspian oil will not resolve Turkey's political and economic crises. Nor can a pipeline route from Baku to Ceyhan be viewed as the most efficacious method of bringing the Caspian's oil and natural gas to world markets. Such an undertaking would face formidable technical[58] and security problems. The Supsa-to-Ceyhan pipeline would run approximately 1,050 miles, cost $3.3 billion, and traverse areas menaced by Kurdish insurgency.[59] The Turkish option, then, owes its currency to ideological factors that are on display in the American debate on the pipeline routes.

There is, first, what might be termed the anti-Russian, "neo-Cold War" option, based on the desire to limit Russian influence. A significant faction within this group, particularly those associated with the petroleum industry, seek an opening to Iran, including a southern pipeline route, as a means of enhancing the U.S. position relative to Moscow and obtaining the most cost-effective means of shipping Caspian resources. This position is the most inimicable to Turkey's interests. But there is, second, an anti-Iranian, "clash of civilizations" approach that favors cooperation with Russia in pursuit of a common "Western" campaign to contain Iran and Islamist movements. This view is more compatible with Turkey's agenda, provided that the anti-Iranian emphasis orients the pipeline westward rather than north to Russia. The third position is the "neoconservative" trend, which also favors the anti-Iranian theme and calls for an explicit and prominent Turkish (and Israeli) role in the pipeline security system, envisioning the creation of an American-Turkic-Israeli "strategic consensus." These three approaches compete for influence within the Clinton Administration, with the latter two being dominant.

The U.S. policy response has been, at one level, to favor the "multiple pipeline" concept[60] as a delaying tactic, an acknowledgement that the "facts on the ground" favor the resuscitation of Russia's northern route.[61] At another level, the Clinton Administration has expressed support for the Turkish pipeline, based on the logic of strategic denial[62] and the political/ideological appeal of the "clash of civilizations" and "neoconservative" rationales, despite the reticence shown by Western petroleum companies, alert to the risks attendant in the Ceyhan route.[63]

The Clinton Administration's view converges with Turkey's, for whom the completion of the Ceyhan project would bring tangible benefits. First, direct access to Azeri oil would help to alleviate Turkey's domestic energy shortfall. With a population of 62 million, half of whom are under 25, Turkey has witnessed rapid demographic expansion, particularly in the urban areas.[64] And, despite its political and macroeconomic problems, Turkey's economy has experienced rapid growth. While Turkey's domestic energy requirements are expected to grow by 200 to 300 percent over the next decade,[65] its indigenous supplies are limited.[66] The prohibitively high exploration and development costs, owing to Turkey's forbidding geography, ensure continued foreign oil dependency. One response has been a shift toward natural gas, which has resulted in contracts and pipeline deals with Russia, Iran,[67] and Turkmenistan.[68]

The Ceyhan route also would gain Turkey transit fees and reduce the cost of energy imports. This is to say nothing of the other economic benefits associated with the establishment of Ceyhan as a world oil terminal. The closure of the Iraqi pipeline is said to have cost Turkey $27 billion.[69]

Then there is the matter of Turkey's long-term role in the Caucasus and the Caspian energy regime. While there had been much speculation that upon the USSR's demise, Turkey would attract the support of the region's Turkic population and establish a preeminent position, Ankara's ambitions have been constrained by Russia's continued primacy, the proliferation of local conflicts, and the inherent limitations on Turkey's ability to project its influence. Its influence over the pipeline's ultimate destination is also limited. Ankara's concerns over environmental damage from excessive tanker traffic through the Bosphorus is not likely to deter the AIOC, but internal disputes within various Turkish governments have diminished Ankara's ability to press its case for Ceyhan.[70] Should, however, the ideological and strategic factors informing American policy triumph, a Baku-Supsa-Ceyhan pipeline promises to substantially deepen Turkey's role in the Caspian.

NATO's Reach

Not the least of Ceyhan's attractions is the promise that a Turkish terminal would bring Azerbaijan's oil within NATO's reach. According to Eisenstat, Turkey is "the link between NATO and the West in this critical region."[71] There is, in fact, a fortuitous convergence between the drive to expand NATO eastward and the desire to protect the Caspian's oil and natural gas reserves. How to safeguard the passage of the oil to world markets, then, becomes a problem not only for the states directly involved, but also for the West, its oil companies, financial institutions, and security structures. In this sense, the Caspian and its pipeline system has become a NATO "out-of-area" concern.

If the central strategic problem in this quest is the management of Russia, deterring Moscow from aggressively asserting its claims in the region while not jeopardizing the West's relations with the Yeltsin government or upsetting its internal balances, the expansion of NATO into central/eastern Europe provides a model that may be applicable to the

Caucasus. The emergence of the neo-liberal government in Russia, the muted internal reaction of Yeltsin's acquiescence to NATO expansion,[72] and the crystallization within the CIS of a pro-Western subsystem comprised of Azerbaijan, Georgia, and Ukraine, has given the United States and its new friends in the Caspian area confidence that the geo-politics of Eurasia can be remade.[73]

Azerbaijan's drift toward NATO is indicative of this trend. Its spokesmen have viewed the alliance as a stabilizing force "not only in Europe, but worldwide," one that plays "an extremely important part from the point of view of balancing forces and interests in the South Caucasus."[74] A balancing mechanism is critical for Azerbaijan in view of its security problems with Armenia and Russia's military relationship with the Yerevan government.[75] For Aliyev, oil and pipelines are compensatory devices, attracting NATO to Azerbaijan in order to offset the Russian-Armenian alliance.

Azerbaijan's association with NATO has become increasingly explicit. Azerbaijan joined the "Partnership for Peace" program[76] and has received visits from NATO officials.[77] For his part, Aliyev argued that Azerbaijan is "situated in the east of Europe," thereby eliminating geographical objections to Baku's incorporation in Euro-Atlantic security structures.[78] Other Azeri officials have suggested that NATO forces be used to protect Azerbaijan's oil facilities and pipelines.[79]

Bilateral American-Azerbaijani security talks have involved U.S. assurances to Azerbaijan that the so-called flank agreement regarding conventional forces in Europe (CFE) would not compromise Azeri security.[80] The United States has also promoted the Nagorno-Karabakh peace process.[81] The *Journal of Commerce* reported that the Clinton Administration had contemplated the use of U.S. forces in peacekeeping capacities.[82] Aliyev's visit to the United States in August 1997 revealed a further deepening of Azeri-American relations. He met with President Clinton and initialed a $10 billion exploration and development contract with Exxon, Chevron, Amoco, and Mobil.[83] Additionally, Aliyev entered into a military cooperation agreement with the U.S.,[84] held talks with several major companies,[85] and received support in the opinion pages of American newspapers in this effort to have removed the sanctions imposed on Azerbaijan as part of the 1992 Freedom Support Act.[86]

As a critical regional ally of the United States and a member of NATO, Turkey provides a mechanism for Azerbaijan and other pro-Western CIS states to integrate into Euro-Atlantic structures. On one level, of course, the Turkish relationship contains difficulties for Azerbaijan. Concern has been raised in Baku concerning Turkish "meddling" in Azeri internal affairs.[87] Turkey's attempts to open relations with Armenia have aroused suspicion[88] and Azeri reticence about the Turkish role was given public expression in September 1997 when Yilmaz chose not to visit Azerbaijan during his first official visit to the region.[89] There exists, however, a general convergence of interests between Turkey and Azerbaijan, bilaterally, in terms of direct Turkish economic investment,[90] cultural influences, and security cooperation,[91] and multilaterally in terms of threat-assessment and operational convergence with Euro-Atlantic structures. Azerbaijan's involvement in NATO military exercises, held in Turkey, constitutes a tangible expression of this cooperation.[92]

Azerbaijan's movement toward the West parallels that of Georgia and Ukraine. As the entry point for the Eurasian corridor, Georgia occupies a privileged place in the West's design for the region.[93] Although Russia retains the ability to destabilize the internal balance in Georgia through its manipulation of the Abkhazian issue, the Georgian president Eduard Shevardnadze has become increasingly bold in his distancing from Moscow, his shift to the West,[94] and his attempt to fashion a pro-Western tendency within the CIS.[95]

Shevardnadze's efforts to integrate Georgia into Euro-Atlantic financial and security structures[96] have coincided with an intensification in Georgia's relations with Turkey. "I think it's time," Shevardnadze observed in March 1997, "for a more active Turkish participation in the settlement of the Abkhazian conflict and other conflicts in the Caucasus region."[97] Shevardnadze has also signaled his approval of the Ceyhan option as the logical termination point for the Baku-to-Supsa pipeline.[98]

A similar pattern is observed with respect to Ukraine. Its leadership has asserted that the deepening of cooperation with the Atlantic alliance is among its "strategic goals."[99] President Kuchma has insisted that Ukraine seeks "integration" in Western "economic structures" and "a deepening of trans-Atlantic cooperation."[100] Then in July 1997, Ukraine signed an agreement linking itself to NATO's multi-lateral security framework[101] and has emerged as an "informal leader of a pro-NATO subsystem in the Caucasus/Black Sea region,[102] an ambition enhanced by Kiev's deepening bi-lateral ties to Azerbaijan[103] and Turkey.

The projection of Western and NATO influence can also be seen in Central Asia. The United States has promoted military cooperation among Uzbekistan, Kazakhstan, and Kyrgyzstan.[104] In September 1997, the Central Asia Battalion joined with 500 American troops as well as Russian and CIS forces[105] in manoeuvres designed, as an American official remarked, "to breed familiarity with NATO and to tie these armed forces into the new NATO idea of peacekeeping."[106]

Both Aliyev and Shevardnadze have expressed interest in "Dayton solutions" for their respective conflicts,[107] holding as they do the prospect of NATO involvement in a peacekeeping capacity. For its part, the United States has betrayed little reticence about taking on such projects. Commenting on the significance of the military exercises in Central Asia, the American commander asserted that its "message" was that "there is no nation on the face of the earth that we cannot get to."[108]

"Market Civilization"

From Turkey's perspective, Russia remains both a threat and an opportunity. Russian statements critical of Turkey and its role in the Caspian are easily found. The depiction of Turkey as a "Trojan horse" of Western influence[109] is neither novel nor irrational. Russian officials have asserted, in one way or another, that "Turkey can have Azerbaijan, but the oil is ours."[110] For her part, Tansu Ciller warned that Turkey "will not allow Caspian oil to go through Russia, since in that case such an important source of energy would fall exclusively under its control and this would mean that the fate of the countries in the region would be in Russia's hands."[111] At another level, Turkey has complained of Russian military aid to Cyprus, Armenia, and the PKK.[112] Conversely, Russia claimed that Chechen guerrillas received assistance and training from Ankara.[113]

But the Caspian question is neither the sole or even the most critical mechanism informing Turkish-Russian relations. Turkey's economic ties to Russia far exceed its economic involvement with its Turkic kinsmen in the Caucasus and Central Asia. Total trade with Russia is $3.3 billion as compared with $800 million for the Caspian region.[114] Much of the Turkish-Russian commerce centers on the energy sector, as Russia is Turkey's primary source of natural gas.[115]

Turkish and Russian interests converge also in their desire to seek a place in the American-dominated neo-liberal "market civilization."[116] The domestic consensus for such an orientation remains problematic in both societies. Yet, Moscow and Ankara appear to be moving in the same direction, toward the "Washington consensus" on questions of political economy and increasing integration into Western financial markets.[117] American and Turkish elites seek comity between Ankara and Moscow, while remaining firm in their determination to deny pipeline dominance to Russia. In a sense, then, the state of Russian-Turkish relations may serve as an index by which the success of the West's bid to end Moscow's hegemony in its southern periphery can be measured

Israel's Assertion

An indication of Turkey's identification with the "dominant order" and its agenda can be found in its security ties with Israel. The existence of a formal strategic relationship between Turkey and Israel became apparent with the signing of a military cooperation and training agreement on February 23, 1996.[118] In the broadest sense, the Turkish-Israeli alliance recalls the Nixon Doctrine's "Twin Pillars" structure. But unlike its pre-decessor, the neo-Twin Pillars strategy is not devolutionist: the United States and NATO continue to play a direct military role in the advance-ment of their regional agendas.

The role of the neo–Twin Pillars concept in the Middle East is the con-tainment of Iran and Syria.[119] In this regard, it was designed to serve as an anchor for a broader regional coalition, involving Arab states willing to join the Israeli-Palestinian "peace process" camp. Operationally, it has been deployed against Syria,[120] whose support for the PKK and Hezbol-lah united Turkey and Israel. Syria is, according to Turkish security offi-cials, "the headquarters of the terrorism that threatens both Turkey and Israel."[121]

Throughout 1996 and 1997, troop mobilizations occurred on both sides of the Syrian-Turkish border.[122] Reports surfaced of bombing inci-dents in Syria, which Damascus attributed to the Turkish government. Syria had hoped that the accession of the Erbakan government would occasion a split in the Turkish-Israeli alliance,[123] but it continued apace. The initial military cooperation and training agreement[124] was expanded and complemented with visits by key political and security officials, including a meeting between Israeli Foreign Minister David Levi and Erbakan.[125] In May 1997, General Cevik Bir and American officials met with Israeli Defense Minister Yitzaq Mordechai in Tel Aviv as part of reg-ular "strategic dialogue."[126] Talks continued throughout 1997 and into 1998 regarding joint American-Israeli-Turkish naval manoeuvres that took place later in the eastern Mediterranean.[127] The alliance with Israel also contributes to Turkey's position with the United States.[128] Mordechai's urging to an American audience that Turkey's military be "given what it is asking for in order to maintain the domestic balance in that country"[129] is indicative of the type of "cover" Israel can offer to off-set the limitations to Turkey's influence within the American political system.

The U.S. position on the utility of the Turkish-Israeli relationship has been unambiguous. "[I]f certain Arab countries don't like that," an offi-cial noted, "that's just tough, because Israel is a country that needs broad support."[130] In fact, the American bearing was so confident that the unusually strong Arab reaction to the Israeli-Turkish alliance and Turkey's anti-Kurdish campaign,[131] which ranged from denunciations to suggestions of a possible rapprochement between Syria and Iraq, left the United States unmoved.[132]

There is also the question of Israel's energy requirements. As with Turkey, Israel is increasing its shift from oil to natural gas.[133] Agreements effected with Egypt were meant to furnish Israel with natural gas shipped across the Sinai,[134] but political difficulties have made this plan problem-atic. Israel has, therefore, sought alternative arrangements with Russia and Turkmenistan.[135] The manner in which Russian or Turkmen gas would be shipped to Israel[136] revealed the technical and political difficulties inherent in Israel's involvement in the Caspian system. Nevertheless, upon his visit to Baku, in August 1997, Prime Minister Benjamin Netanyahu asserted that Israel "was involved in this project."[137] Israel is able to provide Azerbaijan with the type of political support within the United States that it has long furnished to Turkey.[138] With Turkey, Israel is also able to offer a security framework supplemented by the United States designed primarily to counter Iran.[139]

The establishment of open ties between Israel and Azerbaijan, has raised the possibility of Israel and the West employing Azerbaijan and northern Iraq as "strategic enclaves"[140] to support the dual containment policy in the Persian Gulf and deepen Israeli-American-Turkic hegemony in the Caspian system. For its part, Tehran has warned that in courting the "Zionist regime," Baku was "playing a dangerous game."[141] Israeli and American concerns about Iran's non-conventional military capabilities[142] and Russia's military relationship with the Islamic Republic served to heighten Western fears of a joint Iranian-Russian threat to the nascent Caspian security order.[143]

From Turkey's perspective, however, the principal threat to the Azerbaijan-Turkey pipeline ambitions is found not in the confrontation between the American-Israeli-Turkic bloc and Iran, but in its abatement. A shift in the American, Azeri, or Israeli[144] threat perception regarding Tehran, a dissolution of the anti-Islamist strategic consensus, or a movement toward normalization with the new Khatami government would imperil the Ceyhan route.[145]

America's Taliban

America's relations with the Afghan militia, the Taliban, constitutes a source of anxiety for Turkey. Turkey has invested more than $1 billion in the Caspian region[146] and has offered Ceyhan as the terminal for the massive oil and gas supplies from Kazakhstan and Turkmenistan.[147] While the interests of the United States and Turkey fundamentally converge with respect to denying Russian and Iranian primacy in Central Asia, America's relations with the Taliban government also sets up Afghanistan as a possible rival to Turkey as a potential exit route for Caspian and Central Asian oil and gas.

Attempts by Western oil companies to construct oil and natural gas pipelines from Kazakhstan and Turkmenistan south through Afghanistan to Pakistani ports[148] expose a divergence in the American and Turkish agendas. In May 1997, for example, an agreement was reached between Pakistan and Turkmenistan that would involve a pipeline running through Afghanistan to Pakistan.[149] An alliance emerged between UNOCAL, the

Saudi Delta Mir, the Pakistani government, and the Taliban concerning the use of Afghanistan as a transit route.[150] The commercial imperative for these routes was based on the projected energy needs of the Pakistani and Indian markets as well as the logistical advantage of the Persian Gulf ports in moving oil and gas to the Far East.[151]

A rationale developed in the United States for the tacit support of Taliban's effort to consolidate its control over and to "bring stability" to Afghanistan.[152] For a brief period, at the end of May 1997, it appeared that Taliban would succeed, as its militia moved north from Kabul, hoping to unseat Abdul Dostam, an Uzbek and head of the powerful northern alliance. By the end of the month the operation appeared successful, especially after Dostam's ally, Abdul Malik, defected to Taliban's side.[153] "At least for a few hours, a lot of people were hopeful," observed an American official.[154] "If you have one government controlling virtually all the country," he continued, "commercial banks and international financial institutions would have been likely to get involved." For his part, Dostam was forced to flee to Turkey.

In Turkey, Dostam's defeat was viewed with alarm. He had long been a focus of Turkish support in Central Asia.[155] The demise of organized resistance to Taliban meant the possibility that "the hurdle of instability" had been eliminated, permitting the United States, Saudi Arabia, and Pakistan to move forward on their pipeline plans, a scheme that would come at the expense of Turkey's desire to see Central Asian oil and gas move through Ceyhan.[156] Suddenly, Turkey's position in Central Asia seemed precarious, in part as a result of American policy.

Turkey's fears were allayed when the Taliban failed to hold Mazar-i-Sharif after Abdul Malik turned against the Taliban with the support of the Shi'i Hezb-i-Wahdat.[157] There followed a general retreat of the Taliban toward Kabul. However, the Taliban retained control over the areas to be traversed by the Turkmen gas pipeline and in October 1997 the militia entered into contracts with Western petroleum companies.[158] Turkey is likely to continue to expand its Central Asian role in a manner beneficial to the Euro-Atlantic agenda,[159] but its pipeline ambitions could be undermined should commercial enticements attenuate America's animus toward Islamist regimes and movements.

Conclusion

The international system devised by the Clinton Administration for and in and around the Caucasus and Caspian regions is said to be informed by liberal principles. Referring to U.S. policy in these regions, Talbott has insisted that the era of the "Great Game" was over. "What is required now," he has said, "was just the opposite: for all responsible players in the Caucasus and Central Asia to be winners."[160] It is tempting to question this construct, dismissing, as it does, the possibility that substantive conflicts of interest will continue to shape the politics of this area. Skepticism is heightened when the U.S. State Department's spokesman asserted that the United States does not "believe in spheres-of-influence."[161]

Yet, a closer look reveals that American policy, while perhaps motivated by liberal impulses, has a traditional hegemonic aspect. This is suggested by the organizing principle of U.S. policy toward Russia and the former Soviet republics. "Integration," Talbott has stated, "is key to U.S. policy toward Russia."[162] The integration of Russia, via its neo-liberal elite, into the Euro-Atlantic order is the predicate to the "civilizing" of Russia. The ascendancy of this "Atlanticist" or "mondialist" tendency, joined now to the Western epistemic community, is viewed as critical to Moscow's peaceful devolution of authority in the Caucasus.[163] Peacekeeping and the multi-lateral conflict control it exercises in the Caucasus and Central Asia are the corollary to this principle and have the practical effect of establishing a permanent Western presence in the region.

Turkey's ability to convert the American assertion into tangible advantage will be complicated by several factors. First, there is the possibility, suggested in the case of Afghanistan, that U.S. interests may diverge from Turkey's. This may involve siding with a Caspian state at odds with Turkey or siding with Russia, with whom America may have more substantive mutual interests.

Second, there is the prospect of a de-ideologization and even normalization of American-Iranian relations.[164] This would seriously compromise the viability of the Ceyhan route and throw into question the architecture of the anti-Iranian "strategic consensus" from which Turkey has benefited.

Third, it is not clear that the United States or NATO is prepared to engage in a long-term "hegemonic management" in the Caspian region. The dynamics of local conflicts, the proximity of Russia and Iran, as well as the prospects for a Moscow-Tehran counter-systemic trend,[165] combined with the minimal assets deployed thus far by the West in this region, make such a project more an aspiration than a reality. It is also unclear as to the effectiveness of a Turkish security role given the demands placed on its forces by unresolved indigenous conflicts. Not the least of these is the Islamist assertion, which threatens, in Turkey and throughout the Caspian region, to overturn the existing, narrowly based, and often corrupt secular order.[166]

For the moment, though, the hubris expressed in the Clinton Administration's claim that America is the "indispensable nation" cannot be dismissed blithely. Turkey's present leadership understands this and appears anxious, with the Ceyhan prize in view, to make the most of America's ascendancy.

Notes

1. See report by Xinua News Agency, May 5, 1997.
2. Quoted in *British Broadcast Corporation—Summary World Broadcasts*, May 8, 1997 [Hereinafter referred to as "BBC-SWB"]; "Azeri Crude to be Shipped via Turkey, Aliyev," reported by Agence France Presse [hereinafter referred to as "AFP"], May 6, 1997.
3. Quoted in S. Ivanov, "Bloc and Blockade," in *Sovetskaya Rossiya*, reported in *Foreign Broadcast Information Services-SOV-97-099*, May 20, 1997, p. 3.
4. For the use of this term in its contemporary context, see, for example, G. Melloan, "Who's Winning and Losing, the New 'Great Game,'" in *The Wall Street Journal*, March 31, 1997, p. A15; H. Pope, "Great Game II: Oil Companies Rush Into the Caucasus To Tap the Caspian," in *The Wall Street Journal*, April 25, 1997, pp. A1.
5. For a discussion of this concept, see P. Haas, ed., *Knowledge, Power, and International Policy Coordination* (Columbia, South Carolina: University of South Carolina Press, 1996), p. 3.
6. See H. Pope, "Report Says Caspian Sea Oil Deposits May be Twice as Large as Expected," in *The Wall Street Journal*, April 30, 1997, p. A10. For a dissenting view on the widely publicized figures, see D. Hiro, "Why is the US Inflating Caspian Oil Reserves?," in *Middle East International*, September 12, 1997, p. 18.
7. See D. Jehl, "Qatar's Treasure-Trove of Gas," in *The New York Times*, July 23, 1997, p. C1.
8. See S. Levine, "Daubs of Oil and Exultation as Caspian Pipeline Opens," in *The New York Times*, November 13, 1997, p. C2.
9. See "Azerbaijan Nears More U.S. Deals," in *The Wall Street Journal*, June 24, 1997, p. A18.
10. See report by V. Abdullayeva, Itar-TASS (Moscow), reported in *FBIS-SOV-97-225*, August 13, 1997.
11. See D. Ottoway and D. Morgan, "Former Top U.S. Aides Seek Caspian Gusher," in *The Washington Post*, July 6, 1997, p.A1. This pattern is also observable in the United Kingdom, where the former Foreign Secretary Malcolm Rifkin has become the chairman of Ramco. See the report by Kontact News Agency (Tbilisi), September 19, 1997, reported in *BBC-SWB*, September 22, 1997.
12. The role of this community, "a sociological group with a common style of thinking," is critical in its public-education role as well as its issue-framing capacity within the American foreign policy-making environment. See Haas, *Knowledge, Power*, p. 3.
13. See Nemtsov's remarks on NTV (Moscow), November 16, 1997, reported in *BBC-SWB*, November 18, 1997. However, he also asserted that with the opening of Russia's northern route for "early oil," Russia had "already won" the pipeline game. For President Yeltsin's comments regarding the American assertion in the Caspian, see S. Kinzer, "Azerbaijan Has Reason to Swagger: Oil Deposits," in *The New York Times*, September 14, 1997, p. A12.

14. See J. Baker, "America's Vital Interest in the 'New Silk Road,'" in *The New York Times*, July 21, 1997, p. A19; comments by Senator Sam Brownback (R-Kansas), in *FDCH Political Transcript*, July 21, 1997.

15. On the financial aspects of the "New Silk Road" concept, the emergence of a "Caucasus Common Market," and the role of extra-regional powers such as the United States, Japan, Saudi Arabia, as well as Turkey, see S. Shermatova, "Money Recarving Caucasus Borders," in *Moscow News* (Moscow), September 18, 1997, no. 37.

16. For a discussion of U.S. policy implications, see Brownback, *FDCH Political Transcripts*.

17. Prepared remarks by Strobe Talbott, "A Farewell to Flashman: American Policy in the Caucasus and Central Asia," delivered at the Central Asian Institute, The Johns Hopkins School of Advanced International Studies, Washington, D.C., in *Federal News Service* (July 21, 1997).

18. See Eizenstat's remarks before the Senate Foreign Relations Committee, *Federal News Service* (July 22, 1997).

19. Ibid.

20. See D. Morgan and D. Ottoway, "Drilling for Influence in Russia's Backyard," in *The Washington Post*, September 22, 1997, p. A1.

21. To the Clinton Administration's critics, U.S. policy toward Russia was characterized by "reflective deference," if not appeasement. See, for example, F. Gaffney, "Awakening to Our Caspian Concerns," in *The Washington Times*, July 30, 1997, p. A19.

22. During the 1994 conference between Al Gore and Victor Chernomyrdin, which occurred during a Russian assault on Chechnya, Gore asserted that the matter was a Russian internal problem. See M. Gordon, "Gore in Russia Hoping to be Cast in a Different Light," in *The New York Times*, September 22, 1997.

23. See S. Talbott, "The Great Game is Over," in *The Financial Times*, September 1, 1997, p. 14.

24. On the shift in U.S. policy, see C. Clover, and B. Clark, "Oil Politics Troubles Central Asian Waters," in *The Financial Times*, September 23, 1997, p. 9.

25. See S. Talbott, "The Struggle for Russia's Future," in *The Wall Street Journal*, September 25, 1997, p. A22.

26. See Talbott, in *Federal News Service*.

27. See Eizenstat, in *Federal News Service*.

28. For the paradigmatic statement, see A. Lake, "Confronting Backlash States," in *Foreign Affairs*, March/April 1994, pp. 45-55.

29. See, for example, F. Starr, "Power Failure," in *The National Interest*, Spring 1997, pp. 20-31; Z. Brzezinski, B. Scowcroft, and R. Murphy, "Differentiated Containment," in *Foreign Affairs*, May/June, 1997.

30. E. Shirley, "The Iranian-American Confrontation," in *The Wall Street Journal*, May 23, 1997, p. A18. According to the *Journal* Edward Shirley is the pseudonym for a former Iranian specialist in the CIA's Directorate of Operations.

31. See remarks by Richard Cheney, the former Secretary of Defence and now chief executive officer of Halliburton Inc., in which he argued that the

Clinton Administration's dual containment policy was providing Russia more leverage over the Caspian region. Address before the U.S.-Russia Business Council, on C-SPAN, November 19, 1997.

32. See J. Hoagland, "South of Europe," in *The Washington Post*, March 6, 1997, p. A21.

33. See R. Corzine, "US Decision Sparks New Round of Pipeline Politics," in *The Financial Times*, July 31, 1997, p. 4. But see R. Greenberger, "Approval of Pipeline Through Iran Isn't A Signal to Tehran, U.S. Official Says," in *The Wall Street Journal*, July 28, 1997, p. B7.

34. On the practical implications of the Clinton Administration's continued "hardline" regarding Iran, see B. Clark, "US Renews Threat Against Investors in Iran," in *The Financial Times*, October 31, 1997, p. 6. Representative of the anti-Iranian sentiment within the U.S. political system are the views of Senator Alfonse D'Amato. "Iran is," he asserted, "the foremost sponsor of international terrorism and threatens our national security and the interests of our allies." See Hearing of Senate Banking, Housing and Urban Affairs Committee, in *Federal News Service*, October 30, 1997.

35. See Talbott, in *Federal News Service*.

36. See R. Burt, "Are We Losing Turkey?," in *The Wall Street Journal*, June 17, 1997, p. A18.

37. On Turkey's self-perception, see the discussion of its 1996 Defense White Paper in C. F. Foss, I. Kemp, and L. Sariibrahimoglu, in *Jane's Defense Weekly*, September 10, 1997, p. 39.

38. See J. Barham, "Turkish Government Promises Unwavering Secularist Course," in *The Financial Times*, July 14, 1997, p. 4.

39. See "Creeping Coup," in *The Financial Times*, June 23, 1997; "Turkey's Troubles," in *The Wall Street Journal*, June 24, 1997.

40. Quoted in S. Kinzer, "Turkish Military Raises Pressure on Leader of Islamic Government," in *The New York Times*, June 13, 1997, p. A1.

41. For the Turkish view, see M.A. Birand, "The U.S. Arms Embargo Has Been Resumed," in *Sabah* (Istanbul), May 19, 1997, p. 12.

42. See "Turkish Forces Cross Border into Iraq to Attack Kurdish Guerrillas," in *The New York Times*, May 15, 1997, p. A4; comments by Strobe Talbott during his June 1997 visit to Turkey on *TRT* (Istanbul), June 2, 1997, reported in *BBC-SWB*, June 4, 1997.

43. See J. Dorsey, "Confusion About Turkey's Condition Keeps Foreign Investors on Sidelines," in *The Wall Street Journal*, October 3, 1996, p. B4.

44. See J. Dorsey, "Turkey's Leader Faces a Censure Vote," in *The Wall Street Journal*, October 14, 1996, p. A14.

45. Refah's share of the Turkish electorate remained at 20 percent several months after Erbakan's fall. See J. Barham, "Islamic Party Faces Turkish Court Closure," in *The Financial Times*, November 11, 1997, p. 4.

46. For the U.S. reaction, see report by AFP, March 5, 1997, reported in *FBIS-WEU-97-064*.

47. On opposition to the Islamists from the business sector, see S. Kadak, "Businessmen Protest Against Reactionarism," in *Sabah*, May 15, 1997.

48. On the military's presumption to uphold Turkey's secular traditions, see K. Couturier, "Turkish Army Piles Pressure on Islamists," in *The Finan-*

cial Times, June 12, 1997, p. 2, providing details based on the briefings by Major-General Fevzi Turkeri, chief of intelligence in the general staff, concerning the "threat" posed by the Islamists to the Turkish Republic.

49. Quoted in "US Concern over Turkey," in *The Financial Times*, June 14, 1997, p. 2.

50. See S. Kinzer, "Pro-Islamic Premier Steps Down In Turkey Under Army Pressure," in *The New York Times*, June 19, 1997, p.A1.

51. See S. Kinzer, "In a Rebuff to Religious Party, Secularist is to Lead Turkey," in *The New York Times*, June 21, 1997, p. 5.

52. Quoted in K. Couturier, "New Turkish Premier Pledges 'Freedom-Loving' Government," in *The Financial Times*, July 1, 1997, p. 3. On the government's attempt to reassert its control over the education system, see J. Barham, "Young Turks, the Pawns in Education Move," in *The Financial Times*, September 18, 1997, p.2.

53. For Thomas Friedman's view of "The Paradigm," meaning global financial system and the pressure it can exercise over "backlash" forces, such as Erbakan's Islamists, see "Turkey Wings It," in *The New York Times*, July 17, 1996.

54. See J. Barham, "New Turkish Ringmaster Cracks Whip," in *The Financial Times*, July 17, 1997, p. 2.

55. See J. Barham, "World Bank to Lend $1.5 Billion," in *The Financial Times*, July 26, 1997, p. 2.

56. See J. Barham, "Bull Run May be Over for Turkey," in *The Financial Times*, September 17, 1997, p. 38.

57. See J. Barham, "Turkey Still Reluctant to Grasp Inflation Nettle," in *The Financial Times*, October 9, 1997, p. 3.

58. See H. Pope, "Azerbaijani Leader Favors Pipeline Backed by U.S. in Rebuff to Russia," in *The Wall Street Journal*, May 9, 1997, p. A14.

59. See ibid.; M. Lelyveld, "Delays in Caspian Pipeline Shocks Experts," in *The Journal of Commerce*, June 25, 1997, p. 3A. Alternative figures regarding the pipeline's length and cost can be found in "Shell is Seeking Export Alternatives as Caspian Oil Production Rises," in *The Journal of Commerce*, September 18, 1997, p. 10A.

60. See, for example, the White House spokesman's assertion reported by AFP, November 18, 1997, that the United States "has long been in favor of multiple routes for pipeline trafficking."

61. On Russian-Chechen relations and the reopening of the Chechen section of the Novorossiysk pipeline, see B. Clark, "Chechen Chief Takes Struggle to Washington," in *The Financial Times*, November 14, 1997, p. 2.

62. See S. Heslin, "The New Pipeline Politics," *The New York Times*, November 10, 1997, p. A23.

63. See remarks by Charles Pitman, president and chairman of Amoco Eurasia Petroleum Company, in *PR Newswire*, October 23, 1997.

64. See the data in K. Couturier, "Turkey Aims to Satisfy Its Fuel Needs," in *The Washington Post*, October 20, 1997, p. A17.

65. Ibid.

66. See J. Kennedy, "Booming Oil/Gas Demand," in *Oil and Gas Journal*, February 24, 1997, p. 29.

67. See J. Barham, "Turkey Signs $13.5 Billion Deal to Import Russian Gas," in *The Financial Times*, April 29, 1997, p. 4. On Turkey's relations with Iran and Shell Oil's desire to construct a Iran-to-Turkey pipeline, see report in *BBC-SWB*, March 17, 1997.
68. See "US Urges Turkey to Seek Gas Elsewhere," in *Platt's Oilgram News*, February 18, 1997. American officials promoted Turkish-Turkmen ties, asserting that Turkey's gas deal with Iran furnished a "false security."
69. *BBC-SWB*, May 15, 1997.
70. For a discussion of internal Turkish disputes over Caspian policy, see J. Barham, "Turkey Presses Case for Pipeline," in *The Financial Times*, September 5, 1997, p. 4.
71. P. Glachart, "US Caspian," reported by AFP, October 25, 1997.
72. See C. Freeland, "Russia Agrees to NATO Expansion Plan," in *The Financial Times*, May 15, 1997, p. 1.
73. Quoted in C. Freeland, "From Empire to Nation State," in *The Financial Times*, July 10, 1997, p. 11.
74. Report by Interfax (Moscow), May 26, 1997.
75. For commentary on the Russian-Armenian military relationship and arms transfer issues, see reports by Interfax, September 15, 1997, reported in *FBIS-SOV-97-258,* and report by Turan News Agency (Baku), September 15, 1997, reported in *FBIS-UMA-97-258.*
76. Interfax, February 14, 1997.
77. Ibid.
78. Ibid.
79. Ibid.
80. Itar-TASS, March 12, 1997, reported in *FBIS-SOV-97-072.*
81. Talbott, "The Great Game is Over," p. 14.
82. See M. Lelyveld, "Delay in Caspian Pipeline Shocks Experts," in *The Journal of Commerce*, June 25, 1997, p. 3A; M. Lelyveld, "U.S. Use of Troops Weighed to Protect Caspian Sea Oil," in *The Journal of Commerce*, June 5, 1997, p. 3A. For the Russian response, see A. Yurkin, "U.S. Involvement in Caucasus Would Weaken Russia's Position," reported by TASS, June 6, 1997.
83. See P. Baker, "Clinton Courts Head of Oil-Rich Azerbaijan," in *The Washington Post*, August 2, 1997, p. A12.
84. See Turan News Agency (Baku), "Aliyev Signs Military Cooperation Accord in USA," August 1, 1997, reported in *BBC-SWB*, August 4, 1997.
85. See S. Gaines, *The Chicago Tribune*, July 24, 1997, p. 4.
86. See M. Lelyveld, "Foreign Policy, Oil, Azerbaijan: A Battle to Sway Washington," in *The Journal of Commerce*, July 29, 1997, p. 1A. For an example of U.S. lobbying efforts to overturn the embargo, see the remarks by Charles Pitman of Amoco before the Senate Foreign Relations Committee in *PR Presswire*, October 23, 1997.
87. For Azeri complaints regarding Turkey's bid to "lever its way" into oil deals, see *Platt's Oilgram News*, April 14, 1997, p. 2.
88. See "Turkey Makes Bid for Central Asia," in *The Journal of Commerce*, September 9, 1997, p. 12A.

89. See V. Bayramov, "Turkish Entrepreneurs Promoted Visit of Mesut Yilmaz to Kazakhstan—Azerbaijan Could Have Received Share of $500 Million," in *Azadlyg* (Baku), September 12, 1997, p. 6, reported in *FBIS-SOV-97-255*.

90. See, for example, reports by ANS News Agency (Baku), October 15, 1997, reported in *BBC-SWB*, October 24, 1997.

91. For reports of Turkish-Azeri security cooperation, see V. Abdullayeva and V. Shulman, "Azerbaijan and Turkey Sign Security Protocol," reported by TASS, November 4, 1997.

92. See report by Turan News Agency, May 13, 1997, reported in *FBIS-UMA-97-133*.

93. See report by Georgia Radio, Tbilisi, June 16, 1997, reported in *BBC-SWB*, June 19, 1997.

94. Ibid.

95. See, for example, reports from Georgian Radio, November 17, 1997, reported in *BBC-SWB*, November 19, 1997. On Georgian military cooperation with Ukraine, see report from Georgian Radio, November 3, 1997, reported in *BBC-SWB*, November 5, 1997.

96. See G. Duali, "Shevardnadze Goes to U.S. to Enlist NATO Troops," in *Kommersant Daily*, p. 4, reported in *Russian Press Daily*, July 17, 1997. Also see T. Nagdalyan and A. Asatryan, "Topic of the Day: 'We Will Live Like Good Neighbors and Fraternal States,'" interview with Eduard Shevardnadze, in *Respublika Armeniya* (Yerevan), May 3, 1997, pp. 1, reported in *FBIS-SOV-97-102*.

97. See report by G. Ablotia, Itar-TASS, March 4, 1997, reported in *FBIS-SOV-97-063*.

98. See report from Georgian Radio, June 16, 1997, reported in *BBC-SWB*, June 19, 1997.

99. See D.B. Sezer, "From Hegemony to Pluralism: The Changing Politics of the Black Sea," in *SAIS Review*, Winter/Spring 1997, p. 6.

100. Ukraine is the third-largest recipient of American economic aid after Israel and Egypt. See C. Clover and B. Clark, "Oil Politics Troubles Central Asian Waters," in *Financial Times*, September 23, 1997, p. 9. For Russian views on Ukraine's drift toward NATO and Western financial structures, see "Kiev Mother of All Non-Russian Cities," in *Novaya Gazeta*, no.22, June 2-8, 1997, p. 3.

101. See D. Buchan and D. White, "NATO Signs Charter with Ukraine," in *The Financial Times*, July 10, 1997, p. 2.

102. C. Freeland, "Moscow and Kiev Poised to Take Plunge," in *The Financial Times*, May 28, 1997, p. 3.

103. On Ukrainian military support to Azerbaijan, see "Ukraine Helps Azeris Build-Up, Says Armenia," in *Jane's Defense Weekly*, March 12, 1997, p. 3.

104. See C. Clover, "US Military Plans Central Asian Exercise," in *The Financial Times*, July 4, 1997, p. 4. On U.S.-Uzbek security cooperation, see A. Chytov, "USA Stepping Up Military Cooperation with Uzbekistan," reported by TASS, November 4, 1997.

105. Clover, ibid.
106. Ibid.
107. See Duali, "Shevardnadze Goes to US;" E. Shahinogly, "America . . . Conquest is a Long Way Off," in *Azadlyg*, August 9, 1997, p. 1, reported in *FBIS-SOV-97-224*.
108. See H. Pope, "U.S. Plays High-Stakes War Games in Kazakhstan," in *The Wall Street Journal*, September 16, 1997, p. A16.
109. For Gorbachev's statement along these lines, see Sezer, p. 6.
110. See ibid., p. 17.
111. Quoted in F.N. Kovalev, "Oil, the Caspian, and Russia," in *Mezh-dunaradnaya Zhizn* (Moscow), no. 4, April 1997, pp. 33-39.
112. On Turkey's response to Russian arms transfers to Armenia, see report by Anatolia, May 27, 1997, reported in *FBIS-TAC*. On Cyprus, see C. Bellamy, "Turkish Anger Grows at Russian Missile Deal," in *Independent*, January 15, 1997.
113. See report by Interfax, April 29, 1997.
114. See "Turkey and Asian Pipelines," in *Jane's Intelligence Review*, February 2, 1997.
115. See J. Barham, "Turkey Signs $13 billion Deal to Import Russian Gas," in *The Financial Times*, April 29, 1997, p. 4. For further energy deals effected in August 1997, see reports by AFP, October 30, 1997.
116. For a discussion of this terminology, see S. Gill, "Globalism, Market Civilization, and Disciplinary Neoliberalism," in *Millennium: Journal of International Studies*, vol. 24, no. 3, pp. 399-423.
117. For reports on the intentions, if not the outcome, of these reform efforts, see J. Barham, "Yilmaz Plays a Poor Hand with Skill," in *The Financial Times*, September 26, 1997, p. 3; M. Brzezinski, "Yeltsin Calls for Reforms in Economy," in *The Wall Street Journal*, September 25, 1997, p. 16.
118. For a Turkish perspective, see report by U. Akinci in *Turkish Probe* (Ankara), June 7, 1996, reported in *FBIS-WEU-96-113*.
119. "We are two very important countries in this region," asserted an Israeli official commenting on Israel's relationship with Turkey, "nobody is going to mess around with us." R. Usher, "A Potent Combination," in *Time*, June 2, 1997.
120. For the Syrian view of this alliance, see reports by SANA (Damascus), June 2, 1997, reported in *FBIS-NES-97-153*, and by Abu-Zahr, in *Al-Watan Al-'Arabi* (Paris), reported in *FBIS-NES-97-102*.
121. See *Foreign Report*, no. 2446, May 6, 1997.
122. See, for example, B. Doster, "The Turkish-Syrian Dispute as Part of the Middle East Peace Process: The Problem Hinges on Water," in *Nokta* (Istanbul), January 13, 1996, pp. 56-59, reported in *FBIS-WEU-96-032*. For an Iranian view, see A. Salaheddin, "The Dream They Have Dreamed for the Middle East?," in *Keyhan* (Tehran), June 17, 1996, p. 3, reported in *FBIS-NES-96-128*.
123. See F. Tinc, "A Warm Message From a Syrian Minister," in *Hurriyet* (Istanbul), June 24, 1996, p. 16, reported in *FBIS-WEU-96-125*.
124. E. Blanche, "Long Term Turkish Plans May Tighten Israel Ties," in *Jane's Defence Weekly*, February 5, 1997, p. 5.

125. See "Turkish Pragmatism," in *The Jerusalem Post*, April 10, 1997, p. 6.

126. For details of the meeting, see E. Blanche, "USA Teams with Turkey, Israel on Naval Exchange," in *Jane's Defense Weekly*, May 14, 1997, p. 19.

127. See "Joint Naval Exercise by the US, Israel and Turkey Angers Syria," in *The Financial Times*, September 3, 1997, p. 6.

128. See J. Barham, "Israel Seeks to Strengthen Ties with Turkey," in *The Financial Times*, September 24, 1997, p. 2.

129. See reports in *BBC-SWB*, April 4, 1997.

130. Quoted in H. Kutter, "U.S. Backing Israel-Turkish Cooperation," in *The Jerusalem Post*, May 11, 1997, p. 2.

131. See, for example, "The Blatant Turkish Aggression and the Silence of Western Nations," in *Al-Dustur* (Amman), May 30, 1997, p. 1.

132. See H. Pope, "Is a Syrian-Iraq-Iran Alliance a Possibility?," in *The Wall Street Journal*, June 27, 1997, p. A11; report by Radio Monte Carlo (Paris), May 31, 1997, reported in *FBIS-NES-97-152*.

133. See J. Fedler, "Stepping on the Gas," in *The Jerusalem Post*, March 12, 1997, p. 6.

134. Ibid.

135. See D. Harris, "Sharon, Russians to Discuss Gas Deal," in *The Jerusalem Post*, March 14, 1997, p. 15; D. Harris, "Turkmenis Seek to Supply Gas to Israel," in *The Jerusalem Post*, February 24, 1997, p. 1.

136. See G. Simpson and D. Rogers, "White House Got FBI Data on Party Donor," in *The Wall Street Journal*, June 10, 1997, p. A20; M. Lelyveld, "Controversial Oil Entrepreneur Unveils Pipeline Plan," in *The Journal of Commerce*, March 27, 1997, p. 3A.

137. See J. Bushinsky, "PM Discusses Oil Pipeline in Baku," in *The Jerusalem Post*, August 31, 1997.

138. Concerning the formation of an Azeri "lobby" in the United States, the Azerbaijani Foreign Minister Hasan Hasanov stated, "I would also like to note that the Jewish community in the United States has made positive contributions to the development of bilateral American-Azeri relations." See report in *Bakinskiy Rabochiy* (Baku), September 27, 1997, p. 1, reported in *FBIS-SOV-97-273*, September 30, 1997.

139. See S. Rodan, "Azerbaijani Diplomat Predicts Increased Ties with Israel," in *The Jerusalem Post*, June 9, 1997, p. 2; report by Voice of Israel, August 29, 1997, reported in *BBC-SWB*, August 30, 1997; report by Interfax, August 29, 1997, reported in *FBIS-SOV-97-241*. On Netanyahu's view of Iran as a "common enemy," see report by Moscow's NTV, August 29, 1997, reported in *FBIS-SOV-97-241*.

140. See the analysis by S. Bidiwi, in *Al-Sha'b* (Cairo), May 20, 1997, reported in *FBIS-NES-97-148*.

141. See report by Voice of the Islamic Republic (Tehran), August 30, 1997, reported in *BBC-SWB*, September 1, 1997. The Iranian media has complained also of Azeri persecution of indigenous Islamist movements in order to please its American and Israeli allies. See report by Voice of the Islamic Republic of Iran, February 26, 1997, reported in *FBIS-NES-97-039*.

142. See J. Bushinsky, "PM, Albright Discuss Missile Threat," in *The Jerusalem Post*, September 12, 1997, p. 1.

143. On the Israeli perception, see opinion by B. Kaspit, in *Ma'ariv* (Tel Aviv), September 12, 1997, p. 2, reported in *FBIS-TAC-97-255*. Israeli officials spoke of Russia as emerging as Israel's chief security threat. See A. Shumlin, "Israel Threatens to Freeze Cooperation with Moscow," in *Kommersant-Daily*, p. 3, reported in *Russian Press Daily*, August 27, 1997.

144. On discussion regarding Israeli overtures toward Iran, see S. Rodan, "Iran, Israel Reportedly Forging Contacts," in *The Jerusalem Post*, September 9, 1997, p. 1.

145. Turkey's involvement with Iran in the energy sector and other areas is paralleled by Azerbaijan engaging with Iran in various commercial projects as well. See, for example, *Compass*, September 9, 1997.

146. See report by AFP, November 18, 1997.

147. See, for example, report by TRT-TV (Ankara), September 10, 1997, reported in *BBC-SWB*, September 12, 1997.

148. On the geo-politics of this project and the Pakistani perspective, see I. Khan, "Gas Pipeline: The New Great Game," in *The News* (Islamabad), September 11, 1997, p. 6., reported in *FBIS-NES-97-256*.

149. See A. Nizami, "Oil and Gas Pipeline Deal: A Pipedream?," in *The Nation* (Islamabad), May 20, 1997, pp. 1, reported in *FBIS-NES-NES-97-098*.

150. See, for example, T. Emery, "Afghan Oil: U.S. Firm Tries to Woo Taliban Militia," reported by Inter Press Service, August 17, 1997.

151. For background on the pipeline and its rationale, see A. Nizami, "Oil and Gas Pipeline Deal: A Pipedream?," in *The Nation*, May 20, 1997, pp. 1, reported in *FBIS-NES-97-098*.

152. See, for example, "Great Game Endgame?," in *The Wall Street Journal*, May 23, 1997, p. A18. For the Iranian interpretation of America's design, see the report by Xinhua News Agency, July 28, 1997.

153. See F. Bokhari, "Taliban Success Rouses Fears in Central Asia," in *The Financial Times*, May 28, 1997, p. 3.

154. Quoted in S. Levine, "Unrest in Afghanistan is Disrupting Plans for Pipelines," in *The New York Times*, June 5, 1997, p. C4.

155. See M. Veysi, "Turkey's Gamble with General Dostam's Card," in *SALAM* (Tehran), January 7, 1997, p. 12, reported in *FBIS-NES-97-010*.

156. See report by F. Gazel in *Dunya* (Istanbul), May 28, 1997, p. 1, reported in *FBIS-WEU-97-150*.

157. See reports by AFP filed from Hong Kong, reported in *FBIS-NES-97-150*; C. Clover, "Taliban Advance is Reversed," in *The Financial Times*, June 12, 1997, p. 6.

158. See C. Clover, "Taliban Agree to $2bn Afghan Pipeline," in *The Financial Times*, October 28, 1997, p. 6.

159. On this agenda see V. Kuznechevskiy, "Central Asia: Istanbul Train En Route for Moscow," in *Rossiyskaya Gazeta* (Moscow), October 26, 1997, p. 11, reported in *FBIS-SOV-97-211*. For an Iranian view, see "Tashkent

Meeting and Turkey's Concern for Central Asia," in *Iran News* (Tehran), October 22, 1996, p. 2, reported in *FBIS-NES-96-212*.

160. See Talbott, "The Great Game is Over," p. 14.
161. Quoted in V. Kazyakov, "The World Today," on Voice of Russia World Service (Moscow), August 21, 1997, reported in *FBIS-SOV-97-236*.
162. See S. Talbott, "The Struggle for Russia's Future," in *The Wall Street Journal*, September 25, 1997, p. A22.
163. On the internal Russian debates regarding its southern periphery, see M. Mesbahi, "Russian Foreign Policy and Security in Central Asia and the Caucasus," in *Central Asian Survey*, vol. 12, no. 2, 1993, pp. 181-215; A. Gusher and A. Skvokhotov, "Russian National Security Strategy in the South," in *Svobodnaya Gruziya* (Tbilisi), May 24, 1997, p. 2, reported in *FBIS-SOV-97-202*.
164. Central to this process is, of course, a reduction in tensions between Israel and Iran. For signs of such a shift, see report by J. Lobe, Inter Press Service, October 1, 1997; S. Rodan, "Iran, Israel Reportedly Forging Contacts," in *The Jerusalem Post*, September 9, 1997, p. 1.
165. On the prospects for a "counter-axis" ranged against the U.S.-dominated order, see, for example, reports by IRNA (Tehran), September 13, 1997, reported in *FBIS-NES-97-256* and in *Salam* (Tehran), August 14, 1997, p. 12, reported in *FBIS-NEA-97-272*.
166. For a discussion of corruption, power, and anti-Islamist actions in Turkey, see J. Barham, "Turkey's 'State within a State' Survives Scandal," in *The Financial Times*, November 4, 1997, p. 4.

CHAPTER THIRTEEN

The Military Balance in the Caspian Region

Ali Massoud Ansari

Introduction

This chapter examines the potential for military conflict in the Caspian basin, as the independent republics of Azerbaijan, Kazakhstan, and Turkmenistan adjust to the new realities of economic and resource competition in the post-Soviet era.

Prior to the dissolution of the Soviet Union in 1991, any potential military rivalry in the Caspian region involved Russia, in either its tsarist or Soviet form, and Iran; by and large, in that unequal contest a mutually acceptable agreement was reached between the two governments in which the sea was left effectively de-militarized, though the Soviet Union maintained a small flotilla based in Baku. With the disintegration of the Soviet Union, the states of Azerbaijan, Kazakhstan, and Turkmenistan joined the equation, each with some remnant of the Soviet forces.

Economic imperatives often dictate or indeed compel political and military action. As the history of the Persian Gulf region has shown, the emergence of new wealth can breed further instability as rich new states seek to protect and enhance their newfound status and higher international prestige with the purchase of arms. The recent claim made by Turkmenistan on one of Azerbaijan's three offshore oil deposits only reinforces this view.[1] It may be argued that the voracious appetite of the international military-industrial complex is never far behind the anxiety of emerging new states.

This chapter contends that it may be premature to talk about the "militarization" of the Caspian in a traditional sense; there is evidently a healthy potential for this development in the future, however. The Russian assault on Chechnya and the pipeline/depot at Grozny are useful reminders of the use of military force to secure economic objectives.

In the short-term, the prospects of a face-off between the conventional forces of the littoral states is very unlikely. The likelihood of extra-regional military activity in this region is even less. In geo-political terms, the region is important but not vital to Western economic interests,[2] even though in addition to oil and gas reserves there are other resources available for exploitation, such as chromium in Kazakhstan, one of two major sources of this mineral essential to modern high-tech defense industries.[3]

The danger to the stability of the Caspian region will come from the potential for low-intensity warfare conducted by militias, clan, and separatist movements, mainly ethnic, but some sectarian, and such conflicts will emerge largely from intra-state instability as opposed to inter-state rivalries, as standing armies come to be challenged by the "rebirth of the war band."[4] That these conflicts may spill over or draw in state forces cannot be ruled out, but the conflict will remain essentially low-intensity. This situation, however, paradoxically may result in regular and prolonged, if not intermittent bouts of struggle between rival groups.

The Russian Factor

Russia remains the primary military power in the region, not only by virtue of the forces that have remained on its own soil, but also through the deployment of former Soviet regiments and divisions throughout the other three former Soviet Caspian republics and indeed in Central Asia and Transcaucasia as a whole, either because these units have as yet to be recalled home, or because they are deployed on the basis of new bi-lateral defense arrangements with Russia in border areas. There are a reported 1,000 Russian troops guarding nuclear missiles and the Baikonur space facility in Kazakhstan, while another 15,000 Russian troops remain in Turkmenistan along the border with Iran and Afghanistan.

Russia has been active also in ensuring its presence in the "near abroad," in part to prevent any military vacuum being filled by outside powers. For example, much to the satisfaction of the Russian military, the Russian foreign minister in early 1994 concluded that Russia must continue its military presence in the regions where the sphere of its military interests has existed for ages. At the same time, Russia has assiduously cultivated mutually beneficial relations with Iran.

While Russia may be in a position to contribute to peace and stability in the region, there are however limits to its effectiveness. The Russian experience in the Chechen conflict and earlier in Afghanistan cast doubt on its capability to remain in a prolonged guerilla conflict. Moreover, the newly independent republics are reluctant to have exclusively Russian forces on their soil in a "peacekeeping" capacity. Turkmenistan has made it clear that it does not favor any form of collective security arrangement that involves Russia, and it refused to sign the Collective Security Treaty agreed to in Tashkent on May 15, 1992.[5] All parties, however, seem to agree on keeping the non-regional forces from the area.

Indigenization of Forces

The Russian factor also influences the debate regarding the indigenization of the armed forces in Azerbaijan, Kazakhstan, and Turkmenistan, their military doctrines and general correlation of forces in the region. Since the dissolution of the Soviet Union, the out-migration of the Russian professional classes from Azeri, Kazakh, and Turkmen societies has increasingly deprived the economies of these societies of a cadre of skilled, specialist labor, erstwhile mostly subsidized by the Soviet Union.

The shortages in this area are acutely felt in the armed forces, where many local units once were led by Russian officers now eager to leave. The adverse consequences of the Russian military out-migration with respect to Kazakhstan in particular illustrate the point. Only 3 percent of officers in the former Red Army units in Kazakhstan were of native origin. "With 44,000 troops and fewer and fewer officers, the [Kazakh] deputy minister of defence announced at a Parliamentary committee that the country was at risk of losing control of the troops. [The] shortfall of officers had reached over 40%, but though a military cooperation agreement with Russia had been signed a month earlier, curiously, general Pavel Grachev ordered that those who returned to Russia before the end of the year could avail of increased discharge allowances, thereby increasing the exodus."[6]

By contrast, Turkmenistan's problems in the area of indigenization of its forces are less severe in that the ethnic division between the Russians and natives is not so stark, though there are still difficulties implicit in a Russian officer corps commanding a "national" Turkmen army whose ethnic composition is 88.6 percent Turkmen. This overwhelmingly Turkmen composition however has made the issue of dual nationality less problematic for the Turkmen authorities and the official sanction given to dual nationality has meant that many Russian officers have remained. Russia, meanwhile, has agreed to continue training Turkmen troops. Ironically, therefore, despite Turkmenistan's unwillingness to participate in the various defense pacts, its relations with the Russian military on a bi-lateral level are probably more engaged than either Azerbaijan or Kazakhstan.

The loss of Russian officers and non-commissioned officers aside, the quality of the troops in the former Muslim Soviet republics was not as high as those in the Slavic republics, to begin with. This was due in part to discrimination and also difficulty in co-opting the Muslims with a distinct cultural heritage, into what was effectively a "Russian" army. As a rule, recruits from the Muslim republics received poorer training and education and saw little combat experience. Many also found Russian a difficult language to learn.

Kazakhstan

Kazakhstan created its own national armed forces in 1992, appropriating from the Soviet Central Asian Military District the 40th Army, which consisted of one tank division and three motorized rifle divisions. A 2,000-man-strong national guard had already been established earlier in the year. The first minister of defense, Sadagat Nurgambetov, was appointed on May 5, 1992, and, as if to emphasize the main area of concern, the National Border Troops Command was also established. The avowed intention of this new ministry was to reform the armed forces inherited from the Soviet Union and, within the existing financial constraints, to produce a highly trained, highly mobile fighting force able to respond rapidly to local threats.

The reform of the armed forces is clearly a long-term process, and the defense minister, Lieutenant General Mukhtar Altynbayev, noted at his first press conference, in November 1996, that there had been a perceptible, and clearly necessary, improvement in combat readiness and discipline. Noting the increased anxiety on the part of the Russian defense minister, Igor Rodionov, at the apparent lack of cooperation between Kazakhstan and Russia on defense matters, Altynbayev stressed that problems facing the Commonwealth of Independent States in the military field required cooperation and, particularly, the assistance from Russia.[7]

Organizational streamlining and cost savings priorities aside, the Kazakh objective in developing a rapid deployment force is motivated primarily by Kazakhstan's perception of potential security threats along the country's long borders, particularly with China; the military balance of power favors China. Along China's western border regions, China has come to face its own ethnic problems, however, particularly in the province of Xinjiang. Indeed, apart from the Turkmen, all other ethnic groups present in Central Asia are also represented in Xinjiang; the Kazakh are the largest group at 1.1 million. To add to the region's ethnic mosaic, there are some 260,000 Uyghur from the Chinese autonomous province of Uyghur living in Kazakhstan.

China seems to favor the maintenance of the territorial status quo and the avoidance of provoking ethnic confrontations. Despite the occasional talk of a "strategic vacuum" in the region, China has for all practical purposes been content with allowing Russia to continue its stabilizing role, while China pursues its own policy of military cooperation with Kazakhstan. Following the signing of a border agreement among China, Kazakhstan, Kyrgyzstan, Russia, and Tajikistan, in Shanghai, in April

1996,[8] the Chinese chief of the general staff, Fu Quanyou, flew to Almaty for talks with the Kazakh defense minister, Alibek Kasymov, aimed at "build[ing] up confidence in the military field in border areas." Fu alluded to the "traditional friendship" between the armies of the two countries, calling the border agreement "an important step," which, coupled with the visit, would "deepen the mutual understandings" between the two countries.[9]

In October 1996, the Kazakh defense minister visited China. The Chinese defense minister, Chi Haotian, emphasized China's desire to pursue military cooperation with Kazakhstan, and called for a joining of efforts "to improve friendly relations between the countries and the two armed forces."[10]

Despite its concern with China, the official Kazakh military doctrine, as enunciated by President Nursultan Nazarbayev, has prudently avoided naming potential enemies and has concentrated instead on the nature of possible threats, which includes: (1) desire on the part of state(s) to dominate in the region; (2) desire on the part of state(s) to settle disputes by military means; (3) possession on the part of some state(s) of powerful armed forces, some of which are deployed near the Kazakh borders; (4) internal instability; and (5) the threat posed by the potential military expansion of some states in the region. In its current state of military preparedness, Kazakhstan could do little to counter Russia or China and, therefore, for the time being, diplomacy remains the more effective weapon in the Kazakh arsenal.

Turkmenistan

While Kazakhstan's geography and ethnic composition make it, relatively speaking, more reliant on Russia, Turkmenistan, on the other hand, enjoys the luxury of being able to pursue an increasingly independent policy. It is unwilling to participate in any bi-lateral defense pact with Russia, a position that it can afford to take, in part, because it has managed to retain most of its ethnic Russian professionals.

In October 1996, the Turkmen parliament reaffirmed the country's policy of neutrality and ratified the changes to the republic's military doctrine in order to bring it into line with the country's neutral posture. Accordingly, Turkmenistan has announced that: (1) it does not regard any state as its enemy; (2) it does not wish to participate in any military bloc, alliance, or international coalition that may impose rigid obligations on the country; (3) it does not wish to participate in any collective security arrangement; (4) it does not allow foreign military bases on its territory, nor allow deployment of foreign military forces; and (5) it has pledged not to have, produce, or proliferate nuclear, chemical, or bacteriological weapons of mass destruction of any kind.[11]

Despite its resolve to adopt a policy of positive neutrality and its increasing detachment from the Commonwealth of Independent States, there remain some 15,000 Russian troops in Turkmenistan; that Turkmenistan pays for their upkeep allows it to argue that these troops are in fact part of the Turkmen armed forces. Thousands of Turkmen officers are being trained in Turkey, who eventually are to replace the Russian officer corps. Russia, on its part, has agreed to train Turkmen soldiers and pilots in return for Turkmenistan's consent to the "stationing" of Russian troops in Turkmenistan.

The Russo-Turkmen military agreements, however, have been implemented erratically. Russia has complained that in practice it still pays for the 15,000 Russians in the Turkmen armed forces. The 1992 agreement that created a joint Russo-Turkmen military command for a 5-year period was annulled unilaterally by Russia in 1995 and the Russian officers serving thereunder are now in effect on contracts that, on expiry, will result in these Russian officers returning to Russia.

The foregoing notwithstanding, it seems Russia will continue into the foreseeable future to control Turkmen air defense and border installations, for which purposes Russia is likely to establish additional bases, even though the "formal" Turkmen policy may scoff at such presence. Russian officers are expected to continue with the Turkmen military in joint patrols of the Turkmen border areas.

Azerbaijan

Azerbaijan inherited a substantial military organization, infrastructure, and materiel from the Soviet Union. Its armed forces have been estimated at some 56,000-men strong, with reserves of 560,000. It includes 285 main battle tanks, 456 armored combat vehicles, as well as 42 MiG-25s. Despite this respectable strength on paper, Azerbaijan, too, has suffered from the loss of Russian military personnel. Much of the equipment in Azeri possession remains unserviceable and essentially unusable. This is especially true of the air force. The military effectiveness of Azeri troops is also questionable given their performance in the Nagorno-Karabagh conflict.

Among the former Soviet republics on the Caspian, Azerbaijan may have the best chance of modernizing its armed forces; this will depend largely on it being able to export its oil and gas wealth securely and receive sophisticated equipment and training from the likes of the United States and Turkey. Like Kazakhstan, Azerbaijan has opted for military reforms aimed at creating a much smaller, but better equipped and trained 30,000-man armed forces. In the meantime, in order to satisfy its need for security, Azerbaijan has turned to diplomacy.

The southwestern part of the Caspian is by far the most potentially explosive sub-region in the basin. This is owed to the presence of a multitude of ethnic and religious divisions and conflicts, oil, and competition for influence among Iran, Turkey, the United States, and Russia. Initially, upon independence, Azerbaijan turned for assistance to Turkey and played heavily on its ethnic ties with that country. Turkey was more than happy to oblige, eager to cultivate its one natural ally in Transcaucasia. In June 1996, President Aliyev of Azerbaijan received the chief of the Turkish general staff, General Ismail Karadayi, stating that "Azerbaijan favours friendly and versatile relations with Turkey and is prepared to use Turkish experience to strengthen its own defence capability."[12]

The Turko-Azeri military cooperation is tempered by the Turkish government's conscious decision not to exacerbate the Azerbaijan-Armenia conflict. That decision is owed in part to the geographical limitations to Turkey's reach into Transcaucasia and, in part, to being mindful of reactions by Iran and Russia toward such involvement.

The prospects of Russo-Azeri military cooperation appear slim. In part owed to the presence of Western oil companies in Azerbaijan, the United States in particular is interested in seeing as little of Russia in Azerbaijan as possible. To that end, the United States reversed its initial position whereby Russia would act as the region's guarantor of stability. Presently, the United States favors stability through deployment of multi-lateral peacekeeping forces, if necessary.[13] However, in most cases, such forces would be led by Russia. Yet, interestingly, the peacekeeping force proposed for the Nagorno-Karabagh region consists entirely of forces from the Central Asian republics.

In November 1996, the Kyrgyz foreign minister, Roza Otunbayeva, announced that a Central Asian battalion composed of Kyrgyz, Uzbek, and Kazakh troops stood ready to be deployed in peacekeeping operations in the Nagorno-Karabagh region if the parties requested it.[14] Earlier in the year, the battalion had conducted military exercises in the United States as part of the Partnership for Peace Program developed under NATO, with another exercise scheduled for summer of 1997, in which forces from the United States, Ukraine, Russia, Turkey, Denmark, and each of the Baltic states also would participate.[15] In September 1997, the battalion's manoeuvres took place in Uzbekistan and Kazakhstan.

The immediacy of the potential for oil wealth has made conflict a more acute possibility. A wealthy Azerbaijan may become preoccupied in time with its own security but, for the present, its military options will be limited and the existence of factional fighting and "terrorism" would continue to threaten the economic stability of the southwest Caspian.

In January 1996, an organization called Defenders of the Caucasus called for the withdrawal of Russian troops from the region and asked that there be a general suspension of aid to any republic pursuing ethnic claims. "The ultimatum of the 'Defenders of the Caucasus,' in case their demands are not met, is to start terrorist acts on the oil pipelines, which pump fuel from the Caucasus to other countries. The explosions will be carried out on the most important sections of the pipelines including at the junction pump stations."[16]

The Iranian Factor

Iran is the other regional power in the Caspian region, but like China, Russia, and Turkey, Iran has strategic interests elsewhere. Like these powers, Iran's main concern, like China's, is danger of hostilities and instability spilling over from the ethnic and factional conflicts in the former Soviet republics. It favors, therefore, the maintaining of the status quo, showing no inclination—indeed, little capability—to become involved militarily in Transcaucasian or Central Asian conflicts. Having shed its earlier pretensions to ideological militancy in these areas, as of late Iran has focused its resources on securing participation in the economic activities of the region, especially with respect to attracting attention to the country as a viable outbound conduit for the region's oil and gas.

The Iranian diplomacy in the region speaks softly; with respect to conflicts in Tajikistan and Chechnia, it is practically inaudible, even though the Russian direct or indirect intervention there has cost the lives of Moslems. Another example of Iranian diplomacy running counter to intuition involved the country's initial position to support Armenia, a Christian country, in its conflict with Azerbaijan, a Moslem country and Shi'ite, at that. This alignment signaled to Azerbaijan Iran's displeasure at its open courtship of Turkey on the basis of their Turkic ethnic affinity. Yet, when it came to gaining influence in Central Asia, Iran expediently replaced its policy of "influence through religion" by a policy anchored in Central Asia's own Persian lingual and cultural traditions and affinities. Iran's diplomatic and economic offensives in Central Asia have resulted in Turkey being sidelined and Russia being confined to its role as the power shouldering the region's military responsibilities.

The gradual decline in Turkey's influence has had positive implications for Iran-Azerbaijan relations. Iran is now Azerbaijan's largest trading partner and the autonomous republic of Nakhchevan would be hard-pressed without Iranian economic assistance. On the other side of the Caspian, ties with Turkmenistan are being strengthened. Iranian engineers are at work in Turkmenistan, constructing a link between the Iranian and Turkmen pipeline networks. Attempts by the United States to isolate Iran in its own neighborhood have borne little fruit; the economic and diplomatic ties between Iran and its northern neighbors offer the parties a greater advantage than any gain from going along with an intemperate and fickle United States policy based on a vendetta against Iran.

The comparative success of Iran's policy is arguably in part due to the extreme "low intensity" of its military projection. Had it been perceived as a strong conventional military power, like China, Iran's ability to gain influence within the new republics, sensitive to new threats to their independence, would have been even more limited. Suspicious of Iranian intentions, these new republics would have sought refuge or alliances elsewhere. Iran has downplayed its military strength but has emphasized its cultural and political strengths, including its ability, if it so desired, to be a nuisance. That realization, albeit late, did dawn on Azerbaijan soon after Azerbaijan, under pressures from the United States, ignominiously expelled Iran from an Azeri international oil consortium. In order to placate Iran, Azerbaijan hastily arranged for a second oil consortium with European companies and invited Iran to participate in it.

The Caspian Navies

Iran's main naval strength is located in the Persian Gulf, even though, according to a 1994 Russian estimate, it has some 60 combat patrol boats in the Caspian.[17]

In 1992, Azerbaijan inherited a substantial part of the Soviet Union's Caspian Fleet, once centered at Baku. Included in the legacy were 15 warships and patrol boats, more than 10 auxiliaries, the lion's share of stocks, weapons, hardware, and other gear; many Russian officers and ensigns stayed behind to serve with the Azerbaijani Navy.[18]

Redeployed to Astrakhan, the Caspian Fleet spent the next two years recovering from the drain of some 40 percent of its ensigns, who had stayed behind either in Azerbaijan or had sought employment elsewhere. In September 1994, the commander of the Fleet, Vice Admiral Boris Zinin, reported that most of the negative effects of the withdrawal/redeployment had been overcome and that he was satisfied with the sailors' combat readiness. He remarked that the attitude of the officers and ensigns was good, not that the Caspian sailors would complain.[19] The Fleet consists of 2 frigates, 30 patrol craft and minesweepers, and approximately 20 landing craft. There are an estimated 3,000 sailors in its service. While, technically, the Caspian Fleet is divided three ways between Russia, Kazakhstan, and Turkmenistan, however, for all intents and purposes, it is run by Russia. As on land, Russia is the primary naval power in the Caspian.

Initially, the breakup of the Soviet Caspian Fleet had given rise to the fear that safety of shipping in the Caspian would be endangered. According to Zinin, two factors play into this state of potential chaos. First, the lack of maritime borders makes it difficult for the littoral countries to enforce their authority, particularly with respect to poachers and smugglers.[20] Second, the uncertainty as to borders, and limits of environmental and economic zones, has spurred the five coastal states to develop their own navies, thereby increasing the risk of conflict and confrontation.[21]

While Turkmenistan operates two patrol ships jointly with Russia, with mixed crews, Kazakhstan on the other hand is seeking to expand its own independent navy, which it created in August 1996 with assistance from the United States. The United States provided seven cutters, while two more are to be constructed in Kazakhstan. In October 1996, Germany provided Kazakhstan with four coastal defense cutters.[22]

Conclusion

The military threat in the Caspian region is diffuse and it is all the more unpredictable because of ethnic complexities and economic rivalries among the countries of the region. The conflicts in Chechnya and Nagorno-Karabagh are reminders of that, more so than of some deafening ideological or irredentist battle cry.

It is perhaps coincidental that the area of the Caspian region in which oil is being exploited most rapidly has been also the most violent. To be sure, the conflict over Nagorno-Karabagh predates the development of oil interests; however, the conflict in Chechnya, whatever its ethnic determinants, has an obvious economic dimension, in particular, oil. Traditionally, Grozny had been a major point for transportation and distribution of oil and oil products and had served as the fuel center for the entire northern Caucasia.[23] Therefore, the need to crush quickly the bid for Chechen independence was motivated by Russia's need to restore and "retain control over the region's extensive energy infrastructure."[24]

Other conflicts in Transcaucasia will develop and for the most part they may be motivated by or adversely affect the development, production, and transportation of oil and gas. For example, Turkey's support for Azerbaijan in the Nagorno-Karabagh conflict has prevented the construction of a pipeline route from Baku in Azerbaijan to Turkey through Armenia.

The militarization of the Caspian region in the conventional sense is not likely to occur in the short to medium term. Even though oil money may result in the building up of conventional forces, their utility or effectiveness in a region where inter-state warfare is not perceived as the main threat would remain doubtful. In a region where political instability is threatened by local militias, warlords, and guerilla/terrorist groups, a uni-dimensional military response is insufficient. Regional powers are undoubtedly justified in developing rapid response units, trained particularly in low-intensity warfare.

In Caspian's delicate regional balance of power, cooperation and consensus will prove more conducive to internal and external stability than if sides were to impose their hegemony by threat or use of force.

Notes

1. BBC Summary of World Broadcasts (SWB), SU/2830/F1, January 30, 1997 (Interfax News Agency, Moscow).
2. Sergei V. Solodovnik, "Central Asia: A New Geopolitical Profile," in Uwe Halbach, *The Development of the Soviet Successor States in Central Asia* (Koln, Germany: State Institute for East-West and International Studies, 1995), pp. 11-21. "Although this region possesses potential mineral resources, any termination of their supply to the world market could never produce the same devastating effect as the curtailment of supplies of oil from the [Persian Gulf] or copper from South Africa. Its production potential is not of sufficient significance globally to evoke fierce competition from other states or to instigate intervention to maintain stability and security."
3. See T. Kartha, "Central Asia and the Soviet Military Legacy," in *Asian Strategic Revue* (New Delhi: Institute for Defence & Analyses [1994-95]), p. 128.
4. M. Orr, *The Regional Military Balance: Conventional and Unconventional Military Forces around the Caspian* (Camberly, England: Royal Military Academy, Sandhurst, 1995).
5. G. Bondarevsky and P. Ferdinand, "Russian Foreign Policy & Central Asia," in P. Ferdinand, ed., *Central Asia & Its Neighbours* (Chatham House Papers) (London: Royal Institute for International Affairs, 1994), p. 43. Turkmenistan also refused to sign the "Declaration on the Inviolability of Borders," in August 1993, which the remaining Central Asian states signed on to along with Russia.
6. Kartha, p. 145.
7. BBC SWB, SU/2733 S1/4, November 19, 1996 (Kazakh television ch.1, November 15, 1996 broadcast).
8. BBC SWB, SU/2585 S1/1, April 13, 1996 (Xinhau News Agency, Beijing, reported on April 11,1996).
9. BBC SWB, SU/2610 G/3, May 13, 1996 (Xinhau News Agency, Beijing, reported on May 11, 1996).
10. BBC SWB, SU/2752 S1/7, October 25, 1996 (Xinhau News Agency, Beijing, reported on October 23, 1996).
11. BBC SWB, SU/2733 G3, October 3, 1996 (ITAR-TASS News Agency, Moscow, reported on October 1, 1996).
12. BBC SWB, SU/2648 F/3, June 26, 1996 (ITAR-TASS News Agency, Moscow, reporting on TRT-TV Ankara, Turkey's report of June 24, 1996).
13. For more details, see R. Hollis, "Western Security Strategy in South West Asia," in A. Ehteshami, gen. ed., *From Gulf to Central Asia—New Players in the New Great Game* (Exeter, England: University of Exeter Press, 1994), pp. 188-206.

14. BBC SWB, SU/2770 G/1, November 15, 1996 (Interfax News Agency, Moscow, reported on November 12, 1996).

15. BBC SWB, SU/2765 S1/2, November 9, 1996 (Interfax News Agency, Moscow, reported on November 7, 1996).

16. In a statement, a member of Chechnya's Parliament in January 1996 remarked that "without Chechnya's consent, the oil pipeline will not operate," reported by Kommersant-daily, January 19, 1996, p. 4, col. 7/8, in *CIS & Middle East* (Jerusalem: Hebrew University), vol. 21, no. 1-2 (1996), p. 7. See also FSU 15 Nations: Policy & Security, *Pravda*, January 25, 1996.

17. Foreign Broadcast Information Service, SOV (94-188), September 28, 1994, p.11 (interview with Russian vice admiral Boris Zinin).

18. FBIS SOV (94-188), September 28, 1994 (Krasnaya Zvezda).

19. Ibid.

20. Ibid.

21. Ibid.

22. BBC SWB, SU/2745 S1/4, October 17, 1996 (Kazakh Television, ch.1, broadcast of October 14, 1996).

23. Andrei Denisov, "Chechen Oil May Halt War," in *Moscow News* (No.49), December 15-21, 1995.

24. Michael P. Croissant, "Oil & Russian Imperialism in the Transcaucasus," in *Eurasian Studies*, vol. 3, no. 1 (Spring 1996), p.22. In the same journal, see also Ariel Cohen, "The New Great Game: Pipeline Politics in Eurasia," ibid., pp. 2-15.

Part V

Legal Perspectives

CHAPTER FOURTEEN

Evolution of Iranian Sovereignty in the Caspian Sea

Guive Mirfendereski

Introduction

This chapter examines the international legal history of the Caspian Sea and offers a broad conceptual framework for the orderly and progressive development of its legal regime in the aftermath of the 1991 dissolution of the Soviet Union. The future legal regime of the Caspian may well prove to be a microcosm of international law's best legal principles at work, ones that: (1) balance changes in fundamental circumstances with respect for the pre-existing rule of law as a matter of legal continuity; (2) respect the equality of states as to the use of the corpus of the Caspian and its migratory fauna; and (3) ensure the equitable apportionment of its area for the purposes of exploitation of its seabed and subsoil resources. This chapter will conclude that delimitation of the Caspian into areas of national sovereignty is compelled by legal necessity, so as to ensure an orderly exploitation of the Caspian resources, and to provide legal standing for the littoral states to deal with matters falling within their coastal police powers, such as trans-boundary pollution, sanitation and health, immigration, contraband, taxation, navigation safety, fishing, fisheries management, and environmental protection.

Competing Views of the Caspian

The world's largest saltwater lake,[1] the Caspian measures about 750 miles long from north to south, with an average width of 200 miles. With an area of 143,000 square miles, it lies about 93.5 feet below the ocean level. Its maximum depths of about 3,360 feet occur in the south.[2] It is bordered by Azerbaijan, the Russian Federation—itself consisting of the shores of the republics of Daghestan, Kalmykia, and Russia—Kazakhstan, Turk-menistan, and Iran.

Prior to the dissolution of the Soviet Union, the Caspian Sea was deemed an example of an "inland water," enclosed by the territories of Iran and the USSR and governed, in international law, "in accordance with whatever rules these powers impose[d], in accordance with tradi-tional practices and agreements," whereby Iran and the USSR could have made "an arrangement confined to themselves but analogous with that which appertains to the high seas."[3] Nowhere is this better illustrated than in the parties' mutual agreement in the 1940 Iran-USSR Treaty of Commerce and Navigation (CN)[4] to extend the application of the 1926 International Sanitary Convention,[5] already in force between them as to the open seas, to their vessels and ports on the Caspian Sea as well.[6]

This chapter catalogs a myriad of arrangements made by Iran and the RSFSR/USSR with respect to the Caspian, including the 1931 Persian-Soviet exchange of diplomatic notes declaring the Caspian a "Persian and Soviet sea,"[7] reaffirmed as "Soviet and Iranian sea" in 1935[8] and again in the context of the 1940 CN Treaty as a "Soviet and Iranian sea."[9]

By virtue of a number of agreements, by 1957, the Iran-USSR bound-ary had been established. On land, this boundary consisted of a line run-ning from Turkey for 434 miles to the shore of the Caspian and then continuing from the eastern shore of the Caspian for 616 miles to Afghanistan.[10] Because maps of the Caspian showed no Iran-USSR boundary line cutting across the Caspian, the assumption was made by the outside world that "[t]he limits of Iranian and Soviet [territorial] sov-ereignty in the Caspian Sea ha[d] never been officially determined."[11] Similarly, the most authoritative international law treatise of the time contained no reference to the legal regime of the Caspian as an interna-tional lake or inland sea.[12]

By virtue of the 1940 exchange of diplomatic notes declaring the Caspian an "Iranian and Soviet sea," the Iran-USSR international bound-ary in the Caspian became the circumference of the Caspian shoreline at low water-line.

A significant modification in the Iran-USSR circumferential boundary in the Caspian occurred in August 1962, when the Iranian minister for foreign affairs, Abbas Aram, and the Soviet ambassador at Tehran, Niko-lay Mikhaylovich Pegov, signed an agreement whereby Iran and the USSR agreed, among other things, that a line extending straight from Astara on the west coast of the Caspian to Hassan Kiyadeh in the gulf of Hassan Kuli in the east,[13] would mark the Iran-Soviet "boundary,"[14] (hereinafter referred to as the Aram-Pegov line).

The Aram-Pegov line extends some 500 miles and at its farthest point from the Iranian mainland is some 80 miles from the coast. Possibly not intended at the time as a baseline similar to one enclosing a bay or gulf, the Aram-Pegov line may be used by Iran as the point of departure in measuring any Iranian claim to further areas of exclusive jurisdiction in the Caspian.[15] By virtue of the Aram-Pegov line, the Iran-USSR circumferential boundary in the Caspian may be deemed as beginning at the terminus of the Iran-USSR land boundary on the west coast of the Caspian, extending northward along the USSR coastline and down to the terminus of the Iran-USSR land boundary in Hassan Kiyadeh on the eastern shore of the Caspian, and then westward along the Aram-Pegov line to Astara. In 1970, the USSR's ministry of oil and gas demarcated the Caspian into sectors, an exercise on which Azerbaijan, Turkmenistan, and Kazakhstan base their claims of sovereignty to the Caspian's offshore oil and gas deposits.[16]

The analysis here serves as an explanatory note to the two opposing approaches to the issue of national jurisdiction in the Caspian, especially with respect to the exploitation of its petroleum and natural gas reserves.[17] On the one hand, there is the view held by Iran and the Russian Federation that their rights in the Caspian are grounded in treaties, and also in prescription and custom that developed between them as a matter of international law in successive historical periods, first between Persia and tsarist Russia, later the Russian Socialist Federative Soviet Republic (RSFSR) and the Union of Soviet Socialist Republics (USSR). This contrasts with Azerbaijan's assertion of rights in the Caspian, which is effected by means of domestic constitutional legislation,[18] a practice reminiscent of the unilateral appropriation of a portion of the open seas by President Truman's Continental Shelf Proclamation,[19] or by Peru's 200-mile territorial sea declaration,[20] or by Iran's territorial sea law.[21]

On the tradition of the Iran-Soviet legal regime rests the present view of Iran and the Russian Federation that the Caspian is an international lake and therefore petroleum and natural gas deposits in the Caspian Sea's subsoil belong jointly to and should be exploited jointly by the littoral states.[22] By contrast, there is the view taken by the former USSR republics of Azerbaijan, Kazakhstan, and Turkmenistan that the Caspian is a "sea" and should be split into separate territories.[23] At this writing, it appears that Russia and Azerbaijan have agreed to divide up the Caspian into national sectors. This would bring the Russian view closer to that held by Azerbaijan, Kazakhstan, and Turkmenistan.[24]

A Russian Lake, 1723-1917

In the 1723 Treaty of Alliance, Persia ceded to Russia in perpetuity "the cities of Derbent and Baku with all their appurtenances and dependencies, the length of the Caspian Sea; as well as the provinces of Ghilan,

Mazanderan, and Asterabat."[25] Thereby, the Russian flag came to fly "over the southern shore of the Caspian Sea,"[26] making the Caspian a veritable Russian lake.

From 1723 onward, Iranian diplomacy would seek at every opportunity to acquire from Russia and later from the Soviet Union, bit by bit, some measure of sovereignty in the Caspian. By virtue of the 1729 Treaty of Rasht, Russia ceded back to Persia the provinces of Astarabad and Mazandaran, but on the condition that these provinces in no manner be given to another power, in which event they, with their dependencies, would revert in perpetuity to the dominion of Russia.[27] The treaty, however, left in Russian hands Gilan and "the length of the Caspian Sea," which had been ceded in 1723.

Under the 1732 Treaty of Peace, Russia ceded back to Persia Gilan and Astarabad[28] and in 1725 it evacuated Darband and Baku as well, surrendering "the last of the conquests of Peter the Great in this quarter."[29] However, in neither the 1729 nor the 1732 treaty is there any reference to Russia ceding any portion of the Caspian Sea back to Persia. If anything, the description of the lands being ceded by Russia to Persia in 1729 and 1732 were in reference to said lands being on, from, or up to the Caspian Sea, or being located to the left or the right of the Caspian Sea. Because at the time denying the Ottoman Empire access to the Caspian Sea was a pillar of Peter the Great's foreign policy,[30] it is therefore plausible that Russia would have continued to keep the Caspian out of any treaty with Persia and continue to assert control, if not sovereignty, over the Caspian Sea.

One example of Russian control of the Caspian is provided in the details of Russia establishing in 1840 a naval presence on Ashuradeh Island in the southeast corner of the Caspian. In 1836, Persia had requested naval assistance from Russia to subdue the Turkmen "pirates" raiding the coast of Mazandaran. The assistance was followed by Russia occupying Ashuradeh with the intention of establishing a permanent naval station there. When Persia protested the occupation, Russia argued that its presence there would keep the "pirates" in check.[31] Also telling of Russian control of the Caspian was the banning by the Russian government in 1746 of all British trade across the Caspian in response to British subjects assisting the Persian government in developing a naval surveying and operations capability in the Caspian.[32]

The Russo-Persian wars of 1804-13 resulted in the Treaty of Gulistan, which produced a boundary line extending from Odina Basara along the Aras River to the south of Baku on the Caspian Sea. Persia recognized as belonging to Russia the territories on the northern side of the boundary line, among which were included Karabagh, Shirwan, Darband, Baku, part of Talish, Daghestan, Georgia, Armenia, Abkhazia, "and all other lands situated between the boundary line and the Caucasus and confined between the Caucasus and the Caspian Sea."[33]

In regard to matters touching directly on the Caspian Sea, the Treaty of Gulistan provided for the following: "Russian merchant marine vessels shall have, like before, the right to navigate the length of and land on the shores of the Caspian; in case of shipwreck the Persians amicably shall extend help to them. The Persian commercial vessels also shall have, like before, the same right of cabotage along the length of the Caspian coast and to land on Russian shores; in the case of shipwreck, the Russians shall give to them all necessary assistance. As it was before the war as well as it was the case in the time of peace and at all times, only Russian warships have sailed on the Caspian and they now shall have the same exclusive right, accordingly, no power other than Russia shall deploy warships in the Caspian Sea."[34]

A provision similar to the above-cited passage was contained also in article 8 of the Treaty of Turkmanchai, which marked the end of the 1826-28 Russo-Persian wars.[35] In that treaty, Persia ceded to Russia the governorates of Erivan and Nakhchevan north of the Aras River. A boundary was established as extending from the source of the Karasou River eastward to the Aras River, moving eastward along the Astara River to its mouth in the Caspian, whereby parts of Talish and Lankoran lying to the north of the line also became Russian territory. Persia also recognized as Russian "all the lands and islands" situated between the boundary line on the one side, the Caucasus and the Caspian Sea, on the other.[36] On January 18, 1829, the parties' commissioners at Beiramlu initialed a document entitled "Description of the Frontier between Persia and Russia" and finalized the border laid out in the treaty.[37]

The Treaty of Turkmanchai was accompanied by a commercial treaty, which recognized the right of Persian subjects "to import goods to Russia by way of either the Caspian Sea or the land frontier which separates Russia from Persia."[38]

In the push east to reach the Aral Sea, in 1869, the Russians established themselves first at Krasnovodsk and then at Chekishler, near the mouth of the Atrak River on the eastern side of the Caspian Sea.[39] Persia protested this Russian intrusion south into its territories, to which in December 1869 the Russian minister in Tehran, A. F. Beger, replied that Russia recognized Persian dominion up to the Atrak.[40]

In May 1879, Russia and Persia entered into the Telegraph Convention, providing for a telegraph line between Chekishler and Astarabad and for the integration of the Russian communication line from eastern Caspian to Julfa in the west, using the exiting Persian lines from Astarabad to Julfa. The convention provided for Persia's consent to the construction for "that part of the line which may be in Persian territory between Chekishler and Asterabad."[41] The ambiguity in the preceding provision pointed all the more to the necessity of defining the Russo-Persian frontier in the eastern Caspian region.

The Russian proceedings at Chekishler alarmed Britain. Upon inquiries at St. Petersburg, in July 1879, Russia assured that the intended operations were aimed at putting an end to the "depredations" by the Teke Turkmen and that the troops were under strict orders to respect Persian territory. In reply to the inquiry as to whether Russia and Britain were agreed as to the exact limit of the Persian frontiers, Russia replied that much of the land in that area was terra incognita.[42] In September, the Persian government sent an agent to look into the complaints by the Yamoot tribe at Hassan Kuli about the actions of the Russian military occupying the nearby Kari Kara. The Russians arrested the agent and this set off a diplomatic protest by Persia, claiming interference with its territory and subjects, to which the Russian embassy at Tehran responded by claiming that Hassan Kuli lay to the north of the Atrak and was therefore Russian territory.[43]

In 1880 and 1881, Persia agreed to extend the 1879 Telegraph Convention for as long as the Russian government required the arrangement in order to ensure communication with its military stations east of the Caspian.[44] In December 1881, the parties entered into the Boundary Convention, which established the Persian-Russian frontier in the east of the Caspian as beginning at the Hassan Kuli Bay and following the course of the Atrak River eastward.[45] By virtue of a protocol, dated January 30, 1886, the Russo-Persian frontier in the Babadurmaz-Hassan Kuli sector was defined, and by a further protocol, dated March 6, 1886, the line was fixed as reaching the Caspian Sea at the Gudri crossing.[46]

In 1893, Russia and Persia entered into an Exchange of Territory Convention, which amended article 4 of the Treaty of Turkmanchai, whereby, in western Caspian, Russia presently ceded to Persia the district of Abbasabad, including the Hissar village, in exchange for which Persia ceded to Russia the district of Firouzeh in its Khorassan province.[47]

None of the aforementioned Russian-Persian frontier treaties established a boundary between the two countries in the Caspian Sea. The mouth of the Aras in the west and of the Atrak east of the Caspian, respectively, marked the terminus of the land frontier to the left and right of the Caspian Sea. Simply, it defies credulity to assume that in these successive boundary-making exercises both Russia and Persia ignored or remained ambivalent with respect to defining their boundary in the Caspian Sea. The only plausible explanation for not dealing with a boundary in the Caspian would be that the parties labored under the view that either: (1) by virtue of the 1723 Treaty of Alliance the Caspian belonged in its entirety to Russia; or (2) the sea belonged to both; or (3) with the exception of the immediate coastal waters, the rest of the sea was like an open sea. The last two possibilities notwithstanding, the Persian-Russian boundary in the Caspian would have been in all probability viewed by Russia to be located on the Persian coast at low water-line.[48] This would have been without prejudice to Iran's right of navigation on the Caspian, as granted to it in the Gulistan and Turkmanchai treaties. In 1901, Persia and Russia entered into two commercial conventions, which provided for tariff and other regulations as to trade between the two by way of land and by way of the Caspian Sea.[49] In 1907, Britain and Russia agreed to recognize each other's exclusive spheres of interest in southern and northern Persia, respectively. The Anglo-Russian Agreement made no reference to the Caspian as such other than to provide that the revenue generated by Persian fisheries in the Caspian would continue to be used to service Persia's debt to the Russian and British banks operating in Persia.[50]

A Lake Between Friends, 1917-1927

At the outset of World War I, Russia occupied northern Persia, the latter's neutrality notwithstanding. In December 1917, the nascent Bolshevik government annulled the 1907 Anglo-Russian agreement and in January 1918 the People's Commissar for Foreign Affairs informed Persia that all prior and subsequent conventions entered into by tsarist Russia, which limited or diminished Persian freedom and independence, were void.[51]

Until May 1918, Azerbaijan had not existed as an "independent" political entity. From the eleventh century until 1723, and again from 1735 to 1813, the territory now constituting Russia's Azerbaijan had been either under Persian suzerainty or actually part of the Persian Empire. In the periods between 1723 and 1735, and between 1813 and May 1918, the lands and peoples constituting Russia's Azerbaijan had been part of the Russian Empire. In May 1918, the Bolshevist government in Baku gave way to the Central Caspian Dictatorship and its newly formed National Council of Azerbaijan adopted a declaration proclaiming Azerbaijan as an independent republic, with Baku as its capital.[52]

Meanwhile, in June 1918, with the Russian state in the throes of civil war, the British took control of Rasht and the port of Enzeli on the Persian side of the Caspian.[53] Later in the same month, the first Soviet envoy to Persia, albeit unofficially, F. N. Bravin, styled "the Russian Diplomatic Agent of the Bolsheviks to the Court of Persia," stated in Tehran that he regarded the Caspian as "one of Persia's most ancient possessions," and therefore Persian ships could "sail freely on the Caspian Sea."[54] On June 26, 1919, the People's Commissar for Foreign Affairs wrote to the Persian government proposing an outline of an agreement between Persia and the Soviet Russian government, among which "[t]he Caspian Sea, after it has been cleared of the ships of the Imperialist free-booters, the English, will be declared free for navigation by the vessels flying the flag of a free Persia," and also, "[t]he frontiers of Soviet Russia along the Persian border will be fixed according to the free will of the inhabitants of those regions."[55]

In August 1918, the British force at Enzeli responded to Azerbaijan's call for assistance to stave off attacks by the Turkish forces. After a series of land and naval engagements against the Turks and later against the Bolshevist elements in the Russian Centro-Caspian Flotilla, in August 1919, the British forces turned over their fleet to the Russians and relinquished the control of affairs at Baku. In January 1920, the Azerbaijan republic was recognized de facto by the Allied Powers.[56]

The British abandonment of Baku sounded the death knell for the very short-lived "independent" Azerbaijan republic. In April 27-28, 1920, the Soviet Eleventh Army swept into Baku, upon the "invitation" of the local Soviets,[57] forcing the White Russian fleet to flee to Enzeli, where the British disarmed and placed it in protective custody. On May 18, 1920, in "hot pursuit" of the "rebels," the Soviet Fleet bombarded and occupied Enzeli-Ghazian and Rasht.[58] The British forces retreated inside Persia, "leaving the Persian Caspian littoral to the Reds."[59] On June 4, 1920, the indigenous Jangali movement, active in the Persian provinces of Gilan and Mazandaran, declared the "independent" Soviet Republic of Gilan, which was instantly endorsed by Trotsky,[60] and whose effective jurisdiction the Soviets extended also to Mazandaran.[61]

Neither the British challenge to Russian domination over the Caspian nor the fleeting appearance of an "independent" Azerbaijan republic seems to have changed the legal status of the Caspian Sea. By June 1920, the RSFSR had established its authority over imperial Russia's territorial possessions in and around the Caspian, including Baku, and extended by proxy the power to the Persian side of the Caspian.

The mere declaration of Azerbaijan independence in May 1918 could not have created by itself a "state," particularly one that was being declared from the body of an existing parent state, namely, Russia. In international law, the requirements for statehood consist of the entity having a defined territory, with a more or less permanent population, governed by an internationally independent government in effective control of its territory and people, and capable of discharging its international obligations. Even though some may have characterized the new Azerbaijan republic as a "state,"[62] no amount of wishing could create a de jure "state" until the aforementioned requirements were fulfilled and the parent state deemed as having no longer title or sovereignty thereon as a matter of law. That the Allied powers recognized de facto Azerbaijan makes this point all the same.

In August 1920, Soviet Russia decided that its long-term interests lay in evacuating its forces from Persia and stopping its support for the breakaway Gilan Republic.[63] But, insofar as the Caspian was concerned, on November 27, 1920, Lenin had formulated a "draft decision" in which he stressed the necessity of not setting "ourselves the task of conducting any campaign against Georgia, Armenia or Persia," but pursue instead the "main task" of "[guarding] Azerbaijan and [securing] possession of the whole Caspian Sea."[64]

On February 26, 1921, Persia and the RSFSR signed the Treaty of Friendship.[65] As it related to the Caspian, Soviet Russia ceded to Persia "the Ashuradeh Islands and other islands on the Astrabad Littoral" in southeast Caspian; the parties affirmed the Persian-Russian frontier resulting from the 1881 Boundary Convention, and further agreed that they "shall have equal rights of usage over the Atrak River and the other frontier rivers and waterways."[66] Per article 10, Soviet Russia ceded free of charge to Persia the "Port of Enzeli and the warehouses, with electrical, and other buildings."

Article 7 recognized Russia's right of preemptive but conditional self-defense to "the security of the Caspian Sea," in the context of which Persia agreed to sever any subject of a third power employed "in the Persian navy" who might use his position to pursue hostile intentions against Russia. In article 11, the parties agreed to abrogate expressly article 8 of the Turkmanchai Treaty (1828), which "deprived Persia from the right to have a bahriyyeh [navy] in the Caspian Sea."[67] While article 8 of Turkmanchai did not expressly forbid Persia from having naval vessels on the Caspian Sea, this conclusion is simply a matter of logical inference from the provision in said article that recognized for Russia the exclusive right to maintain naval vessels in the Caspian. Regardless, henceforth, under article 11 of the Friendship Treaty, Russia and Persia were to "enjoy equal rights of free navigation on that Sea, under their own flags." This and the mention of third power subjects being employed in the Persian navy on the Caspian amount to an implied recognition of Persia's right to have a navy in the Caspian.

Insofar as the Caspian fisheries were concerned, per article 14 of the Friendship Treaty, Persia recognized the "importance of the Caspian fisheries to Russia's food supply" and promised that upon the expiration of the existing contracts, Persia shall conclude a contract with the Food Service of the RSFSR "relating to fishing and with special conditions to be determined by said time." The "existing contracts" meant the Lianozov concession, which Persia had granted in 1876 to Stefan Lianozov covering the sturgeon fisheries along Persia's Caspian coast. It was renewed in 1906 to last until 1925, but it was annulled in 1918 in favor of another Russian subject. During World War I, the Lianozov vessels and installations at Enzeli, Hassan Kiyadeh, and Babol fell into Russian/Soviet hands.[68] In the spring of 1922, the Persian government granted to the Soviet authorities the exclusive right for one year to buy the entire Persian catch at Enzeli and Hassan Kiyadeh.[69]

The discussions between the Persian and Soviet governments regarding the Lianozov concession and assets in Persia came to naught and, in August 1923, one Martin Lianozov sold to the Soviet government his rights and properties pertaining to fisheries on the south coast of the Caspian.[70] Included in the sale were three trawlers, fishing rights and fishing installations, and real estate at and off Enzeli, the Koulan Gouda Island, and the stretch along the coast from Persia's Astara south to Enzeli.[71] On October 1, 1927, Persia and the USSR entered into the comprehensive Agreement Respecting the Fisheries on the Southern Caspian Coast, whereby Persia vested the concession for catch and processing of fish in the southern shores of the Caspian for the next 25 years to a commercial-industrial company organized, capitalized, and managed on equal basis by the two governments and, the provisions of the agreement notwithstanding, to operate under the laws of Persia.[72] Article 2 defined the area of the concession, which seems to have consisted of the waters lying within the line from where the Persian-Soviet frontier rivers reached the Caspian in the west and east. The area incorporated parts of the Lianozov concession area but excluded certain mouths of rivers flowing into the Caspian from the interior of Persia.[73] Under Protocol No. 4 of the agreement, Persia undertook not to establish any other enterprise that may export fish and fish products from the Persian side of the Caspian, its rivers, and waterways. In order to prevent the company and private fishers from getting in each other's way, the Persian government and the company were to establish, every three years, fishery zones to be exploited by company and private fishermen, and to regulate the conditions by which the private fishermen could fish in areas directly exploited by the company.[74] Moreover, the Persian government undertook to police the concession area against contraband fishing and to maintain the security of the fisheries installations and storehouses.[75]

Under an exchange of notes between the two governments, Persia and the USSR agreed to the ban of chemical and explosives methods of fishing. While the company was allowed to choose any other fishing method, the parties agreed that the company shall cultivate species of fish whose stock are found to be in decline. Furthermore, as a measure to protect the fish and caviar resources, fishing was banned in certain areas and limited to specific periods in other areas.[76]

By its terms, the 1927 Fisheries Agreement expired in 1952, at which point the joint Iranian-Soviet company ceased operations.[77]

Birth of New Soviet Littoral Entities, 1923-1936

On April 29, 1920, the Military-Revolutionary Committee of the Azerbaijan Soviet Independent Republic united Azerbaijan with Soviet Russia. On May 19, 1920, Azerbaijan was declared and constituted as a

Soviet Socialist Republic and this was followed in September 1920 by a series of military, economic, and other treaties between the RSFSR and Azerbaijan SSR. By March 1922, Azerbaijan SSR had become merged in effect with Soviet Russia by virtue of its membership in the Transcaucasian Soviet Federated Socialist Republic, which included also Georgia and Armenia. In December 1922, the TSFSR, Ukraine, Belarus, and the RSFSR ratified the declaration forming the Union of Soviet Socialist Republics (USSR). By April 1923, Azerbaijan was already an administrative unit of the RSFSR with respect to a variety of areas, including post and communications. In January 1924, the Transcaucasian republics ratified the constitution of the USSR.[78]

None of the constituent documents establishing Azerbaijan as a SSR entity contained a reference to the territorial extent of the Republic, only that, according to the 1924 USSR constitution, the territory of a Union Republic may not be altered without its consent.[79]

On April 25, 1923, Persia and the RSFSR entered into the Postal Convention,[80] in which the parties agreed by declaration to extend to the Federation of Transcaucasian Republics.[81] This would have been necessary in order to permit mail to be sent from Persia to Baku in Azerbaijan. Conspicuously lacking was the need to have the convention extend to any other political/territorial entity in the USSR. Similarly, a jointly executed declaration by Persia and Soviet Russia provided for the extension of the Persia-RSFSR Telegraph Convention,[82] to the Federation of the Transcaucasian Republics. This would have affected the application of the convention to the line between Astara, situated in Soviet Azerbaijan, and Enzeli in Persia. However, no similar declaration was deemed necessary for the extension of the convention's regime to the line that ran in eastern Caspian from Krasnovodsk (presently called Turkmenbachy) to Chekishler (both in present-day Turkmenistan), crossing into Persia and ending at Astarabad. For all intents and purposes, Persia and the RSFSR deemed the lands in the eastern Caspian region as part of the RSFSR territorial entity.

On February 20, 1925, the All-Turkmen Congress of Soviets adopted a declaration calling for the independent Soviet Socialist Republic on the territory long occupied by the Turkmen people, consisting of the areas of Poltorak, Mervi, Kerk, Leninsky, and Tashauz, and that "[t]he Territory of the Turkmen Soviet Socialist Republic cannot be changed without its consent."[83] In May 1925, Turkmen SSR was admitted to the USSR.[84] The Kazakh Soviet Socialist Republic was not formed and admitted to the Union until December 1936, at which time article 2 of the USSR Constitution was amended to provide also for the individual accession of Azerbaijan, Georgia, and Armenia to the Union,[85] thus abandoning the fiction of a federative form of the Transcaucasian SSR.

Like the case with Azerbaijan, nothing in the documentation creating the Kazakhstan and Turkmenistan SSR described the territorial extent of these entities, or expressly or impliedly affected the status of the Caspian as between Persia and Soviet Union.

An "Iranian and Soviet Sea," 1931 and Beyond

Persia and the USSR continued to regulate their relations in the Caspian. The Persia-USSR Postal Convention of August 1929 established regular service for postal packets between the parties overland, by air, and "[b]y sea, across the Caspian Sea"[86] and identified the ports of Baku in the USSR and Pahlavi (formerly Enzeli and since 1979 again Enzeli) in Iran as the postal exchange offices between the two countries.[87]

In October 1931, Persia and the USSR entered into the Convention of Establishment, Commerce and Navigation (ECN). It and the attendant exchange of diplomatic notes contained undertakings between the two parties as to the regime of the Caspian. It provided, among other things, for only vessels belonging to the USSR and Persia, employing their respective nationals, to be present anywhere in the Caspian Sea."[88]

Article 17 of the convention provided for the equal treatment of the vessels of one party in the port of the other party, receiving no less of a favorable treatment than accorded to national vessels. Further, the parties declared generally that cabotage be reserved solely for national vessels, but "each party accords to the vessels flying the flag of the other party the right to engage in cabotage of passengers and cargo in the Caspian Sea." And, to each party was reserved an area of up to 10 nautical miles off its coast for exclusive fishing.[89]

The aforementioned provisions of the convention pointed to the special and equal nature of the parties' rights in the Caspian Sea. Any doubt about that would have been dispelled impliedly by article 17(5), whereby "[a]s for all seas other than the Caspian Sea" each contracting party promised to accord the vessels of the other when in its ports and "territorial waters" the most favored nation treatment, not equal or national treatment. And, under article 19, each party agreed to recognize on equal basis the validity of certificates issued by the other as to the capacity and tonnage of the vessels under the other's flag.

Most significantly, the diplomatic notes exchanged between the Persian and Soviet governments show that the parties viewed the Caspian as a jointly owned inland water. On October 27, 1931, the Persian minister for foreign affairs wrote to the Soviet ambassador at Tehran, stating the two governments "considered [the Caspian Sea] a Persian and Soviet sea."[90] The Soviet ambassador acknowledged receipt of the letter, whose contents he repeated, and signed off by stating that he had taken note of it.[91] Upon the expiry of the 1931 ECN Convention, in August 1935, Persia and the USSR entered into a second ECN Convention.[92] Insofar as the relations of the parties in the Caspian were concerned, the provisions of articles 16, 17, 18, and 19 of the 1931 ECN Convention were repeated verbatim in articles 14, 15, 16, and 17, respectively, of the 1935 ECN Convention. The parties reiterated their earlier understanding that the Caspian was a "Persian and Soviet sea."[93]

Compared to the similar provision of the exchange of notes in connection with the 1931 ECN Convention, in 1935, Persia now was called Iran in international diplomatic parlance, and this time it was the Soviet party initiating the exchange on the joint nature of the Caspian. Significant in the 1935 exchange were the following differences from the 1931 version of the note: (1) the Caspian sea was stated to be of special or exceptional value to *both* governments, not just the Soviet government; and (2) *both* parties would ensure that third-party nationals in each party's Caspian vessel and port service would not work against the interest of the other party. These minor adjustments in the 1935 version of the note may be viewed as an expression of mutuality and equality of the parties' rights, interests, and obligations with respect to their relations in the Caspian.

The 1935 ECN was denounced by its own terms with effect from June 22, 1938. However, in March 1940, Iran and the USSR entered into the 1940 CN Treaty. Insofar as the parties' relations in the Caspian were concerned, the provisions of the 1935 ECN Convention were worked verbatim into the new treaty. Article 12 of the CN Treaty provided for equal and national treatment by one party of the vessels of the other, including the right of cabotage for the transport of passengers and cargo in the Caspian Sea. As before, each party also reserved for its own vessels the exclusive right of fishing in its coastal waters up to a limit of 10 nautical miles.[94] In article 13, as before, the parties agreed that "no vessels other than those belonging to the subjects of and the commercial or transport organisations of one of the high contracting parties, flying the flag of Iran or the U.S.S.R., may exist in the whole of the Caspian Sea."[95]

As had been the case with the 1931 and 1935 ECN Conventions, the exchange of notes accompanying the 1940 CN Treaty contained the reference to the Caspian as an "Iranian and Soviet sea."[96] Yet, neither the 1931 nor the 1935 ECN Convention or the 1940 CN Treaty intended expressly to create an Iranian-Soviet "condominium" or "joint sovereignty" over the Caspian Sea. The terms of the three agreements regarding the parties' equal rights and obligations in the Caspian and at the exclusion of any other party raises the likelihood that the parties may have viewed their relations in the Caspian as emanating from a prescriptive form of joint-ownership, a legal relationship analogous to the notion of "joint tenancy" or "common tenancy" in domestic law. The concept provides for each party to have and hold an equal but undivided interest in the whole of the same domain.[97]

The exchange of notes accompanying a series of successive agreements between the Iranian and Soviet governments, in which they described the

Caspian as a "Persian/Iranian and Soviet sea," simply confirmed the parties' common understanding as to the sea as belonging to them. The carving out of zones of exclusivity, such as with respect to fisheries, for example, simply regulated the parties' relations with respect to their mutual and equal rights to use parts of the jointly or commonly owned domain, akin to the exclusive or personal use of a room by one party in a house owned and inhabited jointly or commonly with another party. In this construction, neither Iran nor the USSR would have been required to obtain the consent of the other to use or exploit areas that were not expressly reserved for exclusive use of the other. Under this scenario, therefore, the vessels of one country could have fished by right anywhere in the Caspian other than in the area that the parties had agreed to regard as exclusive to the other party, while the other party could have exploited the Caspian offshore oil deposits in areas to which the other party had no agreed-upon exclusive right.

More Boundary-Making, 1954-1970

By virtue of the December 1954 Frontier Agreement, Iran and the Soviet Union settled their differences with respect to specified land frontiers on the two sides of the Caspian, declaring as settled all questions relating to the entire length of the state frontier between them and having no longer any "territorial claims against each other."[98] The seaward terminus of the boundary line on the western coast of the Caspian was affirmed as being where the watercourse of the River Astara-Chay reaches the Caspian.[99]

The Presidium of the Supreme Soviet of the USSR ratified the agreement on April 25, 1955, while Iran's "constitutional" monarch ratified it on behalf of Iran on March 20, 1955.[100] While the Soviet ratification may have been in conformity with the Soviet constitution,[101] the Iranian ratification apparently bypassed the constitutional ratification process altogether.[102]

In May 1957, Iran and the USSR signed the comprehensive Frontier Agreement, which confirmed the boundary provided for in the 1954 Frontier Agreement, as demarcated by the Mixed Soviet-Iranian Commission in April 1957.[103] However, the agreement would not come into force until December 1962. The delay could not have been divorced from the general U.S.-USSR cold war tension affecting Iran-USSR relations at the time, particularly with respect to the presence of U.S. missiles in Turkey and now possibly being based in Iran.[104] In that context, on September 15, 1962, the Soviet press reported that Iran had abjured the presence of foreign nuclear missiles on its soil.[105] The document that contained that undertaking provided also for the Aram-Pegov line.[106] While the fact of the Aram-Pegov agreement is not in doubt,[107] its precise contents and language continue to be matters of either eyewitness accounts[108] or hearsay.

At the time of inception of the Aram-Pegov line, the extent of Iran and Soviet territorial sovereignty in the Caspian was defined by treaty as consisting of: (1) joint or common tenancy (sovereignty, or ownership) of the "Iranian and Soviet sea" based on the 1940 exchange of diplomatic notes and the equal rights of the parties to cabotage, national treatment, navigation, and other matters according to the 1940 CN Treaty and the 1921 Friendship Treaty; and (2) a 10-mile-wide band of exclusive fisheries zone for each country under the 1940 CN Treaty. The Aram-Pegov line may have intended to affect the existing regime in one of two ways, either: (1) carve out for Iran an area of exclusive sovereignty or jurisdiction without affecting the rest of the joint/common tenancy regime; or (2) create an international maritime boundary between Iran and the USSR, in which case the joint/common tenancy was extinguished altogether in favor of partition.

Evidence to the contrary notwithstanding, it is more plausible that the Aram-Pegov line may have represented a Soviet concession, without a territorial quid pro quo on the part of Iran, whereby the Soviet Union relinquished its joint or common tenancy rights in the area to the south of the line in consideration for Iran's pledge in the Aram-Pegov agreement not to allow its territory to be used against the Soviet Union or as a base for U.S. missiles,[109] leaving the rest of the Caspian Sea subject to joint/common Iran-USSR ownership. For the Aram-Pegov line to have intended a territorial division of the Caspian, it would have meant, in effect, for the memorandum: (1) to invalidate the relevant provisions of the 1921 Friendship Treaty and of the 1940 CN Treaty; (2) to invalidate completely the exchange of diplomatic notes which had considered the sea since 1931 as "Iranian and Soviet sea"; and (3) to the extent that it applied to the Caspian at all, to amend each country's territorial sea legislation in Iran and the USSR as to the breadth of each country's territorial sea[110] without the benefit of additional domestic legislation. Moreover, the Aram-Pegov line, whether viewed as the seaward limit of Iran's internal waters or of its territorial sea, would have created an area of exclusive Iranian sovereignty, within which Iran would have had exclusive fishing rights[111] and this too would have supplanted the 10-mile exclusive fisheries zone off Iran provided for in the 1940 CN Treaty. It is all too unlikely that these substantial alterations would emanate from a document of such relative obscurity and secrecy as the Aram-Pegov memorandum.

An official U.S. manual published as late as 1970 still claimed that the "limits of Iranian and Soviet sovereignty in the Caspian Sea have never been *officially* determined."[112] In view of the historical record in this chapter, however, it is safe to conclude that as late as 1940, the limits of Iran-

ian and Soviet sovereignty in the Caspian had been determined both officially and publicly. What had not been made public by 1970 was the
"fact" of the Aram-Pegov line, which was created ostensibly by an official
international undertaking, albeit in the quiet and regardless of whether it
intended to be an international boundary or simply a baseline enclosing
Iran's internal waters or one capping a historical bay stretching from
Astara to Hassan Kiyadeh.

The secrecy surrounding the Aram-Pegov agreement raises doubts as
to whether the document ever received the required legislative consideration and constitutional ratification in either Iran and/or the USSR. The
public disclosure of the Aram-Pegov agreement would have been more of
an issue for the Iranian government for the fear that it be perceived by the
various political factions as having sold off rights to the vast tracts of the
"Iranian and Soviet sea" and its oil, which the USSR was pumping out of
oilfields off Baku, in return for a measly portion of the Caspian, itself a
catch basin for Soviet pollution. This may explain why the memorandum
is not disclosed internationally through registration with the United
Nations.[113]

Regardless, the Iranian and Soviet national law deficiency in terms of
the ratification of the Aram-Pegov may not be grounds for invalidation of
the agreement,[114] particularly if both states acquiesced in the agreement
and conducted their affair for an appreciable period of time, as if the
agreement was valid and binding, thereby creating as well a customary or
prescriptive rule of law.

The Aram-Pegov line apparently did codify the basis of a custom that
had been all too evident, at least, in the Iranian-Soviet naval proceedings
in the Caspian since 1937. After many years of preventing Iran from
sending its naval assets to the Caspian by way of the Volga waterway,[115]
in 1937, the Soviet Union relented and allowed Iran's 600-ton, Dutch-
made imperial yacht *Shahsavar* to make its way down the Volga to
Pahlavi.[116] In connection with this event, the Soviet Union is said to have
taken the position that the waters south of Astara on the western shore of
the Caspian constituted Iranian waters.[117]

Among Iran's naval acquisitions after World War II had been a
minesweeper bought from the United States government in 1959 and
christened the *Shahrokh*. Commissioned to serve in the Caspian, in 1969,
the *Shahrokh* voyaged down the Volga, touching at the Soviet port of
Baku before continuing on to Pahlavi.[118] Throughout its passage from the
Volga down to a point opposite Astara, the *Shahrokh* was piloted under
the command of Soviet Navy personnel. At the point off Astara, the
Soviet personnel left the ship and placed its command with the Iranian
Navy officers, along with the refrain "from hereon, your waters, your
pilots."[119]

Beginning with the *Shahsavar* and later also with the *Shahrokh,* when either of them required repairs in either Soviet shipyards or abroad in Europe, they would navigate the Caspian to a point opposite Astara, where they would then be boarded as a matter of routine by Soviet pilots and guided by them to the point of exit from Soviet waters. A reverse protocol was observed on the voyages back to Iran.[120]

In the latter part of 1969, the Iranian Navy detained a Soviet ship that allegedly had strayed into "Iranian waters." The undersecretary for the Middle East Section at the Soviet ministry of foreign affairs, Grigoriy Mikhaylovich Vinogradov, protested the seizure to the Iranian ambassador at Moscow and requested the vessel's release. Tehran ordered the release of the vessel, particularly in view of Soviet assurances that the vessel's encroachment was not related to the agitation against the Iranian government at Syahkal.[121]

At the invitation of the Soviet commander of the Caspian Fleet, Admiral Kudelkin, in the summer 1972, the commander of Iran's Northern Navy, Seifallah Anoushiravani, paid a visit to the Soviet naval base and installation at Baku on board the *Shahrokh*. According to his recollection, from a point off Astara, a pilot from a Russian warship came on board to guide *Shahrokh* to Baku, where Admiral Kudelkin accompanied him to his Operations Room, where hung a large map of the entire Caspian Sea depicting a line extending from the gulf of Hassan Kuli to Astara, as to which Admiral Kudelkin stated that the Soviet government viewed this line as the maritime boundary in the Caspian Sea, which no non-Soviet ship or airplane may cross.[122]

The sighting of the "Astara-Hassan Kuli line" on the Soviet admiralty chart in Baku in 1972 by the Iranian naval officers must have been one of the earliest encounters of the Iranian navy with the documented manifestation of what the Iranian-Soviet secret diplomacy had produced in 1962.[123] In any event, the line would have been consistent with "Iran, on occasion, [having] expressed a view that all waters south of a line between the western and eastern junctions of the borders with the sea are Iranian."[124]

Evidence to the contrary notwithstanding, one may conclude that the Aram-Pegov line embodied and codified the custom/practice between Iran and the USSR that considered the waters to the south of the line part of Iran's "internal waters," subject to Iran's exclusive territorial sovereignty. Its characterization as the outward limit of Iran's internal waters is legally compatible with the notion that the line delimited an area of exclusive sovereignty for Iran while not necessarily meaning that the area seaward of the line would belong exclusively to the USSR. However, in the waters lying beyond the Aram-Pegov line, Russia exercised, as a matter of security, the prescriptive right of accompanying Iranian vessels bound for Soviet ports, particularly if said vessels navigated in waters close and along the Soviet coast.

244 • Guive Mirfendereski

Post-Soviet Legal Regime

At the time of the dissolution of the Soviet Union in 1991, the extent of Iranian and Soviet rights in the Caspian Sea were defined, among other agreements, by the 1921 Friendship Treaty, the 1940 CN Treaty, and the Aram-Pegov line. As of 1943, the 1940 CN Treaty, by its own terms, became "ipso facto ... prolonged for an indefinite period."[125] Significantly, the 1940 exchange of notes considering the sea an "Iranian and Soviet sea" in the context of the 1940 CN Treaty therefore may be deemed as having become an indefinite arrangement, continuing to have the force of law. If viewed as a separate arrangement, then the exchange of notes may be deemed as standing on its own, having also the force of law.

In any event, with respect to the Caspian, the provisions of the various Iran-USSR treaties, including the 1921 Friendship Treaty, the 1940 CN Treaty, the 1940 exchange of notes, and the 1962 Aram-Pegov memorandum have the force of law and are made binding on the USSR's successor states as a matter of internal USSR law[126] and international treaty law.[127] A joint property of Iran and the USSR, under international law of state succession, the Caspian is therefore a joint property of Iran and the USSR's successor states bordering the Caspian.[128] Any change in that regime would require the joint consent of all the littoral states, including Iran.

None of the USSR successor states of Azerbaijan, the Russian Federation, Kazakhstan, and Turkmenistan could have inherited from the USSR any degree of territorial sovereignty or exclusive area of national jurisdiction in the Caspian that: (1) the USSR itself did not have exclusive title to in international law; and (2) the USSR had not passed onto constituent Union Republics on the Caspian as a matter of Soviet administrative/ republic boundary delimitation. While the 1962 Aram-Pegov line was an exercise in international law, the division of the Caspian by the USSR's ministry of oil and gas in 1970 was on its face a *prima facie* case of grand theft. It violated the 1940 Iran-Soviet accord that had deemed the Caspian an "Iranian and Soviet sea," presumably an object of joint or, in the minimum, common tenancy. If it can be shown that Iran had acquiesced in the division by virtue of its prolonged and informed silence, or expressly consented to it, then the point may be made that as a matter of prescription first the USSR and later its successor states on the Caspian acquired the demarcated sectors as part of their patrimony in the Caspian.

The dissolution of the Soviet Union could have resulted in only three possible legal situations. First, on the basis of the theory of joint tenancy or joint ownership, as joint tenants, Iran and the USSR could be viewed as having been the owners of undivided shares in the Caspian as a whole and with survivorship between them, meaning that when one died the other would become its sole owner.[129] Assuming this result obtains as a matter of "general principles of law recognized by civilized nations"[130] in respect to joint tenancy, then by analogy, it may be concluded, as a matter of international law,[131] that on the death of the USSR, Iran became the sole owner of the Caspian. Therefore, there would have been no part of the Caspian that technically could have passed onto the USSR's successor states. While technically the most logical of the legal results, this is not politically a tenable view, because no littoral state other than Iran would recognize such a result in practice.

Second, it may be posited that at the time of the dissolution of the Soviet Union, Iran and the USSR had common tenancy or common ownership of the Caspian, both being the owners of undivided shares in the Caspian, without survivorship between them, meaning that when the USSR died its share in the Caspian passed to its successor states on the Caspian.[132] The total share of the littoral successor states could not exceed one-half of the entire Caspian, as the other undivided half would belong to Iran by law of common tenancy.

Third, one may assume there was not at all an Iran-USSR joint or common tenancy, sovereignty, or ownership of the Caspian, or that it had been dissolved by virtue of the Aram-Pegov line and/or by the Iranian acquiescence in the 1970 USSR demarcation of sectors among its Caspian republics. In that event, the death of the USSR would not effect Iran's rights in the section that is deemed by international law as belonging to Iran. The area of the Caspian not belonging to Iran would be deemed inheritable by the littoral successor states, to be partitioned by and among them. In this context, one view may hold that the Aram-Pegov line constituted an international boundary between Iran and the USSR and, therefore, the area seaward of the line constituted the estate of the USSR in the Caspian, to be partitioned among the littoral successor states. Yet, another view may hold that the Aram-Pegov line constituted a line delimiting the landward area as Iran's internal waters. Under this formula, one may argue that the rest of the Caspian should be divided up between the littoral states, including Iran, in the manner consistent with international law.

The notion of common tenancy of the Caspian by the littoral states is the only acceptable foundation for the traditional view of Iran and Russia calling for the joint exploitation of the resources of the Caspian. The same notion is also the point of departure for the view held by Azerbaijan, Kazakhstan, and Turkmenistan calling for the division of the Caspian into territorial sectors. The difference between them is that the Iranian/Russian view clings to the traditions and practices of the past, whereas the view of the other littoral states suggests that the joint or common tenancy is to be no longer and therefore partition in the property is warranted. The right of these states to demand a partition is inherent in the very notion of common tenancy to which Iran and Russia also subscribe, yet this result will obtain in law only by the unanimous consent of all the littoral states or by the order of a competent judicial or arbitral authority.

Delimitation by Necessity

The theoretical underpinning of any development in the legal relationship among the littoral states with respect to the Caspian must begin with the recognition of the legal concept that the corpus of the Caspian is by law a *res communis,* a shared resource, analogous to the corpus of a river, the atmosphere, or the high seas and as such it is legally incapable of being made the private property of any one riparian state. A similar recognition must be the basis for the legal status of the wild animals, including the migratory fish, such as the sturgeon,[133] that inhabit the Caspian. Under the *res communis* doctrine, the riparian state possesses an equal right to use the common or shared resource and therefore each state is entitled to the use of the resource not perceptibly diminished in quantity or quality by another state's use of the same.[134] The obligation to respect the equal rights of a co-riparian is a duty impressed by the principle of customary international law known as *sic utere tuo ut alterum non laedas,*[135] which literally means "[u]se your own property in such manner as not to injure that of another."[136] Therefore, for example, "a state is not only forbidden to stop or divert the flow of a river which runs from its own to a neighboring state, but likewise to make such use of the water of the river as either causes danger to the neighboring state or prevents it from making proper use of the flow of the river on its part."[137] The riparian's co-equal usufructuary rights therefore are regulated by recourse to the standard of reasonable and equitable sharing in the beneficial use of the resource, as determined by taking all relevant factors into account.[138]

As a matter of international environmental law, the *sic utere* principle was recognized by the tribunal in the Trail Smelter Case (1938), which assessed the extent of Canada's admitted liability to the United States with respect to damage to property in the state of Washington caused by sulfur and other fumes drifting from a smelter in Trail, British Columbia.[139] The principle also made an appearance in the 1957 Iran-USSR Frontier Agreement, where the parties agreed to conduct their mining and agricultural activities in the "immediate vicinity of the frontier line" in a manner so as "not to harm the territory of the other party."[140] The principle and its necessary implication for reasonable and equitable utilization of commonly shared resources were evident also in the case of Lake Lanoux, a body of water shared by France and Spain. In that case, the arbitral tribunal determined that the French project to divert water from the lake could not be objected to by Spain because under the project's compensatory scheme Spain was not being deprived of the flow normally occurring from the lake downstream into Spain.[141] The principle was further enshrined in Principle 21 of the 1972 Stockholm Declaration, which provided for a state the responsibility to ensure that "the activities within its jurisdiction or control do not cause damage to the environment of other states or of areas beyond the limits of national jurisdiction."[142]

That the Caspian is polluted is a foregone conclusion.[143] To rely on and be protected by international environmental law, each littoral state of the Caspian must have an implied or express interest within a spatially cognizable area in order to have the standing to complain about a specified injury to its interest or territory caused by the actions of other states. The case of fixing liability purely as a matter of causation aside, without a clearly defined territorial limit, a state would be hard-pressed to make a case in trespass or nuisance against a polluter before the injury manifests itself on its beaches. The questions of who can complain of the contamination as a matter of legal standing and polluter's legal liability and extent of financial responsibility to the injured party would depend on the physical reach and extent of the contamination affecting the state.

If for no other reason, such as policing and fishing, then for the sake of environmental protection, it behooves every state of the Caspian to declare as its internal waters subject to its complete sovereignty an area contiguous to its coast ("exclusive zone"). In the case of Iran, the Aram-Pegov line would serve the purpose adequately. Any contamination reaching this line would engage the responsibility of the offending state, without waiting for effects of pollution to land or wash ashore or lay to waste the waterfowl habitats.[144]

The extent of the exclusive zone for the other Caspian riparian states may be based on a formula whereby a state's area of inland water is delimited by a line connecting the terminal points of its land boundary on the Caspian. In the case of Azerbaijan, that line may run from the terminus of the Iran-Azerbaijan land boundary at Absharan to the terminus of the Azerbaijan-Russian Federation boundary near Darband. A line from this last-named place to the Russian-Kazakh border near Astrachan will mark the Russian zone. A line from this last-named place to the tip of the Fort Sevcenko peninsula and from there to the Kazakh-Turkmen border north of Bekdas would delimit the Kazakh zone. The extent of the Turkmen zone will be marked by a line from the terminus of the Kazakh-Turkmen boundary on the Caspian to the terminus of the Aram-Pegov line at Hassan Kiyadeh in the Hassan Kuli Bay.

While the foregoing exercise would not yield the same square mile of territorial sovereignty per riparian, the principle behind it partakes from the notion that under the previous Iran-USSR regime the USSR could have been entitled to have had an area of the Caspian delimited as its internal waters similar to the exercise that produced for Iran the area landward of the Aram-Pegov line. Within its exclusive zone, each state would have full and exclusive sovereignty with regard to any matter, including exploitation of the seabed, subsoil, and fishing.

The areas of the Caspian not subject to individual exclusive zones may be dealt with in only one of two ways. On the one hand, the states may agree to treat it as their common or joint patrimony, akin to the "high seas," whereby the four freedoms of navigation, fishing, lying of cables and pipelines, and overflight, would be preserved on co-equal basis for all riparian states. The exploitation of the seabed and subsoil in such a regime would be regulated either by a regional seabed authority or by the naked rule of prior appropriation — that is, first come, first served.

On the other hand, the states may agree to partition the remaining area of the Caspian among them. That exercise may well entail the following. First, the states would agree on the geographical or geometric midpoint of the area. Next, straight lines are drawn from the terminal points of each state's land boundaries to the midpoint, creating thereby a pie-shaped zone bound landward by the line delimiting the aforementioned internal waters or exclusive zone of each state.

The foregoing exercise does not lead to equal apportionment of the Caspian among its riparian states, but the exercise is consistent nevertheless with the international legal standard governing the delimitation of offshore areas of jurisdiction. The standard is one of equal division unless another is justified by equitable apportionment.[145] The exception to this rule obtains where the parties agree to another line, or if another line is made "necessary by the reason of historic title or other special circumstances."[146] A similar rule obtains with respect to the delimitation of the continental shelf between two or more states, provided that another line not based on the principles of the median line or equidistance is agreed to by the parties or is justified by the presence of special circumstances.[147]

Conclusion

The foregoing analysis compels the conclusion that one legal consequence that did not result from the dissolution of the Soviet Union is the so-called legal vacuum, somehow implying that the Caspian suddenly became a *res nullius*—that is, an object of law belonging to no state and thus open to unilateral appropriation or occupation by the first comer or taker. However, no amount of auto-interpretation by any of the Soviet littoral successor states could technically deny, negate, or otherwise limit whatever historical, treaty, customary, or prescriptive rights Iran might have in the same sea. To suggest that the dissolution of the USSR has created a legal vacuum is therefore absurd.

While the aforementioned law of the sea conventions and others may not be directly binding on any of the littoral states with respect to the Caspian itself, they do incorporate time-tested principles of international law and practice in delimiting offshore areas of national jurisdiction. The lack of a systemic approach to the issue of competing sovereign rights of the littoral states based on the Iran-USSR conventional and customary law and general international law is bound to result in chaos, if not armed conflict. There are already signs of serious disagreement between the Russian Federation and Kazakhstan with respect to an offshore area in northern Caspian,[148] and between Azerbaijan and Turkmenistan.[149]

The states around the Caspian may feel free to lay claims of exclusive jurisdiction or sovereignty over any extent of the Caspian,[150] yet the legality of such claims or legality of title to the fruits thereof may be tested as a matter of international law by another riparian, including the seeking of judicial attachment of the oil or other products, or proceeds thereof, obtained from the disputed area of the sea.[151] On the other hand, each littoral state is aware that acquiescence by it in another state's rival claim may inevitably constitute recognition and therefore legitimacy for the position of the claimant. To prevent that from happening, each littoral state at one point or another has protested the bidding, awarding, or conclusion of contracts that affect the areas that it claims.[152]

In this "black gold rush" by the littoral states scrambling for advantage in the Caspian, one may bear in mind the lessons of the seminal case of North Sea Continental Shelf Cases[153] as a way to minimize the legal risk inherent in the present oil-grab. There, the International Court of Justice found that the result of the application of the equidistance principle applied by Denmark and the Netherlands in the delimitation of their continental shelf to the detriment of Germany was not binding on Germany because Germany was not a party to the 1958 Geneva Convention on Continental Shelf and also because the principle itself was not a customary norm of international law.[154] Yet, the Court recognized the principle of equitable apportionment, taking into account all relevant circumstances,[155] which the parties may achieve by negotiations, in the course of which they must take into consideration the following factors: (1) the general configuration of the coasts of the parties as well as the presence of any special or unusual features; and (2) so far as known or readily ascertainable, the physical and geological structure and natural resources of the area concerned.[156]

The law abhors a vacuum and for that reason it is axiomatic in international law that no situation, claim, or controversy may be left unresolved solely for the reason that no law exists to cover it. This prohibition against non-liquet allows the law to develop and adapt to changing circumstances, to which binding rules can be brought to bear, in the minimum, by analogy from similar situations.

Note

1. Geographically speaking, the Caspian is a body of water surrounded by land and as such it is a "lake;" it is called a "sea," however, because it is a salt lake. In its proper sense, the term "sea" refers to a collection of salt water, mostly surrounded by land. See J. Onley, *A Practical System of Modern Geography*, 5th ed., (Hartford, Connecticut: D.F. Robinson & Co., 1830), pp. 8-9. That the Caspian is connected to the outside world by rivers and canals is completely irrelevant to its geographical status as a lake. In international law, the Caspian is deemed a land-locked body of water because it cannot be entered without passing through internal waters or a territorial sea. Otherwise, it "will be deemed part of the high seas, and subject to the same regime as the rest of the open sea." D. J. Latham Brown, *Public International Law* (London: Sweet & Maxwell, 1970), p. 98.

2. Aleksey Nilovich Kosarev and Oleg Konstantinovich Leontyev, "Caspian Sea," in Philip W. Goetz, editor-in-chief, *Encyclopaedia Britannica*, 15th ed., (Chicago: Encyclopaedia Britannica, Inc., 1981), Macropaedia, vol. 30, p. 980.

3. Brown, p. 97.

4. Done in Tehran, March 25, 1940, in *British and Foreign State Papers*, (London: H. Majesty's Staionary Office [hereinafter cited as B.F.S.P]) vol. 144: 1940-1942 (1952), pp. 419-430.

5. Signed in Paris, June 21, 1926, in *B.F.S.P.*, vol. 123 (Part 1):1926 (1931), p. 160.

6. Article 15, *B.F.S.P.*, vol. 144, p. 430.

7. See letter from M. A. Foroughi, Persian minister for foreign affairs, to A. Petrovsky, the Soviet ambassador at Tehran, October 27, 1931, and letter from A. Petrovsky, the Soviet ambassador at Tehran, to M. A. Foroughi, the Persian minister for foreign affairs, Tehran, dated October 27, 1931, in *B.F.S.P.*, vol. 134:1931 (1936), p. 1045 [item No.13] and pp. 1045-1046 [item No. 14], respectively.

8. See letter from A. Tchernykh, Soviet ambassador at Tehran, to B. Kazemi, the Iranian minister for foreign affairs, August 27, 1935, and acknowledgement by Iranian minister for foreign affairs, in *B.F.S.P.*, vol. 139:1935 (1948), p. 574 [item No. 5] and p. 574, [item No. 6], respectively.

9. See letter from Matvei Filimonov, the Soviet ambassador at Tehran, to Muzaffar Alam, the Iranian minister for foreign affairs, dated March 25, 1940, and the latter's reply, in *B.F.S.P.*, vol. 144, p. 431, items No. 1 and No. 2, respectively.

10. Harvey H. Smith, *et al.*, *Area Handbook for Iran*, 2nd ed. (Foreign Area Studies/American University) (Washington, D.C.: U.S. Government Printing Office, 1971), pp. 306-307.

11. Ibid., p. 307. According to a 1981 description of the Caspian, with the exception of its southern shores, "six-seventh of the Caspian coast runs through Soviet territory." *Encyclopaedia Britannica*, Macropaedia, vol. 30, p. 980. Technically, a better description would read "Except for its [south] and [southwest] extremities, which are contiguous to Iranian territory, the Caspian Sea is bounded on all sides by the [USSR]." Joseph L. Morse, ed.,

The Universal Standard Encyclopaedia (New York: Unicorn Publishers, Inc., 1955), vol. 4, p. 1455.

12. See L. Oppenheim, *International Law* (H. Lauterpacht, editor), 8th ed., (New York: David McKay Company, 1952), vol. 1, pp. 476-479.

13. Astara is situated in Iran's Eastern Azerbaijan Province and Hassan Kiyadeh is located at 37/24 N and 49/58 E in Iran's Golestan Province. See generally, *Gazetteer of Iran*, 2nd ed., (Washington D.C.: Defense Mapping Agency, 1984).

14. "Bahr khazar yek darya basteh motaaleq beh Iran va Roussiyeh ast" (The Caspian Sea is a closed sea and belongs to Iran and Russia), in *Kayhan* (Farsi language weekly, London), No. 631, November 17, 1996, p. 8, quoting Ahmad Mirfendereski, political director of the Iranian ministry of foreign affairs (1963-1964) and ambassador to the Soviet Union (1965-1971). Mirfendereski recalls the agreement bearing the date of August 15, 1962, and recalls it consisting of three or four articles—one article reaffirming the parties' commitment to the principles enshrined in the 1921 Friendship Treaty, another article containing Iran's commitment not to allow its territory to be used for launching an attack on the Soviet Union, another article providing for a line from Astara to Hassan Kiyadeh. Ahmad Mirfendereski, interview with the author, October 26, 1997.

15. See, for example, Convention on the Territorial Sea and the Contiguous Zone, done in Geneva, on April 29, 1958, which came into force on September 10, 1964, in *United Nations Treaty System* (hereinafter cites as U.N.T.S.), vol. 516:1964 (No. 7477) (1964), pp. 206-224, articles 3-7.

16. See, for example, report by Lawrence Sheets, Reuters, from Tbilisi, Georgia, August 19, 1997.

17. A comprehensive review of the Caspian picture is contained in the survey "Central Asia: A Caspian Gamble," in *The Economist*, February 7, 1998.

18. The Azerbaijan constitution [1995] defines the territory of the republic as including its "inner waters," and "the Caspian Sea (Lake) sector relating to the Azerbaijan Republic." Chapter II, article 11 of Constitution of the Azerbaijan Republic [1995], adopted by referendum, November 12, 1995, in Gisbert H. Flanz, ed., *Constitutions of the World*, Release No. 96-6: Azerbaijan (New York: Oceana Publications, 1996). In contrast, article 1 of the 1989 constitution, which defined the extent of the Azerbaijan SSR, did not refer to the Caspian or any offshore area. Constitutional Law of the Azerbaijani Soviet Socialist Republic on the Sovereignty of the Azerbaijani SSR (September 23, 1989), in Charles F. Furtado, Jr. and Andrea Chandler, eds., *Perestroika in the Soviet Republics* (Boulder, Colorado: Westview Press, 1992), p. 449.

19. Presidential Proclamation 2667, dated September 28, 1945: Natural Resources of the Subsoil and Sea Bed of the Continental Shelf, 10 *Federal Register* 12303 (1945).

20. Brown, p. 95.

21. See law fixing the limits of the coastal waters and area of the sea under supervision of the government, enacted by the Majlis Shora Melli

(National Consultative Assembly) on Tir 24, 1313 (July 1934), article 1 (establishing a 6-mile wide territorial sea), amended by law amending the law fixing the limits of coastal waters and area of the sea under supervision of Iran, enacted by the majlis on Farvardeen 22, 1338 (April 1960), article 3 (fixing a 12-mile wide territorial sea), in Jamsheed Momtaz, *Hoquq-e Daryaha* (Law of the Seas) (Tehran: Tehran University's Center for Advanced International Studies, 1976), pp. 81-85.

22. See, for example, report by David Chance, Reuters, from New York, August 1, 1997.

23. Ibid.

24. See "Russia, Azerbaijan agree to divvy up the Caspian," in *Iran Times* (Washington, D.C.), No. 1377, April 3, 1998, p. 1.

25. Treaty of Alliance between Russia and Persia, signed at St. Petersburg, September 12, 1723, in Clive Parry, ed., *Consolidated Treaty Series* (Dobbs Ferry, New York: Oceana Publications, Inc. 1968) (hereinafter cited as C.T.S.), vol. 31, pp. 425-428, article 2 (emphasis added). The French text of the treaty refers to "le long de la Mer Caspienne" being ceded to Russia. However, historians' reference to this treaty makes no mention of the "length of the Caspian Sea," confining the description of the ceded area to Darband, Baku, Gilan, Mazandaran, and Astarabad. See, for example, Martin Sicker, *The Bear and the Lion* (New York: Praeger, 1988), p. 10; *Cambridge History of Iran*, vol. 7, Peter Avery, Gavin Hambly, Charles Melville, eds. (Cambridge: Cambridge University Press, 1991), pp. 318-319, quoting S. M. Solov'ev, *Istoriia Rassii s drevneishikh vremen*, Moscow, 1963, vol. 9, p. 384; Percy Sykes, *A History of Persia*, 3rd ed. (London: Macmillan & Co., 1951), vol. 2, p. 233.

26. *Cambridge History of Iran*, p. 321.

27. Treaty between Persia and Russia, signed at Rasht, February 13, 1929, in C.T.S., vol. 33, pp. 157-162, article 3.

28. Treaty of Peace between Russia and Persia, signed at Riascha, January 21, 1732, in C.S.T., vol. 33, pp. 445-451, article 2.

29. Sykes, pp. 253-254.

30. Ibid., p. 254.

31. Ibid., pp. 344-345.

32. For details, see ibid., pp. 269-271.

33. Treaty of Peace and Perpetual Friendship between Persia and Russia, signed on the River Seiwa, October 12, 1813, in C.T.S., vol. 62, pp. 435-442, articles 2-3.

34. Ibid., article 5.

35. Treaty of Peace and Friendship between Persia and Russia, signed at Turkmanchai, February 10(22), 1828, in C.T.S., vol. 78, pp. 105-112. Article 8 provided: "Russian merchant marine vessels shall have, like before, the right to navigate freely the length of and land on the shores of the Caspian; in the case of shipwreck in Persia they shall find rescue and assistance. The same rights are accorded to Persian merchant vessels, to navigate as previously (sur l'ancien pied) the Caspian and land on Russian shores, where in the case of shipwreck the Persians shall receive, reciprocally, rescue and assistance. As for men of war, Russian warships, which

have been from old times (ab antiquo) the only warships with the right to navigate the Caspian, shall have the exclusive privilege to navigate the Caspian, by which reason, as reserved and ensured today and hereby, with the exception of Russia no other Power can have warships in the Caspian."

36. Ibid., articles 3-5.

37. See Agreement between Iran and the USSR concerning the Settlement of Frontier and Financial Questions, signed at Tehran, December 2, 1954, in *U.N.T.S.*, vol. 451:1963 (No. 6497), pp. 250-264, article. 2(A).

38. Treaty of Commerce between Russia and Persia, signed at Turkmanchai, February 10/22, 1828, in *B.F.S.P.*, vol.45: 1854-1855 (1865), pp. 865-868, article 1.

39. For details, see Sykes, pp. 356-357.

40. *Cambridge History of Iran*, p. 341.

41. Signed at Tehran, May 24, 1897, enclosed in letter from R.M. Thomson, Tehran, to the Marquis of Salisbury, June 27, 1879, in *B.F.S.P.*, vol. 70:1878-1879 (1886), pp. 1227-1229.

42. Letter from the Earl of Dufferin, St. Petersburg, to the Marquis of Salisbury, July, 16, 1879, in ibid., pp. 1226-1227.

43. For details, see letter from M. R. Thomson, Tehran, to the Marquis of Salisbury, September 22, 1879, in ibid., pp. 1232-1233.

44. See Telegraph Convention between Russia and Persia, signed at Tehran, January 31, 1881, in *B.F.S.P.*, vol. 72:1880-1881 (1888), pp. 1225-1226.

45. Signed at Tehran, December 9 (21), 1881, in *B.F.S.P.*, vol. 73:1881-1882 (1889), pp. 97-100, article 1.

46. See Protocol, dated December 2, 1954, to Iran-USSR 1054 Frontier Agreement, in *U.N.T.S.*, vol. 451, p. 264.

47. Signed at Tehran, May 27 (June 8), 1893, in *B.F.S.P.*, vol. 86: 1893-1894 (1899), pp. 1246-1249.

48. This would have not been an anomaly for Persia, to have a boundary running on its side or banks. For example, in consequence of the Perso-Ottoman war of 1821-1823 and the ensuing Treaty of Erzerum, in 1847, the Perso-Ottoman boundary in the Shatt al-Arab waterway was fixed such that, with the exception of the frontage off Mohammareh and Abadan, the Ottoman Empire gained sovereignty over the entire waterway up to the left (eastern) bank, while Persia was granted the right to navigate the whole river. For details, see Majid Khadduri, ed., *Major Middle Eastern Problems in International Law* (Washington, D.C.: American Enterprise Institute for Public Policy Research, 1972), pp. 88-94.

49. Convention between the Emperor of Russia and the Shah of Persia regulating their Commercial Relations and modifying Article 3 of the Additional Act of February 10 (22), 1828, signed at Tehran, October 27 (November 9), 1901, in *B.F.S.P.*, vol. 96:1902-1903 (1906), pp. 1279-1283, amended by Supplementary Commercial and Customs Declaration between Persia and Russia, signed at St. Petersburg, December 29, 1904, in *C.T.S.*, vol. 197, p. 366.

50. Convention between Great Britain and Russia relating to Persia, Afghanistan and Tibet, signed at St. Petersburg, August 31, 1907, in *C.T.S.*, vol. 205, pp. 404-408.

51. For details, see Sicker, pp. 35-36; Nasrollah S. Fatemi, *Diplomatic History of Persia, 1917-1923* (New York: Russell E. Moore Co., 1952), pp. 255-257.

52. For details see, *Encyclopaedia Britannica*, 15th ed., Micropaedia, vol. 1, p. 698; Constitutional Act on State Independence of the Republic of Azerbaijan, proclaimed at Baku, October 19, 1991, by President of the Republic, Ayaz N. Mutalibov, in *Constitutions of the World*: Azerbaijan (1996).

53. For details see, Sykes, pp. 488-498, 525-526; Sicker, pp. 38-39.

54. Cosroe Chaqueri, *The Soviet Socialist Republic of Iran, 1920-1921* (Pittsburgh: University of Pittsburgh Press, 1995), p. 147, quoting a statement by Bravin reported in *Iran* (Tehran daily), June 19, 1918.

55. Fatemi, pp. 257-259.

56. For details see, Sykes, pp. 494-498, 525-526; *Encyclopaedia Britannica*, 15th ed., Micropaedia, vol. 1, p. 698.

57. See the Address of the Military-Revolutionary Committee of the Azerbaijan Soviet Independent Republic to Soviet Russia, April 29, 1920 (No. 91), in Mikhail Georgadze, ed., *USSR: Sixty Years of the Union* (Moscow: Progress Publishers, 1982), p. 161. The address admitted that the Committee could not withstand the counter-revolutionaries on its own and asked the government of Soviet Russia to enter into a union with it in order to struggle against world imperialism, wherefore "we ask you immediately to render effective assistance by sending Red Army detachments."

58. *Cambridge History of Iran*, p. 345; Fatemi, p. 259; Sicker, p. 40; Sykes, pp. 526, 550.

59. Fatemi, p. 260.

60. Sicker, p. 41; *Cambridge History of Iran*, pp. 345-346. The text of the letter by Mirza Kuchik Khan, the President of the Soviet Republic of Gilan, to Comrade Vladimir I. Lenin, expressing solidarity with Soviet Russia, is reprinted in Fatemi, p. 221. For a detailed analysis of the rise and demise of the Soviet Republic of Gilan (1920-1921), see Chaqueri.

61. Sicker, p. 41.

62. See, for example, Sykes, p. 488 (referring to Azerbaijan as a "state [coming] into being").

63. For details, see Fatemi, p. 261; *Cambridge History of Iran*, pp. 344-346; Sicker, pp. 41-44.

64. Chaqueri, p. 264.

65. Signed at Moscow, February 26, 1921, in *League of Nations Treaty Series* (hereinafter cited as L.N.T.S.), vol. 9:1922 (No. 268), pp. 401-411.

66. Ibid., article 3.

67. The term "bahriyyeh" in Persian means "navy." The League of Nations' English translation of article 11, however, states that article 8 of Turkmanchai "forbids Persia . . . to have vessels in the waters of the Caspian Sea." Article 8 of Turkmanchai actually recognized Persia's right to have merchant vessels in the Caspian.

68. For details, see Fatemi, pp. 299-300.

69. Ibid., p. 300.

70. Ibid., pp. 300-301.

71. For greater detail, see Extract from Contract with Martin Lianozov, August 10, 1923, in *B.F.S.P.*, vol. 126 (Part 1):1927 (1932), p. 564.

72. Signed at Moscow, October 1, 1927, in *B.F.S.P.*, vol. 126 (Part 1), pp. 947-953, articles 1 and 12.

73. Ibid., article 2.

74. *B.F.S.P.*, vol. 126 (Part 1), pp. 955-956.

75. Fisheries Agreement, ibid., articles 18-19.

76. Letter from Ali Gholi Khan Ansari, Persian minister for foreign affairs, Moscow, to L. Karakhan, assistant people's commissary for foreign affairs, October 1, 1927, and reply of the latter of the same date, in *B.F.S.P.*, vol. 126 (Part 1), pp. 956-957.

77. Smith, p. 390. Iran was bound however by article 4 of the agreement itself not to allow any third-power or its subject to fish in this area for an additional 25 years, that was therefore until 1977.

78. Reprinted in *USSR: Sixty Years of the Union*; see the Address of the Military-Revolutionary Committee of the Azerbaijan Soviet Independent Republic to Soviet Russia, April 29, 1920 (No. 91), p. 161 (the address declared Azerbaijan as uniting with Soviet Russia to sovietize the Caucasus and the East, and signed off with "Long live the Azerbaijan independent soviet republic!"); Fundamental Law of the Transcaucasian Federative Soviet Republics, approved by the First Transcaucasian Congress of Soviets, March 2, 1922 (p. 98); Allied Treaty on Forming the Federal Union of Socialist Soviet Republics of Transcaucasia, adopted by the Plenipotentiary Conference of Central Executive Committees of the Transcaucasian Republics, March 12, 1922 (p. 66); Resolution of the First Congress of Soviets of the Union of Soviet Socialist Republics, December 30, 1922, approving the Declaration and Treaty on the Formation of the Union of Soviet Socialist Republics (p. 161); Resolution of the Second Transcaucasian Congress of Soviets on the Constitution (Fundamental Law) of the USSR, approving the Fundamental Law of the USSR, January 5, 1924 (p. 174), The Constitution of the USSR, January 31, 1924 (p. 175).

79. Ibid., article 6.

80. Signed at Moscow, April 25, 1923, in *B.F.S.P.*, vol. 118:1923 (1926), pp. 953-956.

81. The text of the Declaration is found in *B.F.S.P.*, ibid., p. 956.

82. Signed at Moscow, April 27, 1923, in *B.F.S.P.*, vol. 118, pp. 957-960.

83. See Declaration of the First All-Turkmen Congress of Soviets, February 20, 1925, in *USSR: Sixty Years of the Union*, p. 210.

84. See Resolution of the Third Congress of Soviets of the USSR, May 13, 1925, in *USSR: Sixty Years of the Union*, p. 215. Article 2 of the USSR Constitution was amended to include Turkmen SSR as a constituent Republic; while Azerbaijan, Georgia, and Armenia were included in the Transcaucasian Soviet Socialist Republic.

85. Resolution of the Extraordinary 8th Congress of the Soviets of the USSR, December 5, 1936, in *USSR: Sixty Years of the Union*, p. 219.

86. Agreement between Persia and the USSR regarding the Exchange of Postal Parcels, signed at Moscow, August 2, 1929, in *L.N.T.S.*, vol. 109:1930-1931 (No. 2530), pp. 101-113, article 1.

256 • Guive Mirfendereski

87. Supplementary Protocol, dated August 2, 1929, to the Postal Parcels Agreement, in ibid., p. 113, article 1.
88. Signed at Tehran, October 27, 1931, in *B.F.S.P.*, vol. 134:1931 (1936), pp. 1026-1038, article 16.
89. Ibid., article 17, sections 1 and 2, and Final Protocol, done in Tehran, October 27, 1931, in *B.F.S.P.*, vol. 134, pp. 1039-1040 (Ad 17).
90. See endnote 7 above. The entire text read: "Because the Caspian Sea, which is considered by our two governments as a Persian and Soviet sea, is of exceptional value to the USSR, I have the honor of informing that my government will take all necessary measures to which end nationals of third countries in its service in the ports of this sea do not use their stay therein for purposes going beyond the scope of the functions with which they are charged by my government." *B.F.S.P.*," vol. 134, p. 1045 [item No. 13].
91. Ibid., pp. 1045-1046 [item No. 14].
92. Done at Tehran, August 27, 1935, in *B.F.S.P.*, vol. 139:1935 (1948), pp. 554-572, article 18.
93. See endnote 8 above. The letter initiated by the Soviet Ambassador read: "I have the honor to inform you of the following: Given that the Caspian Sea, which is considered by the two governments as a Soviet and Iranian sea, is of exceptional value, it is understood that the two governments will take the necessary measures to which end nationals of third countries in their service in the ports of this sea do not use their stay therein for purposes going beyond the scope of the functions with which they are charged." *B.F.S.P.*, vol. 139, p. 574 [item No. 5].
94. Article 12, section 4, in *B.F.S.P.*, vol. 144, p. 429.
95. The 1940 CN Treaty itself was drawn up in Persian and Russian; its passages quoted here are from the English translation provided in *B.F.S.P.*, vol. 144. The 1931 ECN Convention and the 1935 ECN Convention were drafted in French. The English translation of their passages quoted in this chapter are supplied by the author.
96. See endnote 9 above. The British Foreign Office translated the passage as "[s]ince the Caspian Sea, which the high contracting parties hold to belong to Iran and the Soviet Union (*lit.* to be an Iranian and Soviet sea), has special importance for the high contracting parties." See *B.F.S.P.*, vol. 144, p. 431 [item No. 1].
97. See, for, example, *Black's Law Dictionary*, revised 4th ed., (St. Paul, Minnesota: West Publishing, 1968), pp. 1634-1635.
98. *U.N.T.S.*, vol. 451, pp. 250-266, article 1.
99. Protocol, December 2, 1954, article 2, in ibid., p. 264 [item No. 3].
100. See Protocol, dated May 20, 1955, to 1954 Frontier Agreement, in *U.N.T.S.*, vol. 451, pp. 265-266, p. 250, footnote 1.
101. The Presidium of Supreme Soviet of the USSR, elected by the members of the bicameral Supreme Soviet of the USSR, constituted the highest legislative authority in the USSR. Under article 14 of USSR Constitution [1936], in effect at the time, the jurisdiction for ratification of USSR treaties with foreign states resided in the USSR. *USSR: Sixty Years of Union*, p. 219. Yet, article 18 of the same constitution provided that the territory of a Union Republic "may not be altered without its consent."

102. The Iranian constitutional requirement at the time for parliamentary ratification of international treaties involving Iran's boundaries rested on provisions of Constitutional Law [1906] and Supplemental Constitutional Law [1907]. La Loi constitutionelle du 14 Zigh Adatol-Haram 1324 [30 December 1906] and Additif du Chaban 1325 [7 October 1907] a la Loi constitutionelle, in J. E. Godchot, ed., *Les Constitutions du Proche et du Moyen-Orient* (Paris: Sirey, 1957). Under article 24 of the 1906 law, "Notwithstanding treaties whose secrecy is warranted for the good of the State and the Nation, the conclusion of treaties and conventions . . . must be effected with the approval of the Consultative Assembly [Majlis]" and, per article 46, by the Senate. According to article 3 of the 1907 law, "Iran's frontiers . . . may not be altered unless by law." Per article 52, "treaties kept secret in accordance with [the 1906 law] would have to be made public either when the need for said secrecy has passed or the interest or security of the State so warrants, for which purposes the Emperor shall refer said treaties, with the necessary explanation, to the [Consultative Assembly] and Senate." These constitutional provisions remained in force until the promulgation of Iran's 1979 republican constitution. Article 77 of Iran's republican constitution [1979] provides: "International treaties, protocols, contracts, and agreements must be approved by the Islamic Consultative Assembly." Article 78 provides: "All changes in the boundaries of the country are forbidden, with the exception of minor amendments in keeping with the interests of the country, on the condition that they are not unilateral, do not encroach on the independence and territorial integrity of the country, and receive the approval of four-fifths of the total members of the Islamic Consultative Assembly." See Iran's constitution in Albert P. Blaustein & G. H. Flanz, eds., *Constitutions of the Countries of the World* (Release No. 92-8: Islamic Republic of Iran) (Dobbs Ferry New York: Oceana Publications, Inc. 1992).

103. Treaty between the Government of the Union of Soviet Socialist Republics and the Imperial Government of Iran concerning the Regime of the Soviet-Iranian Frontier and the Procedure for the Settlement of Frontier Disputes and Incidents, signed at Moscow, May 14, 1957, with Protocol and Annexes, in *U.N.T.S.*, vol. 457:1963 (No. 6586), pp. 212-246, part 1, article 1.

104. For details, see Smith, pp. 305, 310-311 and 315-316; Walter Laqueur, *The Struggle for the Middle East* (New York: The Macmillan Company, 1969), pp. 28-29 ("Persia was reminded that according to the treaties of 1921 and 1927 it did not have the right to join any 'anti-Soviet' pact."); J. C. Hurewitz, *Middle East Politics: The Military Dimension* (New York: Frederick A. Praeger, 1969), pp. 98-99 (quoting Khrushchev being nervous about missiles in Turkey).

105. Laqueur, ibid., p. 28, footnote 45 (reported by *Pravda*, on September 16, 1962) and p. 30.

106. See endnote 14 above.

107. See endnote 14 above and, for example, Martin Sicker, *The Bear and the Lion* (New York: Praeger, 1988), p. 93 (reference to Aram-Pegov mem-

orandum [September 15, 1962] regarding Iran abjuring nuclear missiles, citing *Summary of World Broadcasts*, part 4, September 18, 1962, ME/1050/D/1); Laqueur, p. 30 (reference to Iran's pledge not to station nuclear missiles on its soil, quoting the Soviet daily *Pravda*, September 16, 1962).

108. See endnote 14 above.

109. See endnote 14 above.

110. See Smith, p. 307 ("[b]oth countries habitually claim twelve miles as the breadth of their territorial waters").

111. See, for example, 1958 Geneva Convention on the Territorial Sea, *U.N.T.S.*, vol. 450, article 2 ("[t]he sovereignty of a coastal State extends to the air space over the territorial sea as well as to its bed and subsoil").

112. Ibid., emphasis added.

113. Under article 102 of the U.N. Charter, Member States are required to register for publication with the U.N. Secretariat "every treaty and every international agreement." Otherwise, no party to any such treaty or international agreement which has not been registered may invoke that treaty or international agreement before any organ of the United Nations, which also includes the Security Council, the General Assembly, and the International Court of Justice. For the procedural requirements of registration and publication of Member State's treaty and international agreements, see *U.N.T.S.*, vol. 1: 1946-1947, pp. xiv-xxx; *U.N.T.S.*, vol. 76: 1950, pp.xviii-xxviii; *U.N.T.S.*, vol. 846:1972, p. xviii.

114. See, for example, Vienna Convention on the Law of Treaties, in *U.N.T.S.*, vol. 1155:1980 (No. I:18232), pp. 332-353, article 46 (a state may not invoke violation of internal law as invalidating treaty, unless the violation is of an internal law of fundamental importance).

115. For details, see Farajallah Rassai, gen. ed., *Peyk-e Darya: 2500 Sal Rouy-e Daryaha* ("Messenger of the Sea: 2500 Years on the Seas") (Tehran: Imperial Iranian Navy, 1971), pp. 377, 413-419.

116. Ibid., p. 419.

117. Admiral Farajallah Rassai, interview with the author, Alexandria, Virginia, April 24, 1997. Admiral Rassai served as the Chief of the Imperial Iranian Navy (equivalent to the U.S. Secretary of the Navy) from 1959 to 1972.

118. Hameed Ahmadi, *Tarikh-e Neem Gharn-e Nirouy-e Darya-ee Noveen-e Iran* ("The Demi-Centennial History of Iran's Modern Navy") (Tehran: Imperial Iranian Navy, 1976), pp. 144-145.

119. Rassai, interview with author, endnote 117 above.

120. Ibid.

121. Mirfendereski, interview with author, endnote 14 above.

122. Letter from Admiral F. Rassai, to the author, May 13, 1997, with an enclosure by the former Commander of Iran's Northern Navy, Seifallah Anoushiravani, undated, recollecting his voyage to Baku in summer 1971.

123. The earliest instance of the Iranian Navy's curiosity regarding the extent of Iranian territorial jurisdiction in the Caspian appears to have occurred

around the time of Iran's acquisition of the *Shahrokh* and other naval vessels from the United States. According to Admiral Rassai, interview, in the late 1950s and early 1960s, the U.S. military officials in Iran inquired repeatedly from the Iranian Navy the extent of Iran's boundary in the Caspian. Repeatedly, the Iranian Navy referred the question to the Iranian ministry of foreign affairs, which would not give the Iranian Navy a straight answer. Ibid. The inquiries by the United States in Tehran may have had to do with the research being conducted for the first edition of an official guide entitled *Area Handbook for Iran*, which was published by the U.S. government in May 1963 for use by the U.S. military and diplomatic personnel. See Smith, pp. iii-vii.
124. Smith, p. 307.
125. See, article 16, in *B.F.S.P.*, vol. 144, p. 430.
126. See, for example, the various provisions of the USSR Constitution (October 7, 1977), known also as the "Brezhnev Constitution," in effect at the time of the dissolution of the Soviet Union in 1991, in *USSR: Sixty Years of the Union*, p. 297. Article 71 described the USSR as an "integral, federal, multinational state." Article 75 defined the territory of the USSR as "a single entity and compris[ing] the territories of the Union Republics." Under article 73(2), the jurisdiction of the USSR extended to the determination of the state (international) boundaries of USSR and approval of changes in the boundaries between the Union Republics. Article 73(10) provided for the USSR representing the Union Republics in international relations. With the exception of article 72, which recognized the Union Republics' right to secede from the USSR, for all practical purposes, the Union Republics were not subjects of international law, no more than they were ever "sovereign" or "independent."
127. See Agreement Establishing the Commonwealth of Independent States among RSFSR, Ukraine, and Belarus, done at Minsk, December 8, 1991 ("Minsk Agreement"), in *International Legal Materials*, vol. 31:1992, p. 143 (UN Doc.A/46/771 of December 30, 1991), article 12 ("[p]arties undertake to discharge the international obligations incumbent on them under treaties and agreements entered into by the former [USSR]"); Protocol to Agreement Establishing the Commonwealth of Independent States, done at Alma Alta, December 21, 1991, in ibid., p. 147 (UN Doc.A/47/60 of December 30, 1991) (Azerbaijan, Kazakhstan, Turkmenistan, and others joining the Minsk Agreement); Alma Ata Declaration of December 21, 1991, ibid., p. 148 (the sixth declaration, in which Azerbaijan, Russian Federation, Kazakhstan, Turkmenistan, and others undertake to "guarantee in accordance with their constitutional procedures the discharge of the international obligations deriving from treaties and agreements concluded by the former [USSR]."
128. See, for example, Oppenheim, pp. 159-160 (under the principle of *res transit cum suo onere*, treaties of the extinct state concerning boundary lines pass onto the succeeding state), succession takes place with regard to locally connected rights and duties, such as in the case of land and rivers; and pp. 164-166 (same in the case of dismemberment, cession and separation).

129. See, for example, A. James Casner and W. Barton Leach, *Cases and Text on Property*, 3rd ed. (Boston: Little, brown and Company, 1984), p. 255.
130. Statute of the International Court of Justice, article 38, section 1, paragraph (c).
131. See generally H. Lauterpacht, *Private Law Sources and Analogies of International Law* (London: Longmans Green & Co., 1927).
132. See Casner and Leach, p. 255.
133. Among the Caspian's fish stories is one from 1996 when Iranian fishermen landed a 1,100-pound sturgeon with 118 pounds of caviar, worth $107,000. See the Reuters report from Tehran, Iran, September 22, 1997. The sheer size of the fish tells of a fantastic survival tale in a very over-fished sea, or is testimony to mutative or accelerated growth possibly brought on by the sea's pollution.
134. For a brief discussion of the natural law concept of *res communis*, see Guive Mirfendereski, "An International Law of Weather Modification," in *The Fletcher Forum*, vol. 2, no. 1 (January 1978), pp. 53-63.
135. See generally the dissenting opinion of Judge de Castro in Nuclear Tests Case (Australia v. France), Judgment of December 20, 1974, *International Court of Justice Reports* (1974), pp. 388-390, where he stated that the *sic utere* principle was customary international law and it would have been decisive in establishing France's responsibility for nuclear testing in the South Pacific had France not halted the tests and the case had reached its merits before the Court.
136. *Black's Law Dictionary*, p .1551.
137. Oppenheim, pp. 474-475.
138. See, for example, International Law Association, *Report of the 52nd Conference* (Helsinki, 1966), p. 477. This body of aspirational rules at the time were known as the "Helsinki Rules."
139. Trail Smelter Arbitration (1938, 1941), in *United Nations Reports of International Arbitral Awards*, Annual Digest for 1942, vol. 3. The tribunal stated, "[u]nder the principles of international law . . . no State has the right to use or permit the use of its territory in such manner as to cause injury by fumes in or to the territory of another or the properties or persons therein, when the case is of serious consequence and the injury is established by clear and convincing evidence." Ibid., p. 1965.
140. Part 3, article 19, in *U.N.T.S.*, vol. 457, p. 228.
141. Affaire du Lac Lanoux (France/Spain), Award of November 16, 1957, *United Nations International Arbitral Awards*, vol. 12, p. 315-316. For an English translation of the case, see *American Journal of International Law*, vol. 53:1959, p. 156.
142. United Nations Conference on the Human Environment, Final Document, adopted June 16, 1972, in *International Legal Materials*, vol. 11:1972, p. 1416.
143. See generally Siamak Namazi, "The Caspian's Environmental Woes," in this volume. See also, for example, the Reuters report from Tehran, Iran, August 17, 1997. The report quoted the Iranian news agency's report that the Caspian suffers from high pollution levels. Sampling showed some 61,840 cases of microbial pollution in the water. The major pollu-

tants are identified as oil, pesticides, and industrial sewerage. The Iranian report quoted an official calling for the littoral states to work together on measures to protect endangered marine life and reduce the risk to public health caused by eating contaminated fish.

144. Ironically, the Ramsar Convention, one of the earliest international environmental conventions and certainly the first on wetlands and waterfowl, bears the name of Ramsar, an Iranian city on the Caspian shore. Convention on Wetlands of International Importance especially as to Waterfowl Habitat, concluded at Ramsar, Iran, February 2, 1971, entered into force on December 21, 1975, in *U.N.T.S.*, vol. 996:1976 (No. I.14583), pp. 246-250. The convention became binding as between Iran and the USSR upon the USSR's accession thereto as of February 11, 1977, in *U.N.T.S.*, vol. 1026: 1976 (No. A-14583), p. 429.

145. The notion is embedded, for example, in article 12 of the 1958 Geneva Convention on the Territorial Sea, *U.N.T.S.*, vol. 516, p. 212, whereby the boundary between two opposite or adjacent coastal states is deemed to be "the median line every point of which is equidistant from which the breadth of the territorial seas of each of the two States is measured."

146. Ibid.

147. Article 6 of the Geneva Continental Shelf Convention (1958), *U.N.T.S.*, vol. 499, p. 316.

148. See, for example, report by Marat Gurt, Reuters, from Ashgabat, Turkmenistan, September 1, 1997 (quoting the Kazakh foreign minister's objection to the Russia's attempt to tender an area that Kazakhstan considers its area).

149. Turkmenistan claims the offshore areas of Serdar and Azeri as its own and Chirag as partly belonging to it. While Azerbaijan is actively exploiting Azeri and Chirag with the participation of an international consortium, the Azerbaijani plans to develop Serdar, which they call Kyapaz, were dealt a setback when the Russian President Boris Yeltsin annulled the Russian oil company's contract to develop the field. The Azerbaijani President, Heydar Aliev, has admitted that Kyapaz is indeed located on the border of the Azerbaijan and Turkmenistan sectors of the Caspian. For details, see reports by Marat Gurt, Reuters, from Ashgabat, Turkmenistan, September 25, 1997 and September 2, 1997, and by Kynnley Browning, Reuters, from Moscow, August 8, 1997.

150. For example, Turkmenistan holds the position that each littoral state shall have an offshore territorial zone 45 miles wide, while the middle of the sea should be used by all the littoral states. Report by Marat Gurt, Reuters, from Ashgabat, Turkmenistan, September 25, 1997, quoting the Turkmen Deputy Foreign Minister.

151. In consequence of the 1951 Iranian nationalization of the Anglo-Iranian Oil Company, in June 1951, the British party sought in various countries to obtain judicial attachment of Iranian oil shipments. For details, see Fuad Rohani, *Tarikh-e Melli Shodan-e Sanaat-e Naft-e Iran* ("History of the Nationalization of the Iranian Oil Industry") (Tehran: Franklin Publications, 1964), pp. 396-411.

152. In August 1997, one of Iran's vice presidents protested Azerbaijan's production-sharing agreements with four American oil companies to develop the area Azerbaijan considers in its Caspian sector. The area known as the Inam field in southern Caspian lies some 120 miles south of the Azerbaijan capital of Baku at 100 meters deep in the shallow and 200 meters in the deeper waters. See reports by Reuters from Tehran, Iran, August 6, 1997, and from New York, August 1, 1997.

153. North Sea Continental Shelf Cases (Federal Republic of Germany/Denmark) (Federal Republic of Germany/the Netherlands), Judgment of February 20, 1969, *International Court of Justice Reports* (1969).

154. Ibid., p. 46, paragraph 83.

155. Ibid., pp. 53-54, paragraph 101.

156. Ibid., p. 54, paragraph 101.

CHAPTER FIFTEEN

Legal Status of the Caspian Sea

Scott Horton and Natik Mamedov

Introduction

This chapter examines the legal status of the Caspian Sea from the point of view of the Azerbaijan Republic. Azerbaijan follows a policy of monopoly regarding its mineral resources, internal and territorial waters, the continental shelf, flora and fauna within the limits of the territory of Azerbaijan. Under Article 2 of the Economic Independence Law and Article 10 of Property Law, they are the exclusive property of the Republic. Azerbaijan also claims the right of possession, use, and management of resources located within its economic zone within the Caspian Sea. This is the subject of one of the more interesting international legal disputes now pending.

The collapse of the Soviet Union created a new category of international dispute, namely those between the former constituent states of the Union. In the Soviet period their relations were viewed as "international," but being fraternal socialist states, of course they had no disputes or disagreements with one another. With the dissolution of central control in Moscow, however, some of these disputes could no longer be suppressed. One of them is the escalating dispute over rights in mineral resources on the bed of the Caspian Sea. The four former Soviet republics and Iran have developed and articulated positions on this issue.

If anything, the Caspian Sea debate demonstrates that international law—from both the customary and treaty-law perspectives—does not provide ready solutions for territorial issues. However, international legal precedent does narrow and frame the debate considerably. In the end, questions of the ownership of the Caspian seabed and subsoil can only be satisfactorily resolved by mutual agreement among the littoral states, and this resolution is not presently in sight.

Importance of Oil

While a stable and internationally recognized regime in the Caspian is essential to the peaceful exploration and exploitation of its oil and natural gas reserves, the present ambiguities in its legal regime, however, have not deterred foreign investment in the Azeri waters of the Caspian. On September 20, 1994, the "Deal of the Century" was signed between the State Oil Company of the Azerbaijan Republic (SOCAR) and the consortium of Western oil companies, leading to the creation of the Azerbaijan International Operating Company (AIOC) to develop the oilfields of the Caspian shelf.

The $7.5 billion project covers three oil fields located up to about 100 kilometers offshore, namely the Azeri, Chirag, and Guneshly fields. The estimated oil reserves amount to 600 million metric tons (mt) of crude oil to be produced within 30 years. Current estimates quoted in the Azeri press claim that production will peak at 30 million mt (mmt) of oil per annum within 11 years, at which point daily production is expected to rise to 720,000 barrels (bls).[1]

Among other major oil development projects are: (1) joint venture (JV) Baku-Ponder Services, established to rehabilitate more than 2,000 wells, undertake shallow infield drilling and conduct feasibility studies on oil recovery from oil-soaked lands and waters; (2) Chevron, under an agreement with SOCAR, is presently undertaking evaluation of an unexplored region of the Southern Caspian Sea; (3) JV Ansad formed between Attila Doghan (Turkey) and SOCAR to complete development of the onshore Khilli and Neftchala fields; (4) British Petroleum and Statoil are joining in a feasibility study of development of the Dostlug field, with reserves currently estimated at 2 billion barrels and expected production of 200,000 bls/day.[2]

More recently, Exxon Azerbaijan Ltd. signed a memorandum of understanding with SOCAR granting Exxon exclusive rights to negotiate exploration, development, and production sharing agreements covering two offshore exploration blocks located in the Shakh-Deniz oil fields, located 80 to 120 kilometers south of Baku. In April of 1996, Pennzoil sold 5 percent of its 9.8175 percent holdings to Itochu (Japan) for a reported $132 million. In addition to Itochu, a major Japanese trading company, Mitsui, is reported to be close to an investment of about $1 billion in the Azerbaijani oil industry. The level and size of international interest in Azeri oil and gas, however, is also testament to the protection provided for in the Azeri Foreign Investment Law of January 15, 1992, which envisions: (1) participation in joint ventures with legal entities and citizens of Azerbaijan; (2) creation of enterprises, branches, and representative offices wholly owned by foreign investors; (3) purchase by foreign investors of enterprises, real estate, and equity securities; (4) purchase or lease of rights with respect to land and natural resources, as well as other property rights, presumably including concessions and licenses; and (5) conclusion of other foreign investment agreements.

The Foreign Investment Law contains a stabilization provision of a type found frequently in such legislation in the region. It promises protection against subsequent changes in the legal or regulatory regime that has a material adverse effect on an investment. This provision safeguards against "taking;" however, it does not apply if the subsequent unfavorable change occurs in the areas of taxation, military defense, national security, or environmental protection legislation.

Legal Status

International law provides authority to support two general lines of argument with respect to the Caspian: one that it is an "enclosed sea" under Law of the Sea concepts, and the second that it is an international or frontier lake. Following the first theory, the natural resources under the sea would be subject to the jurisdiction of the littoral states following a median line delineation using guidelines furnished by the 1982 Convention on the Law of the Sea; each littoral state would exercise exclusive jurisdiction over the mineral resources underlying its sector.[3]

Taking into account current estimates of petroleum reserves, this resolution would be highly advantageous to Azerbaijan and Kazakhstan, while dealing a more meager hand to Russia and Turkmenistan, among the Soviet Union's successor states.

The position equating the Caspian to an "enclosed sea" has several serious problems, including that the 1982 Convention does not clearly qualify the Caspian as an "enclosed sea" (article 123); that the relevant provisions of the Convention are generally not viewed as a codification of customary law in any event; and that while the Convention applies only to its members, none of the littoral states in question has acceded to the Convention.

The alternative view, that the Caspian is an "international lake," rests most significantly on established custom and usage in the region—in this case, between the Russian Empire and the Soviet Union, on one hand, and Iran on the other. Under this view, the littoral states each have a limited exclusive zone on the rim of the Caspian, while the core region is subject to joint disposition by the littoral states. This view may be supported by some conduct during the Soviet period, but other activities (including some in the petroleum sector) establish a competing usage theory. Moreover, even with respect to "international lakes," the clear trend internationally has been to demarcation down the middle as opposed to some sort of condominium ownership, as with the North American and African Great Lakes, Lake Constance, and Lake Geneva in Europe. In sum, it would seem that international law does not provide a clear resolution, nor do international authorities line up solidly behind any side of the controversy.

The Russian Perspective

Numerous different viewpoints have been advanced on behalf of Russia. Regardless, the participation of Russia's LUKoil company with significant state ownership in several major Azerbaijan-based transactions has been interpreted in some quarters as a sign of Russian assent to Azerbaijani claims. The Russian ministry of foreign affairs has gone to great lengths to rebut such conclusions, however. Assuming the ministry speaks for Russia on the question,[4] the Russian position can be summarized in five essential points:

1. The Caspian is a lake for purposes of international law because it is landlocked, unconnected to the world's oceanic network and therefore not an international body of water subject to rights of navigation for the benefit of the maritime nations. Therefore, the elaborate system of international customary and treaty law that is designed to secure rights of free navigation as well as to define rights in territorial seas, exclusive economic zones, and the continental shelf do not apply to the Caspian.

2. Legally, the Caspian is a joint-use area and issues pertaining to all activities, particularly including the exploitation of mineral resources of the subsurface, should be resolved with the participation and consent of the littoral states.

3. The status of the Caspian must be resolved in a manner consistent with the USSR-Iranian treaties of February 26, 1921, and March 25, 1940. Azerbaijan, Kazakhstan, Russia, and Turkmenistan are successors to the USSR and as such are bound by the treaty commitments made by the USSR. The two USSR-Iranian treaties establish a zone of exclusive economic interest and a core area of joint interest and these principals and the custom and usage created in the course of USSR-Iranian relations establish the customary and treaty law governing the Caspian.

4. The prior treaties with Iran do not state a comprehensive basis for development of the Caspian and that definition of relations between the four littoral successor states of the USSR is needed. However, this can occur only through a formal agreement to be concluded among those states.

5. The unilateral action by any state and efforts to develop the Caspian by any littoral state without the conclusion of a formal treaty regulating such development activities is illegal and will not be sanctioned by the Russian Federation.

The Russian position is hardly an incontrovertible exposition of international law. Even if the Caspian is to be treated as an international lake, it hardly follows that it should be treated in the manner the Russians prescribe; indeed, the strong trend of international custom has been toward a median-line division of international lakes, as in the cases of the Great Lakes between the United States and Canada, Lake Geneva between France and Switzerland, and Lake Constance among Austria, Switzerland, and Germany. Therefore, it is vital to the Russian position to establish the Iran–Soviet Union treaties as the major treaty source governing the Caspian. Custom under those treaties has been quite ambiguous, however. Obviously, the Russian and Iranian sides today can freely offer self-serving characterizations of the relationship. Still, there is ample evidence of a long-established view between the nations that the Caspian is "a sea belonging to the USSR and Iran," to quote an exchange of notes associated with the 1940 Treaty.

In the Soviet Union, certain jurisdictional lines had been established between the Soviet Union and the union republics, which Soviet law considered to be subjects of international law. These lines, established in the relevant industries, provide support to Azerbaijani and Kazakhstani claims to the offshore fields in the Caspian.[5] Hence, the Russian claim that there are no delineations between the Soviet Union's successor states is strained. Russia insists that such delineations must be by formal treaty, which, in fact, is absent. Finally, it should be noted that most formerly constituent republics of the Soviet Union do not view themselves as legal successors bound by the treaty undertakings of the Soviet Union. Moreover, the Transcaucasian and Baltic republics were all, in fact, independent sovereign states prior to their military occupation by the Red Army, and their rejection of succession arguments is therefore particularly persuasive since they can make a claim that they were nations under military occupation who clearly cannot be bound by the treaty undertakings of their occupier. Therefore, it is not all that "clear" that the Iran-Russia treaties should be binding on Azerbaijan, Kazakhstan, or Turkmenistan.

A particularly troubling aspect of Russia's position has been its public stance that no development should occur in the absence of a final treaty reconciling the views of the littoral states. This threat runs sharply counter to Russia's undertakings in the Energy Charter Treaty, a European initiative signed in Lisbon, Portugal, in 1994, which is designed to clarify rights of transportation and access with respect to energy products. Of late, at least in practice, the Russian position may be softening. The Russia-Azerbaijani Oil Transportation Treaty of February 18, 1996, and joint Russian-Azeri efforts to iron out their differences over the pipeline from Baku to Novorossiysk speak of reconciliation between Russia and Azerbaijan. Nevertheless, the Russian ministry of foreign affairs cautioned about the treaty: "[it] should not be construed as changing the

Parties' positions regarding the status of the Caspian Sea." That may be so, but the treaty and LUKoil's participation in Azeri offshore ventures imply the contrary.

Azerbaijan's Position

Azerbaijan's position is driven clearly by the fact that a "median division" theory allows Azerbaijan to claim the lion's share of the Caspian's petroleum resources. The Azerbaijani position is that the Caspian is an international border lake without access to oceans and that, as such, the resources underlying the sea must be divided between the littoral states along median-line principles. In support of this position, Azerbaijan refers to the pattern derived from the division of comparable bodies of inland water, such as Lakes Superior, Huron, Erie and Ontario between the United States and Canada; Lake Chad among Chad, Niger, Nigeria, and the Cameroon; Lake Geneva between France and Switzerland; and Lake Constance among Austria, Germany, and Switzerland.

In the case of the aforementioned lakes, a median-line approach was used and the littoral states ultimately received exclusive zones of jurisdiction with respect to the surface, air, and subsurface of the lakes. Within their zones of exclusivity, the littoral states exercise exclusive rights over fishing, boating, shipping, exploration, and development of natural resources on the floor of the lake and other activities; the permission of another littoral state must be sought only when activities reach into the zone of the other littoral state. Azerbaijan has expressed its willingness to enter into a Caspian Sea treaty, provided that the treaty reflects these principles.

Azerbaijan has rejected Russia's thesis that a resolution of Caspian's legal status precede any development of its offshore resources. The Azerbaijani position states that, in absence of a treaty, the concept of a "median line" governs the division of the lake; Azerbaijan is free to explore and develop the area and grant licenses and concessions therein. Moreover, Azerbaijan points to a series of ministerial acts and laws of the Soviet era that recognized the exclusive authority of the Azerbaijan Soviet Socialist Republic and some of its political subdivisions and state-owned enterprises over an area within the Caspian. More significantly, Azerbaijan points to a bilateral Russia-Azerbaijan Agreement on the Development of Oil and Gas Deposits, dated November 20, 1993, which, in the Azerbaijan's view, recognizes Azerbaijan's authority over part of the Caspian, and particularly the Azeri and Chirag oil deposits. Moreover, Azerbaijan notes that both the Soviet Union and Iran have had a long legacy of unilateral exploitation of Caspian's natural resources, a fact that refutes the basis of their contemporary claim to the Caspian being a "joint interest zone," and that a formal delimitation treaty should precede exploitation/development in the Caspian.

Kazakhstan's Position

Initially, Kazakhstan's strategy on the Caspian was motivated by twin goals of retaining authority over its own Caspian shelf projects and not being confrontational with Russia. Recent trends in Kazakhstan's policy, however, show greater independence. In its most formal statements, Kazakhstan has supported several of the key elements of Azerbaijan's position that all littoral states should exercise exclusive control of the resources of the Caspian Sea located within their territorial waters and continental shelf. Still, Kazakhstan has expressed some interest in and support for various Russian proposed delimitation zones, and has supported Russia's view that the issues should be resolved in a treaty between all the littoral states. On the key question of development pending such resolution, Kazakhstan has taken the position that its economic interests dictate that resolution of the dispute must not impede immediate development of the region, including offshore oil projects.

Turkmenistan's Approach

In the beginning, Turkmenistan's position in the matter had been relatively ambiguous; recently, however, Turkmenistan has questioned the legal validity of Azerbaijan's offshore activities, claiming that some of these infringe upon Turkmenistan's rights. Turkmenistan's Law on National Borders (1992) sets a 12-mile territorial limit in the Caspian Sea, a position that inclines to support the view held by Russia and Iran.

Iran's Position

The dissolution of the Soviet Union created something of an opening for Iran to claim rights to resources that under the erstwhile Soviet Union–Iran regime would have been non-existent. In all recent debates on the Caspian's legal status, Iran has articulated positions that closely follow those of the Russian Federation. On the other hand, Iran's state-owned entities have participated in oil and gas developments and pipeline projects with Azerbaijan, Kazakhstan, and Turkmenistan that, when considered together, signal some willingness to proceed with the development of Caspian's resources without the benefit of a comprehensive treaty resolving the delimitation of offshore boundaries. Iran's position is, however, driven in some measure by Iran's analysis of the United States' close relations with Azerbaijan.

Conclusion

All the littoral states seem to be in agreement that the current controversy would best be resolved by a treaty that delimits clearly the territorial rights of each state and provides a clear path to development of the
Caspian's resources. However, the differences between the states are still
significant and no serious reconciliation of these differences may be in
sight for some time to come. The development of Caspian's resources will
proceed, however, even though some concessions/licenses may be clouded
because of competing claims by the littoral states.

Notes

1. See generally, James P. Dorian and Farouk Mangera, *Oil and War: Impacts on Azerbaijan and Armenia*, East-West Center Working Papers, nos. 16-18 (September 1995).
2. Ibid.
3. Under the United Nations Convention on the Territorial Sea and the Contiguous Zone, *U.N.T.S.* No. 7477, vol. 516 (Geneva, April 29, 1958), p.
2055, "the sovereignty of a [coastal] state extends, beyond its land territory and internal waters to a belt of sea adjacent to its coast, described as
territorial sea." This sovereignty extends to the air space over the territorial sea as well as to its bed and subsoil. The United Nations Convention
on the Law of the Sea (Montego Bay, Jamaica, 1982), which entered into
force on November 16, 1994, establishes that "every coastal state has the
right to establish the breadth of its territorial sea up to a limit not exceeding 12 nautical miles, measured from baselines" (Art. 3). Also, under the
United Nations Convention on the Continental Shelf, *U.N.T.S.* No. 7302,
vol. 499 (Geneva, April 29, 1958), p. 312, which entered into force on
June 10, 1964, each coastal state enjoys over the continental shelf sovereign rights for the purposes of exploring it and exploiting its natural
resources (Art. 2). These rights are exclusive "in the sense that if the
coastal state does not explore the continental shelf or exploit its natural
resources, no one can undertake these activities, or make a claim to the
continental shelf, without the express consent of the coastal state" (Art.
3). Article 1 of the Convention defines the continental shelf to include:
"(a) the seabed and subsoil of the submarine areas of the territorial sea, to
a depth of 200 meters or, beyond that limit, to where the depth of the
superjacent waters admits of the exploitation of the natural resources of
the said areas; and (b) the seabed and subsoil of similar submarine areas
adjacent to the coasts of islands."
4. This summary is based on the most authoritative and thorough statement
of the Russian position, contained in a position paper entitled "Position
of the Russian Federation Regarding the Legal Status of the Caspian Sea,"
which was submitted to the General Assembly of the United Nations on
October 6, 1994. While the Russian Federation has subsequently sug-

gested different negotiating positions, it has never deviated significantly from the October 1994 position.

5. See, for example, USSR Law on Water Space (1970); USSR Law on State Borders (1982); Joint Act of the Azerbaijan SSR Council of Ministers and the USSR Ministry of Oil and Gas (Jan. 18, 1991), all applying notions of strict territorial delineation as opposed to "joint use," and the latter recognizing Azerbaijani authority over properties presently disputed by Russian.

CHAPTER SIXTEEN

Iran's View on the Legal Regime of the Caspian Sea

Mohammad Ali Movahed

Introduction

For thousands of years, the body of water that the Iranians call the Sea of Khazar[1] or Sea of Mazandaran[2] and the rest of the world knows as Caspian[3] has belonged to the peoples and nations living on its coasts. The dissolution of the Soviet Union has not changed that historical fact; what has changed is that newly independent states like Azerbaijan, Kazakhstan, Turkmenistan, and the Russian Federation have been drawn into a debate with respect to the legal status of the Caspian. This chapter provides an examination of the status of the Caspian Sea from an Iranian perspective.

The 1921 and 1940 Iran-USSR Treaties

Iran's view of the legal status of the Caspian is shaped by and rests on Iran's treaty relationship with the former Union of Soviet Socialist Republics, chief among which are the 1921 Treaty of Friendship between Persia and the Russian Socialist Federal Soviet Republic (1921 Treaty)[4] and the 1940 Treaty of Commerce and Navigation between Iran and the Soviet Union (1940 Treaty).[5] Between them, these treaties provided the general premise that: (1) neither party enjoy a preferential status vis-à-vis the other, both parties having equal access and equal entitlement to the Caspian; and (2) the Caspian be closed to all but Iranians and Soviets. More specifically, the exclusive fisheries zone provided for in the 1940 Treaty and its related documents was an illustration of the mutual understanding of the parties in regard to the aforementioned principle of equality. The said zone consisted of an area ten nautical miles wide allocated to each party for its exclusive benefit, thus leaving the vast area in the middle of the sea subject to the common use by both parties. In this manner, the 1921 and 1940 treaties provided the basis of co-ownership of the Caspian Sea by the Soviet Union and Iran.

On December 8, 1991, Belarus, Ukraine, and the Russian Federation entered into the Minsk Agreement, announcing the death of the Soviet Union and birth of the Commonwealth of Independent States.[6] Under article 12 of this agreement, "[t]he High Contracting Parties undertake to discharge the international obligations incumbent on them under treaties and agreements entered into by the former Union of Soviet Socialist Republics." On December 21, 1991, by virtue of the Protocol to the Minsk Agreement,[7] Azerbaijan, Kazakhstan, the Russian Federation, and Turkmenistan, among others, acceded to the Minsk Agreement, becoming state-participants in the Commonwealth of Independent States. On the same day, Azerbaijan, Kazakhstan, the Russian Federation, and Turkmenistan, among others, signed onto the Alma Alta Declaration,[8] whose sixth declaration obligated the declarant to "guarantee in accordance with their constitutional procedures the discharge of the international obligations deriving from treaties and agreements concluded by the former Union of Soviet Socialist Republics."

The general applicable principles of succession of states aside, as a matter of treaty law alone, therefore, the breakup of the Soviet Union cannot be deemed to have altered factually or legally the realities concerning the Caspian Sea as established in the Iran-USSR treaties. It goes without saying that at the time of conclusion of the 1921 and 1940 treaties, prospects for the development of the natural resources of the Caspian's seabed and subsoil did not exist. Therefore, the treaties provided nothing particular in this respect. However, the great deal of interest shown by the foreign companies in the oil and gas resources of this area has resulted recently in intense discussions often colored by excitement and misunderstanding, straining the otherwise peaceful relations among the littoral states.

The Russian View

The present controversy regarding the legal status of the Caspian Sea began with reports that Azerbaijan and a consortium of Western oil companies were negotiating an agreement for the development of the hydrocarbon resources in "the Azerbaijan sector of the Caspian Sea." Because a British oil company was among the parties to the consortium, on April 28, 1994, the Russian foreign ministry promptly sent a note of protest to the British embassy in Moscow, stating that the legal status and regulations of the Caspian Sea are determined by the Soviet-Iranian agreements of 1921 and 1940, and in accordance with the principles and norms of international law, Russia, Iran, and other states on the Caspian are bound by the regulations of these agreements in the absence of other treaties.

A more articulate exegesis of the Russian position is found in the diplomatic note submitted by the Russian government to the United Nations, dated October 5, 1994.[9] Therein, Russia elaborated on its official position regarding the exploration and development of the subterranean mineral resources of the Caspian Sea. According to the note, the standard international law of the sea concepts of territorial sea, exclusive economic zone, and continental shelf did not apply to the Caspian Sea, which has no natural link with the world oceans. This view is buttressed by the 1982 Law of the Sea Convention[10] itself, which in article 122 excludes from the purview of the Convention lakes and other bodies of water having no natural outlet to a sea or ocean. Accordingly, the Caspian Sea is not considered even an "enclosed sea" or a "semi-enclosed sea" under the Convention.

The October 5th note stressed that the 1921 and 1940 Iran-USSR treaties govern the legal status of the Caspian Sea and, accordingly, the coastal states are to abide by the terms and conditions thereof. Nevertheless, the note pointed out that "[t]he legal regime, established in these Agreements, requires further development in view of the changed circumstances, including the establishment of new Caspian States."[11] Hence, the argument proceeds, for now, the coastal states must refrain from any uni-lateral action in relation to the development and exploitation of the Caspian Sea's mineral resources until such time as the coastal states can reach an overall agreement as to a new legal regime in the Caspian. Until such time, the note emphasized, any decision with regard to the development of the mineral resources of the Caspian Sea must be made with the participation of all the coastal states.[12]

The note also emphasized the need for cooperation among the coastal states with respect to fisheries and the environment. After calling attention to the vital significance of the Caspian's resources, the note singled out the particular importance of the sturgeon, in terms of numbers and variety of the specie. The note stressed, therefore, all forms of use of the Caspian's resources, mineral as well as living, must be undertaken through a common and concerted effort so as not to harm the flora and fauna of this unique body of water. To prevent the bringing about of an ecological disaster, the note urged the coastal states to observe the present legal regime of the Caspian and refrain from committing uni-lateral acts, warning that such acts inherently would do violence to the "joint-use" nature of the Caspian Sea, affecting the rights and interests of the other coastal states. "The national claims of one or another Caspian State in relation to the area or the resources of the Caspian Sea," the note concluded, "invariably involve the rights and interests of other Caspian states and cannot be considered justified."[13]

Other Views

The concept of co-ownership or joint use of resources embodied in the aforementioned Iranian and Russian views does not sit well with Azerbaijan and Kazakhstan. The considerable reserves of oil and gas discovered already occur in the sea area adjacent to the coasts of Azerbaijan and Kazakhstan. Accordingly, these states advocate the partitioning of the Caspian Sea into five national sectors, coincidentally in such a manner as to position most of the discovered reserves in their own sectors.

Iran and Turkmenistan have not shown an inclination in becoming involved in the controversy on an international level. Instead, they have expressed their belief and desire that the outstanding issues be resolved through dialogue among the five Caspian states. It is self-evident that sincerity and good faith among the five Caspian states are among prerequisites to reaching any agreement in this respect. Each state must begin with a genuine appreciation of another's legitimate interest, avoiding confrontational posturing; ultimately, the proper development of the Caspian Sea's resources requires stability and confidence, which can only be achieved through close cooperation and understanding.

A Colloquy

In February 1995, an international seminar in London bore witness to a reasoned discussion with respect to the various aspects of the Caspian Sea.[14] The head of the Russian foreign ministry's legal department, Alexander Khodakev, pointed out that the Caspian states must respect each other's legitimate rights and interests. After stressing the need for dialogue in order to resolve the contentious issues related to the Caspian Sea's legal order, he highlighted the importance of creating a system for the coordination of efforts and views among the five littoral states.[15] The foreign minister of Kazakhstan, Vyacheslav Gizzatov, specified that the Caspian Sea's outstanding issues are linked to the vital interests of the five coastal States. Moreover, he stated, peace and stability in the region depended on prudent resolution of issues on the basis of mutual respect for each other's legitimate interests. He expressed confidence that the creation of a regional cooperation body for the development of Caspian's resources could help sort out the differences of opinion among the coastal states.[16]

In the conversation regarding the legal status of the Caspian Sea, certain arguments of obscure nature and ambiguous purport threaten to confuse the issues and complicate the situation, so as to hinder any reasonable resolution of the differences among the coastal states. One such argument is the so-called internationalization of the Caspian Sea regime.[17] According to general international law and the 1921 and 1940 Iran-USSR treaty law, the Caspian Sea is a landlocked sea to be used exclusively by its littoral states. Therefore, it is not clear at all what the suggestion as to internationalization of the Caspian Sea regime has to offer or what the people who entertain such an idea have in mind.

This is a simple point. The international boundaries established by Iran and the Soviet Union, which today form the land frontiers of Iran with Azerbaijan and Turkmenistan, have never been subjected to any doubt by these countries after the breakup of the Soviet Union. Yet, some may go so far as to question the validity and controlling authority of these treaties just so that they may appropriate a part of the Caspian. There is no legal justification, therefore, for Azerbaijan or Turkmenistan to contest the 1921 and 1940 Iran-USSR treaties with respect to their equal and undivided rights in the sea area, to which they have succeeded first as a matter of general international law of state succession and, second, by virtue of their consent to the Minsk Agreement and Alma Ata Declaration.

The Legal Considerations

From a purely legal point of view, there are three questions of crucial importance that require proper examination and analysis. They are: (1) Was there a demarcation line agreed upon by Iran and the Soviet Union partitioning the Caspian Sea between the two countries?; (2) Does the historical conduct of the parties point to an understanding with regard to the partition of the sea?; and (3) If the answers to questions (1) and (2) are in the negative, then could the Caspian Sea be considered a condominium, subject to joint sovereignty of its coastal states?

Some have commented on the existence of a line going straight across the Caspian Sea's southern shore from Astara in the west to the Hossein Kuli Bay in the east. The suggestion is that the division of the Caspian Sea between the Soviet Union and Iran has been accomplished already and the current boundary dispute concerns only the former republics of the Soviet Union bordering the Caspian. This view ignores totally the official Iranian and Russian assertions that there was in fact no such dividing line agreed upon and, therefore, to suggest its existence is a mistake.

The existence of this "imaginary line" apparently originated in a book written by a Soviet writer and now rather conveniently is referred to by some commentators in complete disregard of the official position of both Iran and Russia. For example, in February 1996, the director of international energy policy at the United States Department of State stated: "The Soviet Union and Iran established exclusive zones in the Caspian and an international boundary."[18] The fact of the Soviet Union and Iran having established exclusive fisheries zones in the Caspian is not in dispute. The fisheries zones were established under the 1940 Treaty. The allegation that the two countries established also an "international boundary" between them in the Caspian is not supported by the facts. Equally misinformed, there is the statement that "[a]s regards boundaries, the USSR state boundary in the Caspian was the straight line drawn between the respective Soviet/Iranian land boundaries on either side of the Caspian" and this "boundary between USSR and Iran dividing the Caspian Sea will have been presumed to have divided the sea bed as well."[19]

On the other hand, in 1995, the foreign minister of Kazakhstan, Vyacheslav Gizzatov, attempted to be more specific about the existence of this "imaginary line." "The present status of the Caspian Sea," he wrote," was determined by the Treaty between the RSFSR and Iran (Persia) of 26 February 1921 and by the Treaty between USSR and Iran of 25 March 1940 according to which the waters of the Caspian were divided between the USSR and Iran by the Astara-Gasankouli line."[20] The problem with this assertion is that neither the 1921 Treaty nor the 1940 Treaty refers to a boundary line in the Caspian.[21] As the head of the Russian foreign ministry's legal department pointed out in 1995, "[o]ne sometimes hears assertions to the effect that the Caspian Sea was divided by the so-called Astara-Gassankuly line. Again, this has nothing to do with the legal regime of the Caspian established by the 1921 and 1940 treaties, none of them mention this line."[22]

Some may suggest that if all of the Caspian Sea was the object of joint use by Iran and the Soviet Union, then why was the Soviet Union's exploration for oil off the coast of Azerbaijan SSR in as early as 1949 done without consultation with Iran, never seeking permission from Iran or sharing any of the proceeds with Iran. A cursory review of history will show that this was not the first instance in which the Soviet Union had acted in disregard of its international obligations. Furthermore, this type of rationale is misguided because it is unlikely that Azerbaijan or Kazakhstan will ever consider the treatment they received under the Soviet Union as legitimate or lawful. The breach committed by the Soviet Union is no reason for Russia, Azerbaijan, or other former Soviet republics to deny Iran's rights and interests in the Caspian. Iran's silence or lack of protest in the face of this breach by the Soviet Union is no evidence of Iran's consent or acceptance of the Soviet Union's conduct. For very obvious political considerations, Iran may have maintained silence with respect to the Soviet activities in the Caspian.

Some may invoke Article 3 of the 1927 Iran-USSR Treaty of Guarantee and Neutrality[23] as evidence of Iran and the Soviet Union agreeing in essence to the territorial division of the Caspian. Said article provided that neither party enter into any third-party alliance or agreement directed against the security of the other's "land or waters." Some may suggest that this provision referred to "territorial waters," thus implying that the parties were probably thinking in terms of division of the Caspian into different sectors or territorial waters, as would have been the case with open seas and oceans. This interpretation is not borne out by the text of the provision.[24] Furthermore, the Caspian would not have been the only body of water concerning the contracting parties: Iran, for one, bordered on the Persian Gulf and the Gulf of Oman, the Soviet Union's coastal limits included frontage on the Pacific Ocean, the Black Sea, and the Gulf of Finland.

One may conclude, therefore, with a reasonable degree of certainty, Iran and the Soviet Union never acted in any manner to indicate a delimitation of the Caspian Sea into their respective areas of jurisdiction; they never thought in terms of dividing Caspian's surface or seabed. There being neither evidence of an Iran-USSR demarcation line nor of an Iran-USSR practice or course of conduct with respect to territorial division of the Caspian, the remaining query as to the supposition of a joint Iran-USSR ownership or condominium over the Caspian becomes all the more purposeful. To appreciate the argument in this regard, one must go back to the annals of Iran-Russia relations, to almost 100 years prior to the 1917 Marxist revolution in Russia.

The first treaty between Iran and tsarist Russia with a provision on the Caspian Sea was the 1813 Treaty of Golestan.[25] The treaty was the outcome of a war that had started ten years earlier, in 1803, when the Russian army under the command of General Sissianov occupied Tbilisi and Ganjah, subjecting the inhabitants of the latter town to a three-day massacre. Iran reacted, and in consequence of the ensuing war, it lost a vast area of its possessions in the Caucasus. During the war, the Caspian had proved to be of critical importance to the Russian military, as they attacked Baku from the sea and even attempted but failed to land in Iran's port of Anzali and capture Rasht.

Under the Golestan Treaty, Iran lost Dagestan, Georgia, Baku, Ganjah, Shirwan, and the northern part of Talish. Also, it ceded its right to have warships in the Caspian Sea; only Russian warships were allowed to sail there. Iran, however, retained its right to have merchant ships in the Caspian, both parties being entitled equally to navigate throughout the sea. In 1826, once again war broke out between Russia and Iran, this time lasting for 20 months. The ensuing 1828 Treaty of Turkmanchai[26] repeated the stipulation of the Treaty of Golestan Treaty as to the Caspian.

In consequence of the October Revolution, the Russian revolutionaries made various declarations to the effect that the provisions of treaties imposed by the tsarist regime through coercion and use of force on neighboring nations would be made null and void. On June 26, 1919, the people's commissary for foreign affairs, Lev Davidovich Bronstein, also known as Leon Trotsky, wrote to the Iranian prime minister offering the annulment forever of all treaties and agreements imposed on Iran by Russia through coercion, or incompatible with the principle of independence and security of Iran, or that limited or restricted its freedoms in the territory within its possession and in the adjacent seas.[27] The 1921 Treaty followed: Article 8 of the Turkmanchai Treaty was declared null and void, allowing Iran to maintain warships in the Caspian and providing for the parties to have equal rights and freedom of navigation under their own flags.[28]

The 1921 Treaty provision in respect of the Caspian Sea was amplified and expanded upon later, in the 1927 Iran-USSR Fisheries Agreement[29] and the 1940 Treaty. In 1931, Iran and the Soviet Union signed the Convention of Establishment, Commerce and Navigation.[30] On that occasion, the Iranian foreign minister, while "recognizing the exceptional importance to the Soviet Union of the Caspian Sea, which is considered by our two governments as a Persian and Soviet sea," assured the Soviet ambassador in Tehran that the Iranian government would not allow any third-country expatriate employed in its Caspian ports to engage in any activity injurious or contrary to the interests of the Soviet Union.[31] The Soviet ambassador acknowledged receipt of Iran's assurances, without any objection to the characterization of the Caspian as an Iran-Soviet sea.[32]

It may be stated that in law the presumption is against co-ownership; in other words, co-ownership must be created by a contract. Co-ownership is rarely, if ever, created by a contract. Instead the contract is drawn up to determine the parties' rights to use and enjoyment of a jointly owned property. In other words, there must be a distinction made between co-ownership or a state of community that is a legal fact (*fait juridique*) and the arrangement required for the administration of a common property that is a legal act (*acte juridique*). In any case, to say that the Caspian Sea is owned jointly by its littoral countries is not an impediment to the coastal states to reach a fair and just agreement in the future as to use and enjoyment of the sea.

Moreover, the fact that the Caspian Sea is co-owned does not mean that it must maintain this status forever. If the five Caspian states arrive at the conclusion that the division of the sea into five sections is more in line with the interests of each of the five parties, then the joint ownership of the Caspian Sea can be ended by agreement. Any divided property was once part of a whole and any property can one day be partitioned.

The Gulf of Fonseca Case

In 1992, the International Court of Justice (ICJ) ruled that the Gulf of Fonseca geographically bordering El Salvador, Honduras and Nicaragua was indeed owned jointly by the three countries. The decision was based on the evidence that: (1) the three littoral states were successors to the gulf's erstwhile sovereigns, the government of Spain and the Federal Republic of Central America; and (2) no boundary limits had been specified either before or after breakup of Spanish rule and the Federal Government of Central America. Therefore, the ICJ concluded that the Gulf of Fonseca had its own unique and special status and its waters were subject to co-ownership as a condominium of the three states.[33]

The legal status of the Gulf of Fonseca had been subject to litigation some 75 years earlier before the Central American Court of Justice. On that occasion, El Salvador sued to stop the transfer of a naval base by Nicaragua to the United States, claiming that said grant threatened its security and violated its co-ownership rights in the gulf. In ruling for El Salvador,[34] the court ruled unanimously that the gulf in question had become an area of vital interest to the republics of El Salvador and Nicaragua because of its geographical location and intense economic development activities on its shores. Necessarily, the court argued, any decision with respect to the gulf had to take into account the characteristics of the gulf from the points of view of history, geography, and the vital interests of the littoral states.[35] The court held that the gulf "is a historic bay possessed of the characteristics of a closed sea," and that the parties are agreed that Gulf is a closed sea.[36] By the term "closed sea," the court seems to have meant simply that the gulf was not part of the high seas and its waters thus were not international waters.[37]

The Fonseca case offers a very apt analogy for the future of the Caspian Sea. The Caspian's position, in many ways, is identical to that of the Gulf of Fonseca. Although the land frontier between the former Caspian republics of the Soviet Union and Iran has been specified and agreed upon, the sea boundary between them was never determined. In fact, Iran and the Soviet Union had stated explicitly that the Caspian Sea, in its entirety, belonged to them jointly, equally, and exclusively. The two countries had emphasized repeatedly the equality of their rights in the use of the Caspian. Therefore, just as in the case of the Gulf of Fonseca, the Caspian Sea had never been partitioned in any way by Iran and the Soviet Union. The municipal laws of Iran and the Soviet Union considered the Caspian Sea a "closed sea," meaning a landlocked sea not part of the high sea, exactly in the same sense as the ICJ's characterization of the Gulf of Fonseca.

In the Fonseca case, there had existed a strong community of interest among the three coastal states on the Gulf of Fonseca, similar in many ways to the community of interests among the five states surrounding the Caspian Sea. The community of interest, with its requirements of equality of use, common legal rights, and, more particularly, the absence of any individual preference or privilege in the gulf, represented the embodiment of a "condominium."[38]

Just as the facts in the Fonseca case were *sui generis,* so is the totality of the circumstances surrounding the Caspian Sea. The solutions to problems involved in such a case could not be reasonably expected from patented general rules of international law. The Gulf of Fonseca was not a single-state bay. It was necessary, therefore, to investigate the particular history of the gulf in order to discover the regime resulting therefrom. As the ICJ stated in 1982: "[g]eneral international law . . . does not provide for a single regime, for 'historic waters' or 'historic bays,' but only for a particular regime for each of the concrete, recognized cases of 'historic waters' or 'historic bays.'"[39]

There is no legal presumption against co-ownership as such. In the Fonseca case, Honduras argued against the existence of a condominium on the ground that condominium can only be established by agreement. The ICJ did not find that argument persuasive. The ICJ's finding in the Fonseca case leads to the conclusion that the essential juridical status of the gulf was the same as that of internal waters,[40] since they were claimed "a titre de souverain" and were not, therefore, territorial seas.[41] The same conclusion could obtain naturally and justifiably in the case of the Caspian Sea.

New Arrangement

Any jointly owned property can be divided among its owners. In the Fonseca case, the court recognized this principle when it suggested that the interested parties could replace by agreement the condominium regime with a division and delimitation of the gulf into three separate zones according to international law.[42] An agreed partition of a jointly owned territory and its adjacent waters did in fact take place with respect to the Neutral Zone between Kuwait and Saudi Arabia. Under certain circumstances, the Caspian Sea can also be partitioned. The critical question facing the interested parties is to determine under what criteria the partitioning should be carried out. Should the Caspian states decide to jointly exploit Caspian's subterranean resources, it will be necessary to create a regional body, under the authority of the Caspian states, to manage and to execute the regulations pertaining to the sea. This regional body would apportion certain sectors of the sea to financially and technologically qualified oil and gas companies for exploration and development. It would be the responsibility of this body to control and supervise the work being conducted and to pay each country its share of the proceeds in cash or in kind.

One of the problems with the partitioning of the Caspian Sea into national sectors is that oil and gas resources are not equally or uniformly distributed throughout the sea. The subterranean oil and gas structures are located in specific areas and it is by no means certain that exploration in any part of the sea will result in a commercial discovery. Similarly, oil and gas reserves do not occur neatly in the confines of a specific area of national jurisdiction; they may prove to straddle boundary lines. Therefore, any attempt at the partitioning of the sea must take into account simultaneously: (1) the principle of equality of rights in the use and enjoyment of the Caspian Sea's resources among the littoral states; and (2) the principle of equity and proportionality to smooth over the inequities created by a strict application of norms based purely on equality. Both these principles are enshrined in the 1921 and 1940 treaties. Furthermore, the notion of equitable apportionment of offshore resources is a recognized principle of international law.[43]

A solution may be found in a compromise between the Russian position, supported by Iran, and the position held by Azerbaijan and Kazakhstan—to divide some of the sea into exclusive economic zones and keep the middle portion of the sea a commonly owned area. A precedent for this exists already in the 1940 Treaty, wherein Iran and the Soviet Union each received an exclusive coastal fisheries zone, ten nautical miles wide, with the remainder of the sea left open to equal navigation and fishing by the parties. Under a multi-lateral agreement, each littoral state would receive an exclusive economic zone of an agreed-upon width and the remainder of the sea would be deemed as owned jointly by the littoral states, its exploitation to be supervised by a regional body.

Conclusion

There are problems relevant to the Caspian that do not concern all of the coastal states, even though the resolution or lack thereof would invariably affect the political atmosphere regionally. One such problem area concerns Kazakhstan's right to use the Volga-Don and Volga-Baltic channels, both situated in Russian territory. Kazakhstan's position rests on the following: (1) the Volga waterway system and the channels had been developed under the Soviet regime, toward which all the union republics contributed; and (2) the Akhtupa River, an essential part of the Volga waterway system, itself flows from Kazakh territory. Therefore, the Kazakh position argues, it would not be justified to allocate the entire right of use and benefits of such a waterway exclusively to Russia.

Regardless of the merit of the Kazakh position in the foregoing, the issue represents a controversy to be resolved by and between Russia and Kazakhstan on the basis of equality and equity. Yet, with respect to the Caspian itself, no solution without the consent of all five littoral states would do, for which good faith negotiations, cooperation, and good neighborly relations are required both to create and then to maintain a new legal regime for the Caspian Sea.

Notes

1. The name "Khazar" refers to a people who lived on the northern shore of the Caspian, with their capital city at Itel or Atel located on the west bank of the Volga River. The name "Itil" or "Atel" was the Turkish name of the Volga. See D. M. Dunlop, *The History of the Jewish Khazars*, Princeton Oriental Studies, vol. 16 (Princeton: Princeton University Press, 1954), especially pp. 89-116. See also Arthur Koestler, *The Thirteenth Tribe* (London: Hutchinson & Company, 1976).

2. Mazandaran has been one of Iran's two provinces on the Caspian Sea, the other one being Gilan. In 1996, the Iranian parliament established a third provice on the southeastern corner of the Caspian, to be called Gorgan.

3. The name Caspian derives from the people known as Caspi, who lived on the southern shores of the Caspian, after whom the famous Iranian city of Qazvin too is named. The Caspi or Caspian were the indigenous inhabitants of the area before being replaced by the Aryans around 1000 B.C.

4. Signed at Moscow, February 26, 1921, reprinted as no. 268 in *League of Nations Treaty Series*, vol. 9 (1922), pp. 401-413; *British and Foreign State Papers*, vol. 114 (1924), pp. 901-909.

5. Signed at Tehran, on March 25, 1940, reprinted in *British and Foreign State Papers*, vol. 144:1940-1942 (1952), pp. 419-431.

6. Agreement Establishing the Commonwealth of Independent States among RSFSR, Ukraine, and Belarus, done at Minsk on December 8, 1991, reprinted in *International Legal Materials*, vol. 31:143 (1992) (UN Doc. A/46/771 of December 3, 1991). Hereinafter referred to as the "Minsk Agreement."

7. Protocol to the Agreement Establishing the Commonwealth of Independent States, done at Alma Ata, December 21, 1991, reprinted in *International Legal Materials*, vol. 31:147 (1992) (UN Doc. A/47/60 of December 30,1991).

8. Alma Ata Declaration of December 21, 1991, reprinted in *International Legal Materials*, vol. 31:148 (1992) (UN Doc. A/47/60 of December 30, 1991). Hereinafter referred to as "Alma Ata Declaration."

9. UN Doc. A/49/475, October 5, 1994.

10. Reprinted in *International Legal Materials*, vol. 21:1921 (1982) (UN Pub. E.83.V.5-1983).

11. Ibid.

12. Ibid.

13. Ibid.

14. Seminar on the Caspian Sea, London, February 24, 1995. Hereinafter referred to as "Caspian Seminar (London, 1995)."

15. See Alexander Khodadov, "The Legal Framework for Regional Co-operation in the Caspian Sea Region," in *Central Asia Quarterly Labyrinth* (Summer 1995).

16. See Vyacheslav Gizzatov, "The Legal Status of the Caspian Sea," in *Central Asia Quarterly Labyrinth* (Summer 1995).

17. See, for example, W. E. Butler, "Legal Aspects of the Caspian Sea," a paper presented to the Conference on Oil and Gas Prospects in the Caspian

Region, organized by the Institute for International Energy Studies and the Institute for Political and International Studies, held in Tehran, December 10-11, 1995. Hereinafter referred to as "Caspian Conference (Tehran, 1995)."

18. Remarks by Glen Race at Caspian Seminar (London, 1995).

19. Butler, Caspian Conference (Tehran, 1995).

20. Gizzatov, p. 34.

21. In 1969, the Iranian ministry of foreign affairs published a collection of Iran-USSR bilateral agreements; the collection does not contain an agreement concerning a boundary line in the Caspian Sea.

22. Khodadov, p. 31.

23. *British and Foreign State Papers*, vol. 126 (1927), part 1 (1932), pp. 943-945.

24. According to article 8 of the treaty, ibid., the document was prepared in Russian, Persian, and French; in the case of disagreement, the French version was to be controlling. The relevant verbiage in article 3 refers to the parties safeguarding against alliances "diriges contre la securite du territoire ou des eaux" of one another.

25. Treaty of Peace and Perpetual Friendship between Persia and Russia, signed on the River Seiwa, October 12, 1813, reprinted in Clive Parry, ed., *Consolidated Treaty Series* (Dobbs Ferry, New York: Oceana Publications, 1986), vol. 62 (1813), pp. 435-442.

26. Treaty of Peace and Friendship between Persia and Russia, signed at Turkmanchai, February 10, 1828, reprinted in Clive Parry, ed., *Consolidated Treaty Series* (Dobbs Ferry, New York: Oceana Publications, 1986), vol. 78 (1828), pp. 106-112.

27. The note is reprinted in Nasrollah S. Fatemi, *Diplomatic History of Persia (1917-1923)* (New York: Russell F. Moore Co., 1952), pp. 257-259.

28. The 1921 Treaty, article 11.

29. *British and Foreign State Papers*, vol. 126 (1927, part I), pp. 947-953.

30. *British and Foreign State Papers*, vol. 134 (1931), pp. 1026-1041.

31. Letter from M. A. Foroughi, Persian foreign minister, to A. Petrovsky, the Soviet ambassador to Persia, dated Tehran, October 27, 1931, reprinted in *British and Foreign State Papers*, vol. 134 (1931), p. 1045.

32. See letter from A. Petrovsky, the Soviet ambassador to Persia, to M. A. Foroughi, the Persian foreign minister, dated Tehran, October 27, 1931, reprinted in *British and Foreign State Papers*, vol. 134 (1931), pp. 1045-46.

33. Case Concerning the Land, Island and Maritime Frontier Dispute (El Salvador/Honduras/Nicaragua intervening), Judgment, September 11, 1992, *I.C.J. Reports* (1992), para. 432. Hereinafter referred to as "The Fonseca Case."

34. *Republic of El Salvador* v. *Republic of Nicaragua*, Central American Court of Justice, Opinion and Decision, March 9, 1917, reported in *American Journal of International Law*, vol. 11:674 (1917), p. 730. Hereinafter referred to as "Fonseca (1917)."

35. Ibid., pp. 700-705.

36. Ibid., p. 707.

37. The Fonseca Case, paras 394-95.

38. Ibid., paras 396-417.
39. Case Concerning the Continental Shelf (Tunisia/Libya Arab Jamahiriya), Judgment, February 24, 1982, *I.C.J. Reports* (1982), para. 100.
40. The Fonseca Case, paras. 394-95.
41. Ibid., para. 392.
42. Ibid., paras. 409, 418-420.
43. See, for example, The Fonseca Case, paras 59-61; and North Sea Continental Shelf Cases (Federal Republic of Germany/Denmark, Federal Republic of Germany/Netherlands), Judgment, February 20, 1969, *I.C.J. Reports* (1969), paras. 89-91.

Contributors

HOOSHANG AMIRAHMADI, a Cornell Ph.D., is professor of planning and international development at Rutgers University. He has chaired the Department of Urban Planning and Policy Development and has served as director of the University's Middle Eastern Studies Program. He is the founder of Caspian Associates, Princeton, New Jersey, an organization dedicated to research and analysis of development issues confronting the Caspian region. He is also the founder of the Iranian-American Council (AIC) and the Center for Iranian Research and Analysis (CIRA). Professor Amirahmadi's publications include *Small Islands, Big Politics: The Tonbs and Abu Musa in the Persian Gulf* (St. Martin's Press, 1996), *The United States and the Middle East: A Search for New Perspectives* (State University of New York Press, 1993), *Reconstruction and Regional Diplomacy in the Persian Gulf* (London: Routlege, 1992), *Revolution and Economic Transition: The Iranian Experience* (State University of New York Press, 1990), *Post-Revolutionary Iran* (Westview Press, 1988), *Iran and the Arab World* (St. Martin's Press, 1993), and upward of 150 articles, chapters, and review articles. Dr. Amirahmadi is a frequent contributor/participant at conferences throughout the world, having lectured in no less than 20 countries. He has served as consultant to the United Nations Development Program, the World Bank, the Agha Khan Foundation, multinational corporations, and developing countries.

ALI MASSOUD ANSARI is a Ph.D. candidate in political studies at the School of Oriental and African Studies (SOAS), University of London. He is a Research Assistant at the Geopolitics and International Boundaries Research Center at SOAS, and also assists in the teaching of comparative politics and Middle East politics. He is a member of the British Institute of Persian Studies and the International Institute of Strategic Studies. Mr.

Ansari is a consultant with Control Risks Group Ltd., and his contributions to various publications include the Economist Intelligence Unit's country report on Iran, *World Today,* and *Gulf States Newsletter.* In 1991-95 Mr. Ansari served as editor of Menas Associate's *Iran Focus.* He has worked as a media consultant and researcher, most recently on "The Last Shah," a BBC documentary.

NARSI GHORBAN is Director of Iran Association of Energy Economics, as well as Director of the Energy Department of Asian Pacific, Inc. Dr. Ghorban is a consultant to oil and financial organizations and his clients have included the Long Term Strategy Committee of the Organization of Oil Exporting Countries (OPEC), Mid-Continental Financial Services, Shell, Institute for Political and International Studies, and Institute for Energy Studies. His publications include "Iran's Role in the Export of Oil and Gas from the Caspian Basin," in *Oil & Caviar in the Caspian* (Menas Associates, 1995), "National oil Companies with Reference to the Middle East 1900-1973", in *Oil in the World Economy* (Toutleqo, 1989), and contributions to the Economist Intelligence Unit, CEDIGAZ, *The Iranian Journal of International Affairs,* and *Iran Focus.* He holds a Ph.D. in petroleum economics from the University of London.

HORMOZ GOODARZY is Director of Information Technology Office at Management Sciences for Health, Inc., a private nonprofit primary health and family organization in Boston, Massachusetts. A Ph.D. candidate at the Center for Middle Eastern and Islamic Studies at the University of Durham, England, Mr. Goodarzy holds a Master of Public Administration degree from Harvard University's Kennedy School of Government and a Master of Business Administration degree from the University of Hartford's Barney School. His undergraduate studies concentrated on chemistry and geology. A former Visiting Fellow at Harvard University's Center for Middle Eastern Studies, his main area of research and study has been the Caspian Sea's unique historical, geopolitical, and environmental dimensions.

SCOTT HORTON is a partner with the law firm Patterson, Belknap, Webb & Tyler. Based in New York, Mr. Horton is responsible for the firm's practice in Transcaucasia and Central Asia, representing corporate clients operating in the natural resources sector. He counsels the governments in the region and advises international development banks. The Chair of the American Bar Association's Committee on the Commonwealth of Independent States, Mr. Horton has authored over one hundred articles and monographs dealing with legal transformation in the region.

ALISHER ILKHAMOV is Director of Experts, at the Center for Social Research in Tashkent, Uzbekistan, a private enterprise established in 1991. The center's clients have included the World Bank, the European Community (TACIS Program), Johns Hopkins University, the University of Michigan, U.S. Information Agency, and the International Foundation of Elections System. A Ph.D., his publications include articles on the problems of transition faced by the newly independent states of Central Asia. A recent Visiting Scholar at the Harvard Forum for Central Asian Studies, Dr. Ilkhamov presently is conducting research on the subject of dynamics of social stratification and identities in Uzbekistan.

BIJAN KHAJEHPOUR-KHOUEI is the Managing Director of Atieh Bahar Management Consultancy, based in Tehran, Iran, providing political and economic analysis for foreign and international organizations and multinationals interested in development projects in Iran. He is editor of the monthly newsletter, *Iran Focus,* published in London, and is on the editorial board of the Persian quarterly review *Goftogu* published in Iran. A specialist in Iran's political, economic, and social issues, Mr. Khajehpour-Khouei contributes to international seminars and conferences on Iran. His publications include special reports on Iranian issues published by MENAS Associates (London), such as *Iran's Economy and Stock Market: A Waking Regional Giant?* (January 1997), *Country Profile on Iran* (1996), and *The Elections to the Iranian Majlis* (1996), and articles and reviews published in *Goftogu* and *Iran Focus.* A graduate of Middlesex University, London, with honors, in European business administration, Mr. Khajehpour-Khouei also holds a diploma in economic studies from Reutlingen University, Germany.

FELIX N. KOVALEV was Chief of Working Group on the Caspian Sea at the Russian Federation's Foreign Ministry. Previously, as a foreign service officer of the Soviet Union, he served as first secretary and later counselor in Argentina, deputy chief of the ministry's legal department, representative to the United Nations Conference on the Law of Treaties and the Third Conference on the Law of the Sea, ambassador to Ecuador, and director of the department of diplomatic history at the ministry's archives. In his present position as Ambassador at Large, he concentrates on Russian-Georgian relations. He holds a master's degree from Moscow's Foreign Relations Institute.

NATIK MAMEDOV is an Azerbaijani lawyer and international consultant. His clients have included the New York law firm of Patterson, Belknap, Webb & Tyler, and the National Bank of Azerbaijan. A corporate and finance lawyer by trade, Mr. Mamedov has taught public and private international law courses at Baku State University. His publications include articles on foreign investment and legal reforms in Azerbaijan. A law graduate of Baku State University, he has an LL.M. in corporate finance and law from Widener University School of Law.

BRADFORD R. MCGUINN is a professor of international relations and Middle Eastern studies at Florida International University, where he received his Ph.D. in 1995. Among his many publications is "From Caspian to the Gulf: The Assertion of American Power," which appeared recently in *Middle East Insight,* which was discussed by Stephen Rosenfeld in the *Washington Post.*

MOHIADDIN MESBAHI is Associate Professor of International relations and the Acting Director of the Asian Program at Florida International University in Miami. Dr. Mesbahi teaches in the areas of central Asian studies, international relations theory, and strategic studies. Among his many publications are *Moscow and Iran: From the Islamic Revolution to the Collapse of Communism* (Macmillan, forthcoming), "Tajikistan, Iran, and the International Politics of the 'Islamic Factor,'" in *Central Asian Survey* (Spring 1997), "Russia and Iran: The Emerging Partnership," in *Middle East Insight* (August, 1995), *Russia and the Third World in the Post Soviet Era,* ed. (Gainesville: University Press of Florida, 1994), *Central Asia and the Caucasus after the Soviet Union,* ed. (Gainesville: University Press of Florida, 1994). He is a member of the editorial boards of *Central Asian Survey* (London), *Journal of Political Ideologies* (Oxford), *American Muslim Quarterly* (Washington), and *East-West Review* (Chicago).

GUIVE MIRFENDERESKI is a historian and international lawyer in private practice in Newton, Massachusetts. As a legal consultant in the area of private sector development, he has counseled international organizations and developing countries in privatization, business deregulation, and legal reform projects. As adjunct assistant professor of international law, he has taught development, privatization, and business-related courses at the Fletcher School of Law and Diplomacy, as well as at Brandeis University's Graduate School of International Economics and Finance. His publications include "The Toponymy of the Tonb Islands," in *Iranian Studies* (1996), "The Ownership of the Tonb Islands: A Legal Analysis," in H. Amirahmadi, ed. *Small Islands, Big Politics* (St. Martin's Press, 1996); and numerous other articles and opinion pieces. Previously, he served as a corporate lawyer with the law firm of Gaston & Snow in Boston. A graduate of Georgetown University, he holds a J. D. from Boston College Law School, and a Ph.D. M. A. L. D. and M.A. from the Fletcher School of Law and Diplomacy. He may be reached at Guive@aol.com.

PIROUZ MOJTAHED-ZADEH is chairman of Urosevic Research and Study Foundation, London, and senior research associate at the Geopolitics and International Boundaries Research Center at the School of Oriental and African Studies, University of London. As all expert on boundary issues, Dr. Mojtahed-Zadeh has served as consultant to Middle Eastern governments, international educational establishments, such as the United Nations University, national educational and research establishments, such as the University of Tehran, London University, and the University of Oxford, and to MENAS International Boundary Consultants, London. An international speaker and conference participant, he has lectured extensively on subjects regarding the Persian Gulf, Caspian Sea, and Central Asia in Europe, North America, Middle East and Far East. His publications include *The Amirs of the Borderlands and Eastern Iranian Borders* (Urosevic, 1996), *The Islands of Tunb and Abu Musa* (CNMES, 1995), *The Changing World Order* (Urosevic, 1992), *Political Geography of the Strait of Hormuz* (SOAS, 1991), and upward of 100 articles, in English and Persian. He holds a master's and Ph.D. in political geography from SOAS, London University.

FIROUZEH MOSTASHARI is an adjunct assistant professor of history at Regis College in Weston, Massachusetts. As a post-doctoral Fellow at Harvard University's Center for Middle Eastern Studies, she researched and wrote on Azerbaijan under Russian imperial rule. Her publications include articles and reviews on the Caucasus, Russia, and Iran, the latest of which have appeared in *Harvard Middle Eastern and Islamic Review* and *Journal of Central Asian Studies*. A Ph.D. in history from the University of Pennsylvania, Dr. Mostashari also holds a master's degree in industrial engineering from Cornell University.

MOHAMMAD ALI MOVAHED is an international lawyer in Tehran, Iran. He joined the National Iranian Oil Company in 1954, serving as its coordinator of international agreement, and later as manager of NIOC's legal affairs group, and counsel to the president. A member of the NIOC's board of directors prior to 1979, Dr. Movahed continued his service as the company's legal adviser until his retirement. Also a deputy manager of the enforcement department at the Organization of Petroleum Exporting Countries, he served as a member of NIOC's international negotiating and contracting teams between 1954 and 1979; in 1978, he drafted Iran's petroleum law. Recently, he has served as counsel, arbitrator, and expert at the proceedings of the Iran–United States arbitration tribunal at the Hague. A professor at Tehran University, Dr. Movahed's many publications include *Nafte Ma Va Massael Hoquqi An* (Our Oil and its Legal Problems) (Tehran: Kharazmi Publications, 1971).

SIAMAK NAMAZI is co-founder and director of country analysis at Future Alliances International, a strategic consulting firm based in Bethesda, Maryland. A graduate of Tufts University in international relations, Mr. Namazi has completed the coursework toward his master of science degree in Urban Planning at Rutgers University and presently is pursuing a course of study in International Trade Policy at George Washington University, Washington, D.C. Mr. Namazi is the author of a number of articles on Iran and the Caspian region.

JOHN SCHOEBERLEIN is Director of Forum for Central Asian Studies at Harvard University, which he helped establish in 1993. The Forum coordinates research, curriculum, and information in Central Asian studies at Harvard, and organizes lecture series, publications, and other activities. Dr. Schoeberlein-Engel's research focuses on questions of identity, community, and organization among the Islamic peoples of Central Asia and neighboring regions. As Lecturer on Central Asian Studies at Harvard University, he has taught courses in anthropology, history, and politics. Presently, he is involved with projects that study the impact of national state formation in Central Asia on identity, violent inter-communal conflict in the region, and promotion of community-level participation in economic reform. Dr. Schoeberlein-Engel holds a Ph.D. in social anthropology from Harvard University.

JEAN-FRANÇOIS SEZNEC is an expert in international banking and finance, with a particular emphasis on the Persian Gulf region. He is the founding member and president of Lafayette Group, an investment bank based in Greenwich, Connecticut. An adjunct professor at Columbia University's Middle East Institute, he conducts research on the influence of regional political and social variables on financial markets in the Persian Gulf area, as well as on the politics of oil and petrochemicals in the Persian Gulf region. His publications include a book and articles on international banking and financial markets and Middle East. A graduate of Washington College, he holds an MIA from Columbia University, and a master's and Ph.D. from Yale University.

OTTAR SKAGEN is a petroleum economist, presently serving as project manager with the gas business development division of the Norwegian Statoil. Mr. Skagen advises on Statoil's Azerbaijan project. Previously, he was principal administrator at International Energy Agency, Paris, focussing on Poland, the Baltic states, Ukraine, and countries in Central Asia and Transcaucasia. Between 1988 and 1993, he served as a member of Statoil's corporate strategy unit, analyzing the global oil market and planning risk scenarios. Before that, Mr. Skagen worked for the Norwegian Petroleum Directorate, for GIEK, the Norwegian state export credit insurance agency, and participated as a Norwegian delegate to the Paris Club. Mr. Skagen's publications include IEA's 1994 *Survey of the Energy Policies of Poland* of which he was editor and principal contributor. His work on Central Asian gas issues is due for publication by the Royal Institute of International Affairs, London. Mr. Skagen holds a diploma in economics from the University of Bergen, Norway.

Index